Advanced Multicarrier Technologies for Future Radio Communication

5G and Beyond

Hanna Bogucka

Adrian Kliks

Paweł Kryszkiewicz

This edition first published 2017
© 2017 John Wiley & Sons, Inc.

The right of Hanna Bogucka, Adrian Kliks, and Paweł Kryszkiewicz to be identified as the author(s) of this work has been asserted in accordance with law.

Registered Offices
John Wiley & Sons, Inc., 111 River Street, Hoboken, NJ 07030, USA

Editorial Office
111 River Street, Hoboken, NJ 07030, USA

For details of our global editorial offices, customer services, and more information about Wiley products visit us at www.wiley.com.

Wiley also publishes its books in a variety of electronic formats and by print-on-demand. Some content that appears in standard print versions of this book may not be available in other formats.

Library of Congress Cataloging-in-Publication Data
Names: Bogucka, Hanna, author. | Kliks, Adrian, author. | Kryszkiewicz,
    Paweł, author.
Title: Advanced multicarrier technologies for future radio communication : 5G
    and beyond / by Hanna Bogucka, Adrian Kliks, Paweł Kryszkiewicz.
Description: Hoboken, NJ, USA : Wiley, 2017. | Includes bibliographical
    references and index. |
Identifiers: LCCN 2017016847 (print) | LCCN 2017030272 (ebook) | ISBN
    9781119168911 (pdf) | ISBN 9781119168928 (epub) | ISBN 9781119168898
    (hardback)
Subjects: LCSH: Wireless communication systems–Technological innovations. |
    Multiplexing. | Carrier waves. | BISAC: TECHNOLOGY & ENGINEERING /
    Electrical.
Classification: LCC TK5103.2 (ebook) | LCC TK5103.2 .B64 2017 (print) | DDC
    621.3845/6–dc23
LC record available at https://lccn.loc.gov/2017016847

Cover image: (Main image) © saicle/Gettyimages; (Inset) © nopporn/Shutterstock
Cover design by Wiley

Set in 10/12pt WarnockPro by SPi Global, Chennai, India

Printed in the United States of America

10 9 8 7 6 5 4 3 2 1

To our families.

Contents

Preface

Increasing demand by mobile radio customers (persons and devices) for higher data rates, multimedia services, and more bandwidth, as well as anticipated traffic related to the Internet of Things, creates unprecedented challenges for future mobile communication systems. There seems to be the general consensus on the future Fifth Generation (5G) wireless communication directions and expected key performance indicators to meet these challenges, that is, to aim at achieving significantly higher system capacity, connectivity, energy and spectral efficiencies, while lowering the end-to-end latency for some mission-critical applications. Concerning the network capacity and spectrum usage enhancement, they result from network densification and spectrum aggregation [1]. Spectrum aggregation refers to making use of possibly discontinuous frequency bands and, thus, larger amounts of electromagnetic spectrum. It is known to be possible through a technique called Carrier Aggregation (CA), which has been proposed for the Long Term Evolution Advanced (LTE-A) standard in order to achieve the throughput of 1 Gbps in the downlink for the Fourth Generation (4G) systems in a 20 MHz channel [2]. Although CA applied in LTE-A is a step toward spectrum aggregation, its flexibility in aggregating any kind of spectrum fragments is limited, and the proposed protocols do not allow for dynamic spectrum access and aggregation.

New multicarrier transmission techniques using noncontiguous subcarriers are known to be capable of flexible spectrum aggregation [3, 4] and allow for flexibility of various kinds, specially at adaptive physical and medium access control layers. By applying cognitive spectrum sharing using these techniques in both licensed and unlicensed frequency bands of the future heterogeneous networks, more spectrum can be effectively used, and interference among cells and nodes can be avoided. Dynamic aggregation of potentially noncontiguous fragments of bands in a wide frequency range poses a number of challenges for the baseband processing, antenna and Radio Frequency (RF) transceiver design, particularly in the dynamically changing radio environment. In our book, we present these promising technologies and answer how to meet the mentioned 5G challenges with noncontiguous multicarrier technologies and novel algorithms enhancing spectral efficiency, interference robustness, and reception performance. It is apparent that the deployment of future flexible radios and spectrally agile waveforms has received and is still receiving the necessary scientific recognition.

Multicarrier modulation and multiplexing are a form of Frequency-Division Multiplexing (FDM), where data are transmitted across several narrowband streams using different carrier frequencies. The most known example is the Orthogonal Frequency-Division Multiplexing (OFDM). In the recent years, however, increasing research effort has been focused on some other forms of multicarrier modulation and multiplexing, which enhance the properties of OFDM or employ nonorthogonal subcarriers, use discontinuous frequency bands, and apply subcarrier shaping. In our book, we focus on new multicarrier transmission techniques using noncontiguous subcarriers such as PNon-contiguous Orthogonal Frequency-DivisionMultiplexing (NC-OFDM), its enhanced version, Generalized Multicarrier (GMC) multiplexing, or its special case, namely the Non-contiguous Filter-Bank Multi-Carrier (NC-FBMC) technique. These are techniques capable of flexible spectrum aggregation, flexible transmission and reception methods achieving high spectral efficiency or energy efficiency toward meeting the 5G radio system challenges. We believe that in the coming years, the work on novel multicarrier technologies will be at a height of culmination for application in future radio communication systems (5G and beyond).

In Chapter 1, we discuss the challenges and bottlenecks of the future and 5G radio communication technology based on the spectral agility of waveforms and on the flexibility and efficiency of spectrum usage. The need for practical solutions and implementation based on novel multicarrier technologies are emphasized.

Chapter 2 entitled *Multicarrier technologies in radio communication systems* presents the state of the art in multicarrier technologies for radio communication. We present the principles of multicarrier schemes, OFDM, as well as other known multicarrier techniques. In that chapter, we also address the key advantages and issues in designing multicarrier systems, such as nonlinear distortions, Peak-to-Average Power Ratio (PAPR) reduction techniques, transmission parameter adaptation, reception techniques, and synchronization.

In Chapter 3 on *Noncontiguous OFDM for future radio communications*, we introduce the principles of NC-OFDM as a well-suited technique for future 5G radio communications, able to aggregate discontinuous spectrum bands. Efficient NC-OFDM transmitter and receiver designs are discussed. Moreover, key techniques for enhanced NC-OFDM communications are addressed: reduction of the Out-of-Band (OOB) power to aggregate the fragmented spectrum and to limit and control interference generated to the adjacent frequency bands, spectrum aggregation dynamics, PAPR reduction, signal reception, and the particularly difficult problem of synchronization in the face of the reduced number of used subcarriers and possible interference from frequency-adjacent systems.

In Chapter 4 on *Generalized multicarrier techniques for 5G radio*, we introduce the idea of Generalized Multicarrier (GMC) modulation. It is shown that it encompasses all existing multicarrier techniques, as well as all theoretically imaginable multi- and single-carrier waveforms. Some interesting features of this flexible and generalized waveform description are discussed, showing its potential for the application in future 5G (and beyond) radio communications and flexible programmable transceivers. Moreover, key issues for GMC communications are addressed, such as higher PAPR and increased complexity of the GMC transceivers, including adaptive transmission and reception algorithms.

Chapter 5 entitled *Filter-bank-based multicarrier technologies* presents the principles of Filter-BankMulti-Carrier (FBMC) modulation, which has been recently heavily researched worldwide and is being proposed for some of the 5G radio interfaces. In this technique, the OOB power is filtered on the per-subcarrier basis. Efficient Offset Quadrature Amplitude Modulation (OQAM)-based FBMC transmitter and receiver design with reduced computational complexity is discussed. The prototype-filter design and related receiver techniques are addressed. Other challenges of this technique are also covered, as well as other filter-bank-based techniques recently proposed: filtered OFDM, Cosine-Modulated Multitone signaling, Filtered Multi-Tone (FMT), Universal Filtered Multicarrier (UFMC), or Generalized Frequency Division Multiplexing (GFDM).

Chapter 6 on *Multicarrier technologies for flexible spectrum usage* discusses Dynamic Spectrum Access (DSA) and sharing options for the future multicarrier technologies meeting the desired features of 5G communications. Some interesting DSA methods based on game theory, spectrum pricing, and the so-called *coopetition* are discussed. The issue of the required information signaling is confronted against the required spectral efficiency. Coexistence of the new cognitive radio technologies with the incumbent (licensed) systems is considered. In particular, spectrum aggregation using NC-OFDM and NC-FBMC in the real-world scenarios in the presence of Global System for Mobile Communications (GSM) and Universal Mobile Telecommunications System (UMTS) system base stations and terminals is discussed and evaluated.

Finally, the book is summarized in Chapter 7, presenting *Conclusions and Future Outlook*. This chapter summarizes the key observations obtained from the totality of the presented work. The chapter also includes the discussion of the future outlook for presented technologies in terms of their greater industrial realization, hardware practicality, and other challenges.

Poznań
April 10, 2017

List of Abbreviations

2D	Two-Dimensional
1G	First Generation
2G	Second Generation
3G	Third Generation
4G	Fourth Generation
5G	Fifth Generation
3GPP	3rd Generation Partnership Project
A/D	Analog-to-Digital
ACE	Active Constellation Extension
ACIR	Adjacent-Channel Interference Ratio
ACLR	Adjacent-Channel Leakage Ratio
ACS	Adjacent-Channel Selectivity
ADSL	Asymmetric Digital Subscriber Line
AIC	Active Interference Cancellation
AM/AM	Amplitude/Amplitude
AM/PM	Amplitude/phase
AMC	Adaptive Modulation and Coding
AS	Active Set
ASA	Authorized Shared Access
AST	Adaptive Symbol Transition
AWGN	Additive White Gaussian Noise
BB	Baseband
BEP	Bit Error Probability
BER	Bit Error Rate
BFDM	Biorthogonal Frequency-Division Multiplexing
BLAST	Bell Laboratories Layered Space-Time
BRB	Basic Resource Block
C–F	Clipping and Filtering
CA	Carrier Aggregation
CBRS	Citizen Broadband Radio Service
CC	Cancellation Carrier

CCA	Clear Channel Assessment
CCDF	Complementary Cumulative Distribution Function
CDMA	Code Division Multiple Access
CE	Constellation Expansion
CF	Crest Factor
CFO	Carrier Frequency Offset
CLT	Central Limit Theorem
CM	Cubic Metric
CMT	Cosine-Modulated Multitone
COFDM	Coded OFDM
CP	Cyclic Prefix
CQI	Channel Quality Indicator
CSA	Co-Primary Shared Access
CSI	Channel State Information
CSMA	Carrier-Sense Multiple Access
CR	Cognitive Radio
D/A	Digital-to-Analog
DAC	Digital-to-Analog Converter
DC	Data Carrier
DD	Decision-Directed
DF	Digital Filtering
DFT	Discrete Fourier Transform
DGT	Discrete Gabor Transform
DMT	Discrete Mutlitone
DSA	Dynamic Spectrum Access
DWMT	Discrete Wavelet Multitone
DVB-T	Digital Video Broadcasting-Terrestrial
EAIC	Extended Active Interference Cancellation
EC	Extra Carrier
EGF	Extended Gaussian Function
EVM	Error Vector Magnitude
FBMC	Filter-Bank Multicarrier
FCC	Federal Communications Commission
FD	Frequency Domain
FDM	Frequency-Division Multiplexing
FDMA	Frequency-Division Multiple Access
FEC	Forward Error Correction
FIR	Finite Impulse Response
FFT	Fast Fourier Transform
FM	Frequency Modulation
FMT	Filtered Multitone
FPGA	Field-Programmable Gate Array
GFDM	Generalized Frequency-Division Multiplexing

GIB	Generalized In-Band
GMC	Generalized Multicarrier
GPS	Global Positioning System
GS	Guard Subcarriers
GSM	Global System for Mobile Communications
HARQ	Hybrid Automatic Repeat Request
HIC	Hybrid Interference Cancellation
HPA	High-Power Amplifier
HSDPA	High-Speed Downlink Packet Access
HSPA	High-Speed Packet Access
IBO	Input Back-Off
IC	Integrated Circuit
ICI	Intercarrier Interference
IDFT	Inverse Discrete Fourier Transform
IF	Intermediate Frequency
IFFT	Inverse Fast Fourier Transform
IMD	Intermodulation Distortion
INP	Instantaneous Normalized signal Power
IOTA	Isotropic Orthogonal Transform Algorithm
IQ	In-Phase and Quadrature
ISI	Intersymbol Interference
ISM	Industry–Science–Medicine
LAA	Licensed Assisted Access
LO	Local Oscillator
LTE	Long-Term Evolution
LTE-A	Long-Term Evolution – Advanced
LTE-U	Long-Term Evolution – Unlicensed
LSA	Licensed Shared Access
LU	Licensed User
LUISA	Licensed-User Insensitive Synchronization Algorithm
LUT	Lookup Table
MAC	Medium Access Control
MC	Multicarrier
MCS	Multiple-Choice Sequences
MIMO	Multiple Input, Multiple Output
MLSE	Maximum-Likelihood Sequence Estimator
MMSE	Minimum Mean Square Error
MSE	Mean Squared Error
N-OFDM	N-continuous OFDM
NBI	Narrowband Interference
NC-FBMC	Noncontiguous Filter-Bank Multicarrier
NC-OFDM	Noncontiguous Orthogonal Frequency-Division Multiplexing
NL	Noise-Like

NOFDM	Nonorthogonal Frequency Division Multiplexing
OCCS	Optimized Cancellation Carrier Selection
OFDM	Orthogonal Frequency-Division Multiplexing
OFDMA	Orthogonal Frequency-Division Multiple Access
OOB	Out-of-Band
OQAM	Offset Quadrature Amplitude Modulation
P/S	Parallel-to-Serial
PA	Power Amplifier
PAM	Pulse Amplitude Modulation
PAPR	Peak-to-Average Power Ratio
PCC	Polynomial Cancellation Coding
PHY	Physical Layer
PIC	Parallel Interference Cancellation
PL	Power Loading
PSD	Power Spectral Density
PU	Primary User
PW	Peak Windowing
QAM	Quadrature Amplitude Modulation
QoE	Quality of Experience
QoS	Quality of Service
QPSK	Quadrature Phase-Shift Keying
QSP	Quasi-Systematic Precoding
RAT	Radio Access Technology
REM	Radio Environment Map
RF	Radio Frequency
RP	Reference Preamble
RRM	Radio Resource Management
RSS	Reference Signal Subtraction
RX	Receiver
S&C	Schmidl&Cox
S/P	Serial-to-Parallel
SAS	Spectrum Access System
SC	subcarrier
SDR	Software-Defined Radio
SEM	Spectrum Emission Mask
SIC	Successive Interference Cancellation
SINR	Signal-to-Interference plus Noise Ratio
SIR	Signal-to-Interference Ratio
SLM	Selective Mapping
SNR	Signal-to-Noise Ratio
SOR	Spectrum Overshooting Ratio
SP	Spectrum Precoding
SSA	Static Spectrum Allocation

SSIR	Signal-to-Self Interference Ratio
SSPA	Solid-State Power Amplifier
SSS	Subcarrier Spectrum Sidelobe
STFT	Short-Time Fourier Transform
SU	Secondary User
SVD	Singular-Value Decomposition
SW	Subcarrier Weighting
TD	Time Domain
TDD	Time-Division Duplex
TDMA	Time-Division Multiple Access
TF	Time – Frequency
TR	Tone Reservation
TWTA	Traveling-Wave-Tube Amplifier
TX	Transmitter
UE	User Equipment
U-LTE	Unlicensed Long-Term Evolution
UFMC	Universal Filtered Multicarrier
UMTS	Universal Mobile Telecommunications System
USRP	Universal Software Radio Peripheral
VLSI	Very Large Scale Integration
VSB	Vestigial Sideband
WBI	Wideband Interference
WCDMA	Wideband CDMA
WIN	Windowing
WiFi	Wireless Fidelity
WLAN	Wireless Local Area Network
ZF	Zero Forcing

1

Introduction

Intensive development of mobile radio communication systems can be observed since the deployment of cellular telephone systems, which revolutionized communication in modern society. The pioneer working cellular system was the analog First Generation (1G) Nordic Mobile Telephony (NMT) system deployed in 1981, first in Scandinavia, and then, in some European and Asian countries [5]. In the following years, there have been other analog, and then digital, Second Generation (2G) cellular systems introduced with increasing continental coverage, that is, European Global System for Mobile Communications (GSM) and American IS-95, followed by the global Third Generation (3G) systems, that is, Universal Mobile Telecommunications System (UMTS) and IMT-2000 [6]. For these systems, a number of high-data-rate transmission schemes and improvements have been implemented, leading to the Fourth Generation (4G) standards, significantly increasing the 3G-systems' capacity, coverage, and mobility. Apart from the mobile radio communication standards, the local Wireless Local Area Networks (WLANs) are constantly improving to support very high data rates, for example, the IEEE 802.11.n standard using the Multiple Input, Multiple Output (MIMO) technology allows transmission from 54 to 600 Mbit/s (at a maximum net data rate) [7].

Future wireless networks are challenged with keeping up with the constantly increasing demand by mobile devices for higher data rates, multimedia services support, and ever more bandwidth. In Figure 1.1, contemporary wireless communication standards are presented, reflecting the dependence of their achievable data rates and mobility.

Note that the upper-right corner of the picture in Figure 1.1 remains empty. This is because high mobility usually means high velocity and high dynamics of the radio communication channel. This entails limitations on the possible data rates. At the same time, it is visible that subsequent generations of mobile communication systems have been aiming at higher mobility and higher data rates. The trend toward Fifth Generation (5G) (and beyond 5G) mobile communication seems to have even more demanding assumptions.

Advanced Multicarrier Technologies for Future Radio Communication: 5G and Beyond, First Edition.
Hanna Bogucka, Adrian Kliks, and Paweł Kryszkiewicz.
© 2017 John Wiley & Sons, Inc. Published 2017 by John Wiley & Sons, Inc.

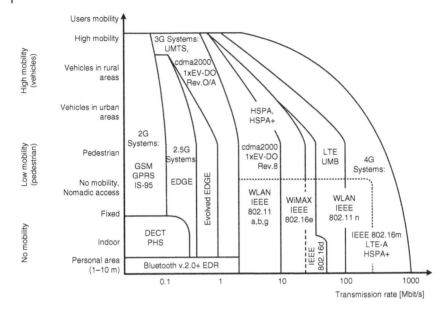

Figure 1.1 Contemporary wireless standards. (Based on [8].)

1.1 5G Radio Communications

According to the Cisco predictions, "annual global IP traffic will surpass the zettabyte (1000 exabytes) threshold in 2016, and the two zettabyte threshold in 2019. (…) By 2019, global IP traffic will pass a new milestone figure of 2.0 zettabytes per year. (…) Traffic from wireless and mobile devices will exceed traffic from wired devices by 2019. (…) Globally, mobile data traffic will increase 10-fold between 2014 and 2019" [9]. According to the recent Ericsson Mobility Report [10], the number of mobile subscriptions at the beginning of 2016 totaled 7.3 billion, while it is predicted that in 2019, this number will be 9.3 billion. Moreover, communication of billions of machines and devices that are expected to comprise the Internet of Things poses even greater challenges, never encountered before. That is why 5G wireless communication is the focus of research and industry interest, aiming at achieving 1000 times the system capacity, 10 times the energy efficiency, data rate, and spectral efficiency, 25 times the average mobile cell throughput, and significantly lower latency compared with today's 4G [11]. The paradigms for future 5G systems provided in [12] are ultrahigh capacity (1000 times higher per square-kilometer), ultralow latency (lower than 1 ms), massive connectivity (100 times higher), ultrahigh rate (up to 10 Gbps), ultralow energy consumption. Although these performance targets do not need to be met simultaneously, they provide the basis for the Gbps user experience for 5G networks [12]. In the practical real-world scenarios, 5G networks should support data rates exceeding 10 Gbps in the indoor and dense outdoor environments, and several 100 Mbps in urban and suburban environments, while 10 Mbps should be accessible almost everywhere, including rural areas in both developed and developing countries [13].

It is also anticipated in [13] that 5G networks will not be based on one specific radio-access technology. Rather, 5G communication system will consist of a portfolio of access and connectivity solutions addressing the demands and requirements of mobile communication beyond 2020. The specification of 5G will include the development of a new flexible air interface, which will be directed to extreme mobile broadband deployments, and target high-bandwidth and high-traffic-usage scenarios, as well as new scenarios that involve mission-critical and real-time communications with extreme requirements in terms of latency and reliability [13]. Apart from the *extended mobile broadband* and *mission-critical* communication, other use cases considered for 5G radio are *massive machine-type* communication, *broadcast/multicast* services, and *vehicular* communication.

Similar vision on 5G capabilities is presented by the European experts of 5G Infrastructure Association in [14]. According to this vision, by increasing *key performance indicators* (data rates, data volume, reliability, mobility, energy efficiency, density of served devices, inverse of the end-to-end latency, and service development time) indicated in the radar diagram in Figure 1.2, the 5G communication technologies will be an economy booster, paving new ways to organize the business sector of service providers, as well as fostering new business models supported by advanced information and communication technologies. They will provide user-experience continuity, the Internet of Things (machine type of communication), and mission-critical (low-latency) services.

To support these 5G radio communication system requirements, in particular for the mobile broadband use case, new spectrum is required. The World Radiocommunication Conference (WRC) in 2015 took a key decision that it will provide enhanced capacity for these kind of systems, that is, to allocate the 694–790 MHz frequency band in ITU Region-1 (Europe, Africa, the Middle East, and Central Asia) for the mobile broadband radio services. Full protection has been given to the incumbent systems operating in this frequency band (e.g., digital television broadcasting). Moreover, WRC-15 decided to include studies in the agenda for the next WRC in 2019 for the identification of bands above 6 GHz. Higher frequency ranges in the millimeter-wave frequency bands are also studied widely to use there unlicensed bands for 5G transmission, as well as cognitive radio technologies that allow for spectrum sharing and, thus, higher efficiency of frequency-band utilization.

Technologies that look promising for 5G include massive MIMO antenna systems, energy-efficient communications, cognitive radio networks, and small cells (pico- and femtocells) including extremely small, mobile femtocells.

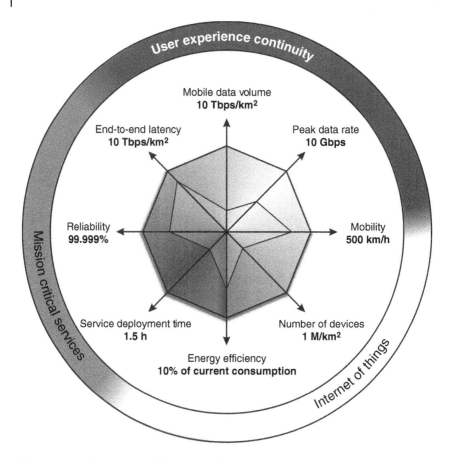

Figure 1.2 5G disruptive capabilities. (Based on [14].)

In part, these technologies will tackle the problem of the relatively poor wireless service inside buildings. Wireless devices are used indoors about 80% of the time, yet today's cellular architecture relies mostly on outdoor base stations. Such indoor wireless communications will increase the energy efficiency of wireless systems, because by separating indoor traffic from outdoor traffic, the base station would face less pressure in allocating radio spectrum and could transmit with lower power. For mobile users in cars, trains, and buses, mobile femtocells can also improve service quality [11]. There is the general consensus on the forecast that the increased traffic will be handled by heterogeneous networks, which will use different types of network nodes equipped to handle various transmission power levels and data processing capabilities and support different radio-access technologies [15]. These technologies are supported by different types of backhaul links. Thus, their interoperability is understood more broadly than the interworking of wireless local area networks and cellular networks.

Low-power micro nodes (base stations and mobile terminals) and high-power macro nodes can be maintained under the management of the same operator, sharing the same frequency band. Thus, joint radio resource and interference management needs to be provided to ensure the coverage of low-power nodes. Moreover, the nodes can use discontinuous bands and aggregate fragmented spectrum. For this purpose, new transmission techniques are being considered for future heterogeneous communications [3, 4]. Enhanced Orthogonal Frequency-Division Multiplexing (OFDM) and Filter-Bank Multi-Carrier (FBMC) are multicarrier technologies that can be mentioned as examples of application in the future 5G radio interfaces. By applying intelligent spectrum sharing methods, strong interference among cells can be avoided, as well as interference originating from coexisting systems. The objective of heterogeneous networks targets the improvement of overall capacity as well as cost-effective coverage extension and green radio solution by deploying additional network nodes within the local area range, such as low-power nodes in micro-, picocells, home-evolved Node-Bs (HeNBs), femto nodes, and relay nodes. Moreover, future 5G networks will support device-to-device (D2D) communications omitting intermediary base station, allowing devices at close distance to each other to communicate without going through the main network infrastructure. This kind of communication is viewed as very effective for traffic offloading and for improving spectrum reuse in densely populated areas.

The topic of 5G heterogeneous networks has gained much momentum in the industry and research community very recently. The 3rd Generation Partnership Project (3GPP) Long Term Evolution Advanced (LTE-A) has started a new study item to investigate heterogeneous network deployments as an efficient way to improve system capacity as well as effectively enhance network coverage [16]. It has attracted the attention of IEEE 802.16j standardization. There have been a number of special issues in leading scientific journals and magazines focused on 5G communications and heterogeneous network key issues and prospective performance (e.g., Refs [17–20]), as well as dedicated events and workshops at major scientific conferences. Undoubtedly, a need is still recognized by both industry and academia to better understand and elaborate more on the technical details and performance gains that can be made possible by future heterogeneous networks. To support the heterogeneity of radio-access technologies and networks, radio resource sharing, and infrastructure sharing in the 5G communication context, virtualization of networks and their control is intensively researched.

1.2 Challenges for Future Radio Communications

Future heterogeneous networks come with the challenges, and there are important technical issues that still need to be addressed for successful deployment and operation of these networks. In theory, the overall capacity scales with the number of small cells deployed in a unit area. By reducing each cell's radius and by introducing more cells in a given area, more capacity can be offered, and spectrum reuse can be increased. However, as cells get closer, the hyperdensification of networks is challenged in many ways. Let us consider the following equation based on the capacity of an additive white Gaussian noise (AWGN) channel. The throughput R of a user in a cellular system is upper-bounded by [21]:

$$R < C = m \, \frac{B}{n} \log_2 \left(1 + \frac{P}{\sigma_I^2 + \sigma_N^2} \right), \tag{1.1}$$

where C denotes capacity, B denotes the base station signal bandwidth, n (load factor) denotes the number of users sharing the given base station, m (spatial multiplexing factor) denotes the number of spatial streams between a base station and a user device, and P denotes the desired signal power, while σ_I^2 and σ_N^2 denote the interference and noise power, respectively, observed at the receiver. The signal bandwidth B can be increased by using additional spectrum, if it is available. The load factor n (≥ 1) can be decreased through cell splitting, which involves deploying a larger number of base stations and ensuring that user traffic is distributed as evenly as possible among all the base stations. Spatial multiplexing factor m can be increased using a larger number of antennas (with suitable correlation characteristics) at the base station and user devices.

Cell splitting has the favorable side effect of reducing the path loss between a user device and base stations, which increases both desired and interfering signal levels P and σ_I^2, effectively lowering the impact of thermal noise σ_N^2. As a result, interference mitigation is paramount for link efficiency improvement in modern cellular systems. This requires a combination of adaptive resource coordination among transmitters and advanced signal processing at the receivers. The aforementioned parameters for wireless capacity enhancement may be viewed under a common umbrella of *network densification*. Network densification is a combination of spatial densification (which increases the ratio m/n) and spectrum aggregation (which increases B) [21]. Spectral aggregation refers to using larger amounts of electromagnetic spectrum, spanning from 500 MHz to the millimeter-wave bands (30–300 GHz). Aggregating potentially noncontiguous fragments of bandwidth across such disparate frequency bands poses numerous challenges for antenna and Radio Frequency (RF) transceiver design, which need to be overcome in order to support spectral aggregation.

Overall, noncontiguous multicarrier techniques, such as Non-contiguous Orthogonal Frequency-Division Multiplexing (NC-OFDM) Non-contiguous Filter-Bank Multi-Carrier (NC-FBMC), have been recognized as suitable candidates for 5G transmission due to their potential for achieving spectrally efficient communications by aggregating and exploiting fragmented unoccupied spectrum while simultaneously achieving high data rates [3]. Both these techniques possess the ability to efficiently use fragmented spectrum opportunities as well as perform spectrum shaping in order to suppress interference that may affect nearby wireless transmissions. To counteract the potential for significant interference resulting from NC-OFDM Out-of-Band (OOB) power emission, several techniques have been proposed in the literature, which are designed to significantly suppress these sidelobes in order to make coexistence between systems utilizing adjacent spectrum feasible [4]. On the other hand, NC-FBMC, which applies subcarriers filtering, handles the OOB power emission at the required level. This comes at the expense of increased computational complexity. One can envision that the OOB level can be adjusted by adaptively modifying the filter characteristics, that is, by adaptive waveform design. However, so far, there has been no published work on this kind of pulse-shape adaptation. Given the possible constraints of limited computational and energy resources available via a user equipment and other elements of the radio-environment context, a practical approach to this problem that achieves a balance between the OOB interference mitigation efficiency and its associated costs is needed. Moreover, there are many issues of such spectrum aggregating, noncontiguous multicarrier signal reception. One of them is synchronization in the presence of self-interference (among subcarriers) and external in-band (especially narrowband) interference, the other being reception/detection quality.

As shown earlier, in a hyperdense deployment, not only desired signal strength but also interference from other cells increases. Increasing other-cell interference needs to be mitigated, and mobility management mechanism is required as the mobile users see cell edges more frequently. Furthermore, as some privately owned small cells implement restricted access schemes, they can generate/receive strong uncoordinated interference to/from external cells sharing the same radio resources [22]. The deployment of small cells is mostly unplanned, so a network self-organizing mechanism needs to be developed. The self-organizing features of small cells can be generally classified into three processes: (i) self-configuration, where cells are automatically configured by downloaded software; (ii) self-healing, where cells can automatically perform failure recovery; (iii) self-optimization, where cells constantly monitor the network status and optimize their settings to improve coverage and reduce interference [23]. The promising performance gain by deploying more cells can only be achieved by successfully addressing these problems. It can be noted that there have been some achievements in recent years, in the development of enhanced intercell interference coordination in 3GPP LTE-Advanced systems [24]. Still, intelligent incentive schemes are needed to motivate privately owned small cells to open up for wider access.

Besides the issue of meeting the overwhelming traffic demands, network operators around the world now realize the importance of managing their cellular networks in an energy efficient manner and reducing the amount of CO_2 emission levels [25]. As a result, the terminology of "green cellular network" has become very popular recently, showing that the energy efficiency as one of the key performance indicators for cellular network design [26, 27]. Although the deployment of small-cell networks is seen to be a promising way of catering to increasing traffic demands, the dense and random deployment of small cells and their uncoordinated operation raise important questions about the implication of energy efficiency in such multitier networks. Besides introducing small cells into existing macrocell networks, another effective technique is to introduce sleep mode in macrocell base stations and offload the traffic to smaller or more energy-efficient cells [28, 29]. Moreover, in some cases of dense small-cell deployment, sleep mode can also be possible and advantageous for open-access picocells to reduce energy consumption in the area [30]. Proper traffic balancing between cells of diverse coverage and networks will also result in higher Quality of Experience (QoE) for the end users by lowering the probability of blocked calls [31].

In the 5G Era, networks, systems, and nodes will need to be context-aware, utilizing context information in a real-time manner based on networks, devices, applications, and the user and his/her environment. This context awareness will allow improvements in the efficiency of existing services and help provide more user-centric and personalized services. For example, networks will need to be more aware of application requirements, QoE metrics, and specific ways to adapt the application flows to meet the QoE needs of the user. The context-based adaptations of various transmission and network parameters will have to take into account the following context information: device-level context, application context, user context, environment context, and network context [32, 33]. This context information itself consists of different parts/components, each of which affects the individual steps of the decision-making process in a different way, as shown in [34]. There are two important performance aspects related to the use of context information: signaling overhead and information reliability. They directly relate to key performance metrics such as network capacity and energy efficiency.

1.3 Initiatives for the Future Radio Interface Definition

The common understanding of future 5G communication and considered frequency bands of hundred of GHz implies that using one radio interface to address this wide range of frequency bands is not a good approach. This is because propagation characteristics, implementation aspects, and compatibility issues are different for different frequency ranges. Therefore, the overall 5G wireless-access solution will most likely consist of multiple well-integrated radio-interface solutions [13]. Nevertheless, suitable signal waveforms to meet 5G communication requirements have been researched, especially the ones possessing parametric definition (and, thus, design flexibility) and potentials for spectral agility for dynamic spectrum access, spectrum aggregation, and spectrum sharing. For example, the EU Horizon 2020 project FANTSTIC-5G studies flexible air interfaces for 5G systems operating in the frequency bands below 6 GHz [35]. Moreover, a new common radio interface is envisioned in the millimeter-wave band. The vision presented in [13] is that 5G networks will incorporate LTE access (based on OFDM)along with new air interfaces in a transparent manner toward both the service layer and users.

OFDM is already a mature technique, successfully applied in a number of wireless standards. Therefore, there are many propositions to use similar techniques for the future radio interface. Many of them originate from the telecom industry and are serious candidates for 5G waveforms. For example in [12], the so-called *filtered* OFDM is proposed as flexible waveform technology to support multiple access schemes, frame structures, application scenarios, and service requirements. It can also facilitate the coexistence of different systems efficiently. In this approach, groups of OFDM subcarriers are filtered. These groups may have different subcarrier spacings, symbol durations, and guard times. According to our categorization in Chapter 5, this proposed method can be viewed as a number of independent filtered OFDM waveforms or a more flexible version of Universal Filtered Multicarrier (UFMC).

To avoid filtering in the Dynamic Spectrum Access (DSA) networks, allowing for the dynamic spectrum aggregation, enhanced OFDM waveforms have been studied, for example, in [3, 4, 36–43]. The enhancement bases on some signal processing and optimization methods for required spectrum shaping, where the design (and redesign) of spectrum shaping filters in the dynamically changing radio communication environment is not possible. Details of this group of multicarrier transmission schemes are given in Chapter 3.

FBMC waveforms have also been intensively studied for the application in the future radio-access networks. They have some prominent features resulting from per-subcarrier filtering. Since the subcarriers spectra are shaped individually, FBMC transmitter still possesses a good degree of flexibility in theoretically aggregating any kinds of fragmented frequency resources and, thus, the spectrum sharing. A number of European 6th and 7th Framework Programme projects have focused on this type of waveform design with the objective of physical layer design for the future DSA and cognitive radio networks, for example, URANUS, PHYDYAS, EMPhAtiC, METIS, or 5GNOW. Chapter 4 presents details on the generalization of multicarrier waveforms based on filter banks for per-subcarrier filtering, while Chapter 5 is focused on FBMC systems as they are more commonly understood.

Let us also mention the recent proposition for 5G Radio Access Technology (RAT) based on Generalized Frequency Division Multiplexing (GFDM) being considered to be a flexible version of OFDM, with the option of the subcarriers to be not orthogonal to each other. This RAT has been described in [44–47]. GFDM, FBMC, as well as two other multicarrier waveforms, namely UFMC and Biorthogonal Frequency Division Multiplexing (BFDM), have been the subject of investigation in the EU 7th Framework Programme project 5GNOW [48]. Some less-detailed description of these schemes can be found in Chapter 5, where some categorization of the contemporary proposed filter-bank-based transceiver structures is provided. However, in fact, GFDM scheme is close to the Generalized Multicarrier (GMC) scheme, which has been proposed earlier in [49, 50] and researched within the EU 6th Framework Programme project URANUS [51, 52]. We devote Chapter 4 to this kind of generalized waveform.

2

Multicarrier Technologies in Radio Communication Systems

Multicarrier modulation is a form of Frequency-Division Multiplexing (FDM) [6, 53], where data are transmitted across several narrowband channels located at different carrier frequencies. As opposed to conventional FDM systems, where narrowband subcarrier signals are separated by guard bands in the frequency domain, multicarrier modulation allows for a potential overlapping of adjacent subcarriers under a certain set of operating conditions, thus making this form of data transmission spectrally efficient. The parallelization of data symbols across several simultaneous subcarriers yields relatively long symbol duration when compared with the encountered duration of a time-dispersive channel impulse response. As a result, communication systems employing multicarrier modulation can efficiently handle the effects of intersymbol interference due to multipath propagation.

Multicarrier modulation implementations have been approached in a number of manners, depending on how the data is demultiplexed to substreams modulating parallel subcarriers. In general, these approaches can be categorized into two classes of multicarrier modulation, namely [3, 53]:

- *Discrete Fourier Transform (DFT)-based multicarrier modulation*: applies DFT and harmonic basis functions for subcarrier modulation; It can be efficiently implemented using the Fast Fourier Transform (FFT) algorithm, for example, radix-2 FFT with $N \log_2(N)$ complexity (in terms of the number of operations) for N subcarriers (for N being the integer power of 2). Numerous commercial network standards employing it include Orthogonal Frequency-Division Multiplexing (OFDM) and Discrete MutliTone (DMT) modulation [7, 54–61].
- *Filter-Bank Multi-Carrier (FBMC) modulation*: applies band-pass filters at the transmitter to shape the spectra of modulated subcarriers prior to combining them in the multicarrier transmit signal. At the receiver, the bank of band-pass filters is applied in order to separate the received data for distinct subcarriers [49, 62–64]. Several examples of FBMC include [3]: (i) *complex-exponentially modulated filter banks* and cosine-modulated

Advanced Multicarrier Technologies for Future Radio Communication: 5G and Beyond, First Edition.
Hanna Bogucka, Adrian Kliks, and Paweł Kryszkiewicz.
© 2017 John Wiley & Sons, Inc. Published 2017 by John Wiley & Sons, Inc.

filter banks that modulate a prototype low-pass filter using complex exponentials and cosines, respectively; (ii) *transmultiplexers* that can be considered to be the functional dual of subband coders; (iii) *perfect reconstruction filter banks* designed to eliminate Inter-Carrier Interference (ICI) under ideal channel conditions; (iv) *oversampled filter banks* that employ a sampling factor higher than the total number of subcarriers; or (v) *modified DFT filter banks* with offset-QAM mapping that delays either the real or the imaginary components of each subcarrier signal with respect to each other to minimize ICI.

Thus, there exists a wide range of multicarrier signals that can be classified based on several characteristics, including whether the subcarriers are filtered or nonfiltered, orthogonal or nonorthogonal, and precoded or nonprecoded. All multicarrier signals can be described by the so-called Generalized Multicarrier (GMC) representation, which will be discussed in Chapter 4.

The popularity of the multicarrier transmission utilizing orthogonal waveforms (subcarriers), as in OFDM, is caused by its excellent properties. First, if the coherence bandwidth of the considered channel is significantly larger than the inverse of the orthogonality time of the OFDM signal, then the channel can be treated as flat-fading in a subchannel defined around each subcarrier frequency. This, in consequence, allows for relatively straightforward and simple equalization of the received symbol and correction of the influence of the transmission channel distortions. Moreover, in practical designs, the duration of each OFDM symbol is much smaller than the coherence time of the channel; thus, the channel characteristic does not change within one symbol. In order to mitigate the influence of the intersymbol interference, the guard period is added in front of each data block. The price paid for the lack of intersymbol interference is the degradation of the spectral efficiency. To deal with these limitations, and to improve the properties of the OFDM-like signals, various additional techniques have been proposed for application in orthogonal multicarrier systems, just to mention, adaptive modulation and coding methods [53, 65–68] or efficient resource management techniques [69–73]. Furthermore, OFDM technique is already applied in Fourth Generation (4G) systems [58, 59] and recommended for application in Fifth Generation (5G) radio communications. It is also widely considered for the cognitive-radio scenarios, mainly in its noncontinuous form [74–77].

Besides the indisputable advantages of OFDM-based systems, this kind of multicarrier modulation suffers from two serious disadvantages, mainly:

- high out-of-band signal emission caused by the processing of high-amplitude samples in the nonlinear devices (such as power amplifiers), corresponding to a high peak-to-average power ratio [6, 53, 78, 79],

- high sensitivity to frequency offsets and synchronization errors, due to the fact that the orthogonality between the subcarriers is lost in a mobile radio channel [53].

Apart from these disadvantages, it is worth mentioning that the spectral efficiency of the OFDM is limited [80–82] due to the existence of the Cyclic Prefix (CP) and the orthogonality requirement between the transmit pulses, which determines the frequency distance between subcarriers. Such a limitation together with other drawbacks of multicarrier systems using orthogonal subcarrier waveforms has led toward implementation of new sophisticated ways of signal modulation. These in particular allow for even strong overlapping between the neighboring pulses on the time–frequency plane [81, 82], thus increasing the total number of bits that can be transmitted in a certain time slot and frequency band. Moreover, modification of the time–frequency characteristic of the basic transmit pulse (waveform) enables higher spectrum utilization in the context of cognitive networks, due to the possibilities of, first, out-of-band emission control and second, efficient subcarrier allocation within the given frequency band [75].

There have been several attempts to generalize the structure of a multicarrier transceiver [49, 51, 52, 83]. Consequently, a class of data transmission signals can be defined that is based on a GMC waveform characterized by two generic requirements [3]:

- the number of subcarriers is greater than 1, and
- all transmit waveforms are created (i.e., translated in time and modulated in the frequency domain) based on the unique, predefined pulse (called also mother function or prototype pulse shape); such waveforms can constitute both complete or overcomplete basis set [49, 84, 85].

Based on these assumptions, each transmission scheme can be defined by proper selection of the set of parameters such as the following: the shape of the transmit pulse, the FFT size, the number of subcarriers, the length of CP . Such generic description of the transmit signal has been proposed in the literature (mainly in the work by Giannakis *et al.* [49, 83, 86, 87]), where it has been assumed that the subcarriers modulated by the signal pulses are mutually orthogonal. However, a more generalized approach to multicarrier signals can be considered, that is, when the assumption of orthogonality does not have to be fulfilled. Such an option has also been investigated in the past years [81, 82, 84].

As it has already been mentioned, although the idea of proper shaping of the transmit and receive pulses is not new, the renaissance of this technique has been observed in the last few years. The reason for this interest is its potential application in future radio communication systems, for example, in cognitive radio or future 5G communication systems. Various international European research projects have addressed this issue, just to mention a few: URANUS (Universal RAdio-liNk platform for efficient User-centric accesS) [84], NEW-COM, NEWCOM++, and NEWCOM# (Network of Excellence in Wireless Communications) [88], PHYDYAS (Physical Layer For Dynamic Spectrum Access and Cognitive Radio) [89], ACROPOLIS (Advanced coexistence technologies for radio optimization in licensed and unlicensed spectrum) [90], EMPhAtiC (Enhanced Multicarrier Techniques for Professional Ad-Hoc and Cell-Based Communications) [91], or COST ACTION IC0902 (Cognitive Radio and Networking for Cooperative Coexistence of Heterogeneous Wireless Networks) [92].

A number of papers that show the possibilities of the application of the nonorthogonal signals for multicarrier transmission [80–82, 85, 93] have been published. They provide the analysis of various modulation cases including biorthogonal basis and the so-called Weyl–Heisenberg frames. Application of these modulation schemes in real-world scenarios has also been addressed [3, 4, 37, 94]. Moreover, some of these papers deal with advanced transmission techniques in the context of the Weyl–Heisenberg-based waveform transmission. It is worth highlighting that as in the existing multicarrier systems (e.g., OFDM, Filtered Multi-Tone (FMT), DMT), the GMC signals suffer from high variation of the time-domain signal envelope (measured by means of the Peak-to-Average Power Ratio (PAPR) metric). However, due to the specific features of the GMC signal, the typical (existing) algorithms for PAPR minimization cannot be applied in a straightforward manner. Moreover, the spectral efficiency in the case of GMC transmission can be improved by application of link adaptation techniques. Furthermore, the signal reception methods should also be adapted for the generic, in particular nonorthogonal transmission case. Finally, it is also worth mentioning that from the implementation point of view, the parameters of the GMC transmission scheme influence not only on the link-level but also on the system-level simulations of the considered systems [95]. All of these aspects can improve the overall system performance, and that is the reason why these issues are addressed in this book.

In consequence, based on the indicated advantages of various multicarrier modulation schemes, we are motivated to present various forms of multicarrier technologies for their application in future 5G wireless communications with an opportunistic access to radio resources, flexibility and interoperability, and aiming at high spectral efficiency at the reasonable computational cost. We focus on new methods of transmission and reception of the advanced multicarrier signals that can improve the performance of the wireless systems when compared with the existing solutions.

Here, we start the discussion with the most popular one, namely OFDM, that is already applied in many wireless standards and is also considered as a good candidate waveform for the future 5G radio communication systems.

2.1 The Principles of OFDM

As mentioned earlier, OFDM technique allows for parallel transmission of multiple data streams using orthogonal subcarriers. It has a number of advantages that makes it well suited for radio communication, as well as some flaws and challenges that need to be addressed in the telecommunication-system design process. This is why it has been described in thousands of scientific papers and tens of books, for example, in [96–102], just to name the examples of related comprehensive works on OFDM in wireless systems.

The multicarrier signal resulting from modulation of N subcarriers of f_n frequencies (where $n = 0, \dots, N - 1$) by the data symbols can be expressed as

$$\tilde{s}(t) = \sum_{n=0}^{N-1} \left[\mathfrak{R}\left\{ d_n^{(p)} \right\} \cos(2\pi f_n t) + \mathfrak{I}\left\{ d_n^{(p)} \right\} \sin(2\pi f_n t) \right] \tag{2.1}$$

for $pT_\mathrm{B} \le t < (p + 1)T_\mathrm{B}$, where $\mathfrak{R}\{d_n^{(p)}\}$ and $\mathfrak{I}\{d_n^{(p)}\}$ are the in-phase and quadrature component, respectively, of the n-th data symbol modulating the n-th subcarrier at the p-th modulation interval, while T_B is the time duration of this interval. Let us note that $\tilde{s}(t)$ is the real part of the $s(t)$ signal:

$$\tilde{s}(t) = \mathfrak{R}\{s(t)\} = \mathfrak{R}\left\{ \sum_{n=0}^{N-1} [d_n^{(p)} \exp\ (j2\pi f_n t)] \right\}, \tag{2.2}$$

where $j = \sqrt{-1}$, $d_n^{(p)} = \mathfrak{R}\{d_n^{(p)}\} + j \cdot \mathfrak{I}\{d_n^{(p)}\}$ is the complex data symbol, and $\exp\ (j2\pi f_n t) = \cos(2\pi f_n t) + j \cdot \sin(2\pi f_n t)$ is the complex subcarrier of frequency f_n. The orthogonality of subcarriers is ensured, if the distance between neighboring subcarrier frequencies equals

$$\Delta f = \frac{1}{T}, \tag{2.3}$$

where $T = N\Delta t$ is the orthogonality period, and Δt is the sampling interval of signal $\tilde{s}(t)$. Assuming that $f_0 = 0$, the modulated subcarrier frequencies equal $f_n = n\Delta f$. If the sampling frequency equals $1/\Delta t = N\Delta f$, the m-th sample of an p-th OFDM symbol resulting from modulation of the n-th subcarrier equals

$$s_{n,m}^{(p)} = d_n^{(p)} \exp\ (j2\pi f_n m\Delta t) = d_n^{(p)} \exp\ \left(\frac{j2\pi nm}{N} \right). \tag{2.4}$$

The OFDM digital signal samples are composed of $s_{n,m}^{(p)}$:

$$s_m^{(p)} = \frac{1}{\sqrt{N}} \sum_{n=0}^{N-1} \left[d_n^{(p)} \exp\ \left(\frac{j2\pi nm}{N} \right) \right] \tag{2.5}$$

for $m = 0, \dots, N - 1$. Note that (2.5) is the formula for the Inverse Discrete Fourier Transform (IDFT) multiplied by $\sqrt{N}$. (The scaling by $\sqrt{N}$ is introduced in order to obtain the same power of the signal at the output of the modulator as that at its input.) The major conclusion from these formulas is that the bank of modulators and summation necessary to form an OFDM signal can be implemented as IDFT, in particular, using its fast (low-complexity) algorithms IFFT. Moreover, demodulation at the receiver can be implemented using DFT and its low-complexity algorithm (FFT scaled by $\sqrt{N}$ in order to maintain the same input and output power).

It is right that the fast IFFT and FFT algorithms as well as the progress in the design and production of the Very Large-Scale Integration (VLSI) chips implementing them that made the OFDM commercial application in the telecommunication systems possible. The terminology used for signals and their processing in an OFDM system reflects the application of Inverse Fast Fourier Transform (IFFT) in the OFDM transmitter (and FFT in a receiver). Digital data symbols processed before IFFT at the transmitter and appearing at its inputs are called *frequency-domain* symbols, while the signals at its output – *time-domain* samples.

Let us now describe the signal processing at the OFDM transmitter. Because of the presence of the frequency-selective fading and noise, in a radio communication channel, error-correcting coding and interleaving are usually applied in the OFDM transmitter. (Coded OFDM (COFDM) has been described in many papers, e.g., in Refs [103–105].) Before modulating the subcarriers, the sequence of Forward Error-Correction (FEC)-coded data is usually extended by the pilot symbols and control symbols, which are used for channel estimation and carry information related to transmission parameters. A set of some predefined symbols may be transmitted periodically as preamble symbols in order to be used at the receiver for synchronization algorithm. The zeroed guard symbols are often added at the edges of the utilized band in order to adjust the sampling frequency and obtain the sequence adjusted to the used IFFT order. The IFFT block is the central element of the OFDM transmitter implementing multicarrier modulation using orthogonal subcarriers.

The sequence of OFDM signal samples at the output of IFFT is extended by the CP in front of the sequence, which consists of a number of the repeated samples taken from the end of the sequence. This CP is introduced at the transmitter and omitted at the receiver for the received signal at the input of the FFT to be a cyclic (not linear) convolution of the transmitted signal and the channel impulse response. Therefore, this results also in the Inter-Symbol Interference (ISI) removal (there is no overlapping of the received symbols), although the symbols can be still distorted by the channel. Thus, the CP duration (in samples) should be equal to the channel impulse response (in samples) minus one sample. Additionally, it is possible to apply a cyclic postfix and windowing to the transmitted sequence of samples in order to limit the power of the OOB spectrum components.

Finally, the transmit signal is Digital-to-Analog (D/A) converted and fed to the Radio Frequency (RF) stage, which may include signal filtering in order to satisfy the Spectrum Emission Mask (SEM) requirements. The block diagram of a typical OFDM transmitter is presented in Figure 2.1.

The OFDM receiver is presented in Figure 2.2. Reception of an analog signal involves RF filtering, low-noise amplification, demodulation to the Base-Band (BB) (RF/BB conversion), and Analog-to-Digital (A/D) conversion. The OFDM signal samples together with CP are used for time and frequency synchronization algorithms. Next, the CP is discarded, and the time-domain samples of the received signal are demodulated in the FFT block. Due to the application of the CP at the transmitter and its removal at the receiver, ISI is eliminated, and the OFDM equalizer consists of N one-tap equalizers that are supposed to equalize the distortions introduced by the frequency-selective channel to distinct subcarriers. Further signal processing consists in demapping of symbols, deinterleaving of bits, and FEC decoding.

The major effect of the application of multiple carriers to parallel data streams is the extended duration of an OFDM symbol. In the appropriately designed OFDM system, the duration of the channel impulse response is significantly shorter than the OFDM symbol duration, so that the multipath

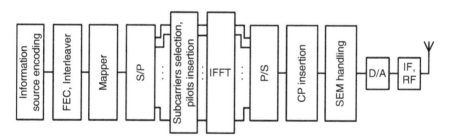

Figure 2.1 The diagram of a typical OFDM transmitter. S/P - serial-to-parallel conversion, P/S - parallel-to-serial conversion.

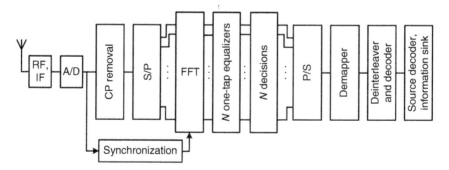

Figure 2.2 The diagram of a typical OFDM receiver.

signal components (echoes) appear within the guard interval consisting of CP. As mentioned earlier, omitting the CP at the receiver has the effect of the linear convolution of the transmitted block of samples (OFDM symbol) with the channel impulse response being equivalent to the cyclic convolution of the two vectors of samples and makes the channel equalization much less complex at the receiver than in case of single-carrier transmission.

The OFDM transmission has higher spectral efficiency than single-carrier transmission without any spectrum shaping. However, in practical applications, it is usually required to apply some spectrum shaping methods, either to protect a single-carrier system from extensive ISI or to respect the SEM requirements. Such practical systems, providing the same data rates, have comparable spectral efficiencies.

2.2 Nonlinear Distortions in Multicarrier Systems

One of the fundamental problems of Multi-Carrier (MC) systems is the occurrence of high-amplitude samples that can be present in Time Domain (TD) signal at the output of the modulator, resulting in a high value of the PAPR. The high value of PAPR in connection with the presence of nonlinear elements in the transmission chain, for example, power amplifiers, results in the in-band and out-of-band distortions, as well as in Bit Error Rate (BER) degradation at the receiver. Thus, efficient methods to minimize the PAPR value are required.

The problem of high PAPR value concerns MC signals rather than single-carrier ones. The vector of data symbols (in frequency domain, i.e., at the input of the IFFT at the transmitter) **d** consists of random variables (data symbols) that are usually assumed as mutually independent and identically distributed, so a time-domain signal sample can be treated as a weighted sum of independent variables. Based on the central limit theorem [106], when the length of the input vector **d** is high, the real and imaginary parts of the output samples after IFFT block can be approximated as the Gaussian distributed random variables. Thus, the envelope of the output samples has the Rayleigh distribution with the mean value equal to $\sigma\sqrt{\frac{\pi}{2}}$ and variance equal to $\frac{4-\pi}{2} \cdot \sigma^2$, that is, the probability density function of variable s being the signal sample at any sampling moment equals

$$f_s(s) = \frac{s}{\sigma^2} \cdot \exp\left(\frac{-s^2}{\sigma^2}\right), \tag{2.6}$$

where σ is the standard deviation of the Gaussian distribution. The real and the imaginary parts of the TD samples are assumed to have their mean value equal to zero and the variance of σ^2. The instantaneous signal power is chi-squared distributed with two degrees of freedom [107–110]. It can be concluded that the most probable absolute values of the complex output samples are rather low, but very high peaks may also occur with lower probability.

When the PAPR value is high, some adverse phenomena can come into play that distort the transmitted signal. If a nonlinear element is placed in the transmit processing chain (i.e., a power amplifier or a digital-to-analog converter), high-amplitude samples are severely distorted or even clipped. The resulting in-band distortion and the rotation of the signal constellation cause the increase of the BER at the receiver. The example tests show [103] that for the coded OFDM case applied in the DVB-T, clipping of around 0.1% of the time leads to BER degradation of around 0.1–0.2 dB in terms of related value of $\frac{E_b}{N_0}$, where E_b and N_0 are the energy per bit in a transmit signal and the noise power spectral density (PSD), respectively. At 1% of clipping, BER degradation is 0.5–0.6 dB. However, BER degradation may not be a decisive factor of system performance in the presence of nonlinear distortions. Notably, clipping, as a nonlinear operation, is the source of out-of-band radiation. The emission of energy outside the nominal transmission band (illustrated in Figure 2.3 as the signal) causes distortion to the neighboring wireless systems and reduces the spectral efficiency. In Figure 2.3, the Clipping Ratio is defined as the ratio between the clipping-threshold value and the square root of the average signal power [111]. In a Digital-to-Analog Converter (DAC), clipping noise occurs due to the limited dynamic range of the converter. However, the problem of high samples already appears in the IFFT block because of the round-off errors and clipping in fixed-point computations. Finally, high-power amplifier is a source of clipping due to its nonlinear input–output characteristic.

Many PAPR definitions can be found in the literature depending on the conditions if the considered signal is finite or infinite and if it is continuous or discrete [112]. Here, let us adopt the PAPR definition as the ratio of the maximum squared absolute value to the mean squared value of the amplitude of TD samples. Thus, PAPR for any vector of signal samples, for example, for OFDM-symbol samples, is defined as

$$\text{PAPR} = \frac{\|\mathbf{s}\|_\infty^2}{\mathbb{E}\{\|\mathbf{s}\|_2^2\}}, \tag{2.7}$$

where $\mathbf{s}$ is the vector of TD samples, $\mathbb{E}\{\cdot\}$ denotes the expectation. Moreover, two mathematic norms that occur in (2.7) are the infinite norm and second-order (Euclidean) norm and denote the maximum absolute value of vector $\mathbf{s}$ and the power of vector $\mathbf{s}$, respectively:

$$\|\mathbf{s}\|_\infty = \max_m |s_m|, \tag{2.8}$$

$$\|\mathbf{s}\|_2 = \sqrt{\sum_m |s_m|^2}. \tag{2.9}$$

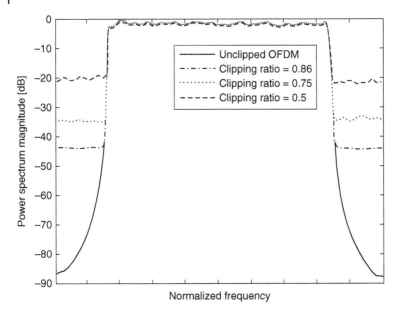

Figure 2.3 The PSD plot of clipped OFDM signal (the effect of out-of-band radiation is also presented).

The PAPR metric defined by (2.7) gives information of how many times the maximal instantaneous power observed in the transmitted data block is higher than the average transmitted power. When the average power is normalized to unity, the PAPR metric is equal to the maximal amplitude of the transmitted sample. Moreover, in the case of normalized power, it is easy to calculate the maximal value of the PAPR metric for the OFDM signal with N subcarriers. In such a case, $\|\mathbf{s}\|_2^2 = 1$, and

$$\text{PAPR} = \|\mathbf{s}\|_\infty^2 = \max_m \left| \frac{1}{\sqrt{N}} \cdot \sum_{n=0}^{N-1} d_n \cdot \exp\left(\frac{j2\pi nm}{N} \right) \right|^2$$

$$\leq \frac{1}{N} \max_m \left| \sum_{n=0}^{N-1} d_n \cdot \exp\left(\frac{j2\pi nm}{N} \right) \right|^2 \leq \frac{1}{N} \cdot N^2 = N \tag{2.10}$$

where d_n is the Frequency Domain (FD) symbol at the n-th input of the MC modulator (IFFT block).

It is often useful to consider the Complementary Cumulative Distribution Function (CCDF) of PAPR, that is, the probability that PAPR of any p-th transmitted symbol is equal or exceeds a certain value – the argument of CCDF:

$$\text{CCDF}(\text{PAPR}_0) = \Pr(\text{PAPR} \geq \text{PAPR}_0) \tag{2.11}$$

as well as the maximal value of PAPR, which can occur in multiple considered blocks of samples:

$$\text{PAPR}_{\text{max}} = \max_{p} \text{PAPR}_{p}, \tag{2.12}$$

where $\text{PAPR}_{p}, p \in (0, 1, \ldots, \infty)$ is the PAPR value in the p-th block of samples. The PAPR metric is one of the most frequently used metrics in defining the system sensitivity to nonlinear distortions. However, when interpreting the actual PAPR value or its CCDF, one cannot extract any information concerning peaks occurring in the transmitted signal that have high amplitudes, that is, above some predefined threshold, but yet below the amplitude of the highest peak. Thus, PAPR metric reflects possible distortions caused by the highest peak only and ignores consequences of the appearance of other peaks. Such an approach can be the reason of drawing wrong conclusions concerning the level of the occurred nonlinear distortions. However, one can find another metric proposition for defining the system sensitivity to nonlinear distortions. The 3GPP forum promotes the Cubic Metric (CM) as a better indication of possible distortions caused by a nonlinear Power Amplifier (PA) than is provided by the PAPR metric [107, 113–115]. It has been observed that the PA voltage gain characteristic can be approximated by means of the third-order polynomial $f(t)$:

$$f(t) = G_1 \cdot s(t) + G_3 \cdot s^3(t), \tag{2.13}$$

where the G_1 and G_3 are the power amplifier model parameters and $s(t)$ is the amplifier input signal. The third-order of the polynomial guarantees the incorporation of the nonlinear terms (i.e., third-order nonlinearities) in the simulation process. To assess this kind of nonlinearities, the CM has also the cubic exponent in their definition:

$$\text{CM} = \frac{\text{RCM} - \text{RCM}_{\text{ref}}}{K_{\text{emp}}}, \tag{2.14}$$

where K_{emp} is a real value that should be empirically defined for each wireless systems (e.g., for Wideband CDMA (WCDMA), it is equal to 1.85, and for the Long Term Evolution (LTE) signal – as a representative of multicarrier signals – it is equal to 1.56 [113]), and RCM is the so-called *Raw Cubic Metric*, defined as follows:

$$\text{RCM} = 20 \cdot \log_{10}\left(\text{rms}\left(\left(\frac{|\mathbf{s}|}{\text{rms}(|\mathbf{s}|)}\right)^3\right)\right), \tag{2.15}$$

while

$$\text{rms}\,(\mathbf{s}) = \sqrt{\frac{1}{N}\sum_{m}|s_m|^2}. \tag{2.16}$$

The RCM$_{ref}$ in (2.14) is the reference value of the RCM defined for WCDMA voice signal equal to 1.52 dB. In the CM definition, all peaks of the transmitted signal that have their amplitudes appropriately high are taken into account. The higher the amplitude of the sample, the higher its impact on the CM value. Let us stress, however, that the RCM incorporates all TD samples in its definition. Thus, to underline the influence of the high samples and to approximate the presence of the third-order nonlinear distortions (resulting due to the processing in the High Power Amplifier (HPA)), the amplitude of the input samples is cubed.

Another metric proposition can be found in [116], where the Instantaneous Normalized signal Power (INP) is used to reflect the nonlinear distortions caused by the PA. It reflects possible distortions caused by a number of samples of the power exceeding the predefined threshold. Finally, the performance of each system, where some algorithms to avoid nonlinear distortions have been applied, can be quantified by means of the so-called *Total Degradation* (TD) metric. This metric measures the degradation of the signal due to the processing through the nonlinear device such as PA. It is then possible to find the optimal value of Input Back-Off (IBO) to minimize the total distortions of the transmit signal.

2.2.1 Power Amplifier Models

An ideal power amplifier should amplify any input signal in linear manner without modifying its phase and without causing any other distortions. In practice, however, the processing characteristic of a PA and especially HPA is not linear. (The processing characteristic of HPA is the relation between the input signal amplitude and the output signal amplitude or output signal phase, denoted as AM/AM and AM/PM, respectively.) Typically, it can be divided into three regions with regard to the value of the amplitude of an input signal A_{in}: the *linear* region (in which the input signal is amplified ideally with the full gain), the *suppression* region (in which the input signal gets distorted and the signal gain is not maximal), and the *compression* region (in which all of the input signal values are clipped to the certain level A_{sat}). The exemplary input–output characteristic of a PA is illustrated in Figure 2.4; also the characteristic of the ideal linear amplifier is presented as the reference. In order to increase the energy efficiency of the PA, its operating point should be as close to the compression region as possible.

Since nonlinear devices (mainly PA) have significant impact on the transmitted signal, efficient and accurate approximation of their input–output characteristic is of high importance. One of the applied models is the so-called *soft limiter* or *ideal clipper* that does not change the phase of the amplified signal and ideally amplifies the input samples if their amplitudes fall within a certain range. If the amplitude of the input sample is higher than the upper

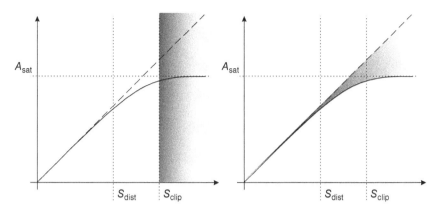

Figure 2.4 Typical AM/AM characteristic of a power amplifier.

bound of this range, the sample will be ideally clipped. In other words, there is no compression region in the soft limiter. Its AM/AM characteristic is presented in Figure 2.5. The soft limiter does not approximate the practical PA because the true characteristic of a PA is not linear even in the range below the saturation level. One of the most common models of practical PA is the so-called Traveling Wave Tube Amplifier (TWTA) [117]. Its example input/output characteristics is illustrated in Figure 2.6. Although the TWTA is widely applied in satellite communications, its practical usage in wireless terminals is limited, mainly due to the nonlinear AM/PM characteristic.

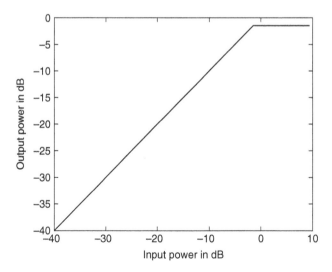

Figure 2.5 The AM/AM characteristic of the *soft-limiter* power amplifier.

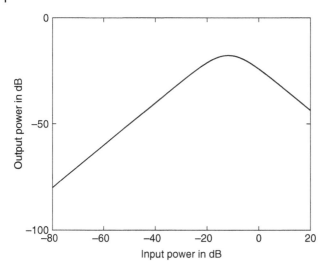

Figure 2.6 The AM/AM and AM/PM characteristic of TWTA.

Usually, another class of PAs are used in wireless terminals, that is, Solid State Power Amplifier (SSPA), whose AM/AM characteristic is almost linear and whose influence on the phase of input data is negligible. In the considerations and experiments discussed later in this book, the *Rapp* model [118, 119] of the SSPA model has been used, whose AM/AM and AM/PM characteristics are defined as follows:

$$g_{\text{AM-AM}}(s) = v\frac{s}{\left(1 + \left(\frac{v|s|}{A_{\text{sat}}}\right)^{2p}\right)^{\frac{1}{2p}}}, \tag{2.17}$$

$$g_{\text{AM-PM}}(s) \cong 0, \tag{2.18}$$

where A_{sat} is the saturation level of the PA, v is the gain of the considered PA, and p is the Rapp-model parameter. Often, p is equal to 2. The AM/AM characteristic of the SSPA for $p = 2$ is presented in Fig. 2.7. When $p \to \infty$, SSPA tends to the ideal clipper. The comparison between various PA classes is presented in [120]

It is worth mentioning here that the PA are often characterized by the so-called IBO parameter, defined as [110]:

$$\text{IBO} = \frac{A_{\text{sat}}^2}{P_{\text{in}}}, \tag{2.19}$$

where P_{in} is the power of the input signal to the PA. Interpretation of this value is as follows: if IBO is high, the nonlinear distortions are negligible (input power is far below the saturation level squared). This also means that the operating

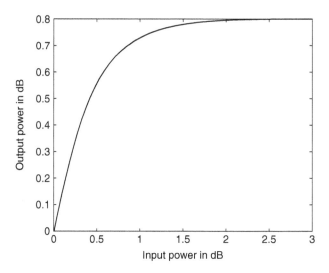

Figure 2.7 The AM/AM characteristic of SSPA using the Rapp model; $p = 2$.

point is far from the optimum and the power efficiency of the PA is low. Contrarily, when the PA operates close to its suppression region, the power efficiency is high; however, the IBO value is low, and the degradation of the signal can be significant.

2.3 PAPR Reduction Methods

Many PAPR reduction methods can be found in the literature [78, 111, 121–123]. Most of them have been designed specifically for OFDM systems. From all of the well-known PAPR reduction methods, some of these techniques can be considered as preferable ones, because they do not require transmission of the so-called side information, containing the information necessary to restore the original data signal intentionally distorted at the transmitter by a PAPR reduction algorithm. Another preference for a PAPR reduction method is to choose the scheme that should not require any modifications in the reception algorithm. These constrains are the consequences of the following pragmatic approach: future multistandard terminals should operate correctly with the legacy standards. This means that the signal received by any wireless standard receiver must be processed as if the standard transmitter generated the transmit signal waveform. There are also other constraints upon the choice of the PAPR reduction scheme, for example, minimization of the computation complexity or the energy consumption. Obviously, any PAPR reduction method can be applied in a multistandard terminal if the application of such a method is explicitly indicated in the standard. Hence, let us focus on solutions that do not require any modifications of the receiver structure. The following methods seem to have potential for the application in future 5G systems:

- Clipping and Filtering (C–F) [110, 124–126],
- Peak Windowing (PW) [127–129],
- Reference Signal Subtraction (RSS) [121]
- Active Constellation Extension (ACE) [79, 108, 109, 130, 131],

In what follows, these methods are shortly described and compared. Additionally, the Selective Mapping (SLM) method has also been presented as a representative method from another group of PAPR reduction techniques (requiring the side information) in order to compare its efficiency with the multistandard suited methods.

Clipping and Filtering

C–F is a very effective technique of PAPR and CM reduction, since very large peaks occur with relatively low probability. The idea of clipping is very simple – all TD samples that have their amplitude higher than some predefined threshold ξ_{tr} are clipped. This operation can be mathematically described as follows [110, 125, 126]:

$$\hat{s}_m = \begin{cases} \xi_{tr} \cdot \exp \left(\jmath \phi_{s_m} \right) & \text{for} \quad |s_m| \geq \xi_{tr} \\ s_m & \text{for} \quad |s_m| < \xi_{tr}, \end{cases} \tag{2.20}$$

where ϕ_{s_m} is the phase of the complex sample s_m. One can see that the phase of the time-domain sample remains unchanged during the clipping operation – only the amplitude is decreased. The process of clipping can be illustrated as multiplication of the original signal by one or a sequence of narrow (spanning over one sample) rectangular pulses. However, clipping of the analog signal is a nonlinear operation that causes the broadening of the signal spectrum [110, 126]. This is because multiplication of signals in the time domain is equivalent to the convolution of their spectra in the frequency domain. Since the rectangular pulse has a very wide spectrum, the result of the aforementioned spectra convolution is much broader than the spectrum of the original transmit signal. The Out-of-Band (OOB) frequency components of the clipped signal are much stronger than those of the original signal. The effect of the OOB radiation should be minimized in order to satisfy the standard requirements concerning the allowed level of signal power in the adjacent frequency bands. One can consider application of clipping in the frequency domain, which preserves the bandwidth of the original signal after clipping. But in such an attempt, the clipping noise would distort the transmitted signal in its nominal band and could not be removed or reduced [111, 132, 133]. To alleviate the distortions introduced by the clipping operation, filtering of the clipped signal is introduced. The FD signal has to be upsampled (additional zero samples are added in the middle of the data block before IFFT) to approximate the analog signal. The upsampling rate should be at least four. After the upsampled signal is clipped, the OOB distortions are filtered (either in the time [126] or in the frequency domain [132]).

Unfortunately, C–F method has some significant drawbacks. First, it degrades the transmitted signal because some of the frequency-domain symbols are shifted on the constellation plane, deteriorating the system performance. Moreover, filtering after clipping can cause peak regrowth, which decreases the efficiency of this method in PAPR reduction. This is why clipping and filtering should be repeated until the required PAPR level is reached. Finally, filtering, especially in the time domain, is relatively computationally complex.

Peak Windowing

To avoid the effects of clipping in a nonlinear device and the out-of-band distortions, another PAPR reduction method has been proposed called PW [127–129]. This technique has relatively low computational complexity. Whereas in the C–F method, the signal is multiplied by narrow (spanning over one sample) rectangular windows in the high peak locations, the main idea of PW is to multiply TD peaks by specially predefined function, for example, Gaussian function. In the peak windowing method, the window shape is selected in such a way that its bandwidth is as narrow as possible to reduce the broadening of the transmitted signal spectrum. The major drawback of this method is its strong influence on the BER characteristics at the receiver. Regardless of this adverse property, this method is often used in commercial systems, for example, in WiFi.

Reference Signal Subtraction

The idea of RSS is to subtract a predefined reference function from the transmitted signal [121]. Because the subtraction operation executed in the TD is equivalent to subtraction of spectra in the FD, the transmit signal spectrum is not broader as in peak clipping and windowing, where multiplication of signals is equivalent to the convolution of their spectra. The amplitude of the subtracted signal is one of the parameters describing the method. The shape of the reference function should be chosen in such a way as to minimize the BER augmentation. One of the possible pulse shapes is the raised cosine function multiplied by the *sinc* function.

Active Constellation Extension

The ACE method originally proposed for OFDM systems is based on the amplitude predistortion of some selected data symbols at the input of IFFT to

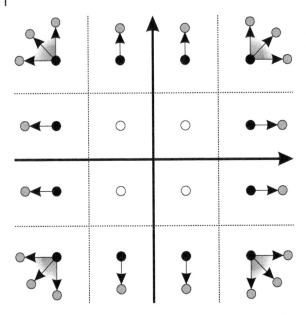

Figure 2.8 The illustration of predistorting capabilities for 16QAM constellation diagram.

● Active constellation point   ○ Fixed constellation point

decrease the PAPR at its output. In this method, only the outer constellation points can be predistorted in order to maintain fixed minimum distance between the constellation points (see Figure 2.8).

In order to find a set of data symbols to be predistorted, the special vector of metrics μ_{nm} has to be calculated. The value of metric μ_{nm} reflects the contribution of the data symbol d_n at the input of the IFFT to the output sample s_m, which has the amplitude higher than the predefined threshold. The threshold level depends on the PA characteristic and should be chosen to minimize the nonlinear distortions introduced by the selected PA and computational complexity of the PAPR reduction method. The algorithm of ACE is the following [108] :

1) Find the set **B** of output samples that have amplitudes higher than the threshold ξ_{tr} and the set **I** of indices of these samples s_m.
2) For every input data symbol $d_n, 0 \leq n \leq N - 1$ and every sample from the set **B**, compute the metric $\mu_n = -\sum_{m \in I} \mu_{nm}$, where $\mu_{nm} = \cos(\phi_{nm}) \cdot |s_m|^{P_{\text{ACE}}}$, where ϕ_{nm} is the angle between the sample s_m $(m \in \mathbf{I})$, and the nth input symbol contribution to the output sample value: $d_n \cdot \exp\left(\frac{j2\pi nm}{N}\right)$. Moreover, $|s_m|^{P_{\text{ACE}}}$ is the weighting function, and p_{ACE} is the parameter of the ACE method.

3) Find S_{ACE} input symbols that have the highest values of the computed metrics (these symbols when predistorted have the strongest influence on reducing the amplitudes of the samples from set **B**), and multiply them by the scaling parameter α (α must be greater than 1).
4) Compute IFFT of the input vector including predistorted symbols to obtain the TD signal waveform with lower PAPR.

In summary, the algorithm finds samples, which after the IFFT operation, are in the opposite phase related to the samples from set **B**. The optimal number of predistorted symbols can be determined with numerical methods.

The drawback of the ACE method is a small increase of the transmit power, so energy effectiveness is decreased, although this power can be easily controlled by the number of distorted symbols S_{ACE} and parameter α. On the other hand, the ACE method has very strong advantage – it does not negatively distort the transmitted signal. Moreover, if the minimum distance between the inner constellation symbols is maintained constant, the predistorted symbols are less sensitive to the channel disturbances (e.g., noise), because the Euclidean distance between them and their neighboring symbols is increased, so BER is lower when compared with the regular OFDM transmission, even in the absence of nonlinear distortions. This, of course, occurs at the expense of power increase.

In order to increase the accuracy and efficiency of the ACE method, the procedure described earlier can be iteratively repeated. Typically, however, only one iteration is assumed. It is worth mentioning that the ACE method has been proposed for application in the new DVB-T2 standard, released in 2009 [55].

Selective Mapping

Finally, let us also briefly describe the SLM technique [134–136]. Its main idea is to represent the same transmitted block in U different forms and choose the representation with the lowest PAPR value for the transmission. In the SLM method, different representations of the same block of data are obtained by multiplication of the transmitted block (before the IFFT operation) by a number of specially defined sequences that usually distort only the phase of the symbols maintaining their amplitudes unchanged. As a result of this operation, different U representations of the same block are created that differ in the phase values of the FD symbols. The SLM method requires transmission of the side information to the receiver about the chosen data representation. It also requires modification of data processing at the receiver. Based on the side information, the received data block is multiplied by the appropriate sequence (the one that has been chosen at the transmitter). The process of SLM is shown in Figure 2.9.

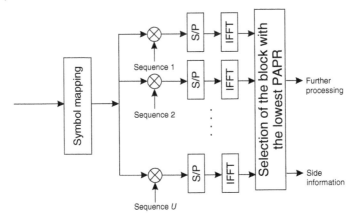

Figure 2.9 The diagram of the Selective Mapping method.

Example Results

Let us examine these standard PAPR reduction methods applied to the OFDM transmission signal. For this purpose, the Rapp model with $p = 2$ of SSPA [118] has been assumed, and the transmission of an OFDM signal with $N = 128$ and QPSK modulation, upsampled four times and interpolated in order to better approximate the analog signal. The results presented in Figures 2.10 and 2.11 show the CCDF of PAPR and of the CM for a number of discussed PAPR reduction methods. For the ACE method, the number of predistorted symbols S_{ACE}

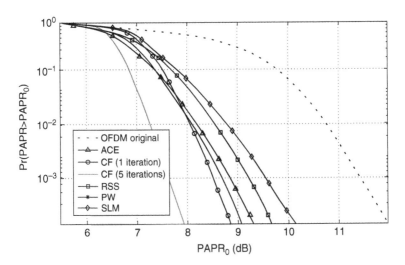

Figure 2.10 The comparative plot of CCDFs for different PAPR reduction methods for PAPR metric.

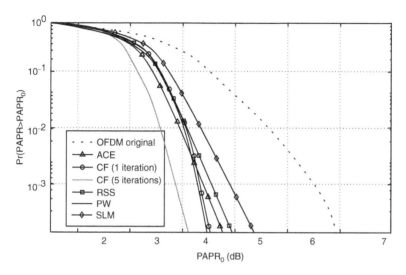

Figure 2.11 The comparative plot of CCDFs for different PAPR reduction methods for Cubic Metric.

has been optimized (S_{ACE} has been equal to around 15%–20% of N). The results indicate that the effectiveness of each method is different with regard to either CCDF of PAPR or CCDF of CM.

Here again, let us stress that the first four presented methods (C–F, PW, RSS, ACE) do not require sending additional information as well as any modification at the receiver, while the SLM method does, and this is why it cannot be considered as particularly suitable for the future generic multistandard terminals that will possibly use the generalized multicarrier waveforms. All the presented PAPR reduction methods, as well as many other ones, have been well described in the literature. The advantages and drawbacks of the presented methods are summarized in Table 2.1.

2.4 Link Adaptation in Multicarrier Systems

Adaptive techniques are being applied in wireless communications to improve the broadly understood efficiency of a telecommunication system. They serve better utilization of available resources (time, frequency, or power) while guaranteeing the Quality of Service (QoS). Some of these techniques are successfully applied in wireless systems, for example, Hybrid Automatic Repeat Request (HARQ) in High Speed Downlink Packet Access (HSDPA) or Adaptive Modulation and Coding (AMC) in LTE, HSDPA, IEEE 802.11 [7], WiMAX [58], or TETRA2 [137, 138]. Adaptive modulation in single-carrier systems is based on the optimal allocation of transmitted power and modulation

Table 2.1 The summary of the selected PAPR reduction method.

Method	Distortion less	Power maintaining	Throughput maintaining	Additional operations in the transmitter	Additional operations in the receiver
Clipping & filtering	×	✓	✓	clipping & filtering	×
Peak windowing	×	✓	✓	Peak windowing	×
Reference signal subtraction	×	✓	✓	Peak detection and subtraction	×
Active constel. Extension	✓	×	✓	IFFT	×
Selective mapping	✓	✓	×	$U \times$ IFFT	Side info. processing

constellation to maximize the transmission data rate, given the instantaneous channel gain and the total transmit power constraint. In general, the optimal bit and power assignment should optimize at least one figure of merit such as the link throughput, bit error rate, energy consumption, or fairness figure. The solution of this classical constraint-optimization problem for independent subbands defined within a given frequency band (e.g., for the OFDM subcarriers) can be found in many books on the information theory, for example, in [66].

Adaptive modulation for frequency-selective channels, originally proposed and applied in OFDM systems, is based on the adaptation of the number of bits assigned to each frequency bin (OFDM subcarrier) with regard to the instantaneous channel characteristic and to the constraint of the total transmit power P_{tot}. In order to find the optimal bit assignment, the available power has to be appropriately distributed over all subcarriers. The solution of this problem leads to the well-known *water pouring* (or *water filling*) principle [66]. In such an approach, the power $P(f_n)$ assigned to the n-th frequency bin in the signal frequency band is equal to the difference between the so-called water-line (water level) W_{level} and the ratio of the noise PSD (which for the Additive White Gaussian Noise (AWGN) equals $\mathcal{N}(f_n) = N_0/2$ for any subcarrier frequency f_n) to the squared absolute value of the channel characteristic at each nth subcarrier: $\frac{\mathcal{N}(f_n)}{|H(f_n)|^2}$. If this difference is negative, the assigned power is set to zero. The water-filling principle of power loading in the frequency domain is illustrated in Figure 2.12.

Let us stress that for bit-and-power loading scheme, with a defined target Bit Error Probability (BEP), the appropriate power assignment should be found

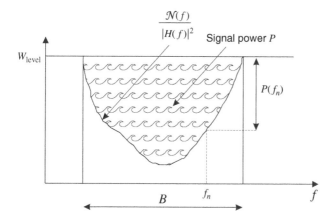

Figure 2.12 Illustration of the idea of water pouring (water filling).

as the difference between the computed water-line W_{level} and the ratio of the noise PSD over the absolute value of the channel characteristic squared and multiplied by the coefficient ρ: $\frac{\mathcal{N}(f_n)}{\rho|H(f_n)|^2}$, where ρ accounts for the assumed BEP.

After finding the optimal power distribution over the available frequency band, the number of assigned bits per each subcarrier (bit loading) has to be calculated according to the following formula:

$$\log_2 [\mathcal{M}(f_n)] = \log_2 \left[1 + \frac{\rho P(f_n)|H(f_n)|^2}{\mathcal{N}(f_n)}\right] \tag{2.21}$$

where $\mathcal{M}(f_n)$ is the constellation order adopted to the n-th subcarrier (of frequency f_n), ρ can be determined for a given constellation scheme, for example, $\rho = -1.5/\ln (5\text{Pr}_b)$ for the QAM constellations, $\mathcal{M} > \in$ and $0 < \text{SNR} < 30$ dB, and Pr_b is the BEP [66].

The aforementioned adaptive technique is called *fast* adaptive modulation. Note that, in this case, the receiver must reliably and rapidly estimate the instantaneous Signal-to-Noise Ratio (SNR), and the appropriate feedback must be sent to the transmitter. To reduce this feedback, model-based channel prediction algorithms can be adopted [139]; however, the model parameters need to be frequently recomputed in the dynamically changing mobile radio channel. This is often impossible in highly mobile environment, and therefore, the pilot symbols are usually transmitted to allow for the Channel State Information (CSI) estimation. On the other hand, the number of transmitted pilots and their energy are limited, which has an impact on the accuracy of the channel estimation (and thus, the quality of CSI and the signal-detection performance). Thus, in practical systems, when the channel gains are estimated based on transmitted pilots, link adaptation schemes are based on incomplete and imperfect CSI. The imperfect CSI is often used for the purpose of the adaptive transmission due to the fact that there is no means to remove an error from its estimation.

As an alternative to the mentioned fast adaptation, *slow* adaptive modulation scheme has been analyzed [140], where modulation parameters are adapted based on the BER performance averaged over a longer period of time. With this technique, the constellation size is changed, tracking the slow variations of the channel, mainly related to the propagation shadowing. Therefore, it requires a slower feedback rate. In [141], the slow QAM adaptation is considered in the mobile scenarios, in which the reception diversity, ambiguous CSI, and energy constraints have to be taken into account.

The choice of the transmission parameters is usually not limited to the symbol-constellation order. In practical systems, the most popular and advantageous adaptation scheme is the AMC. There, apart from the modulation, the error-correcting coding schemes and parameters (rates) are adaptively chosen. The values of parameters representing the options for adaptivity are usually derived off-line from the theoretical analysis of the given transmission scheme performance, and assemble the look-up table called the *code-book*. (Less practical solution, sometimes considered in the literature, would be to define the goal function and to implement the optimization procedure every symbol period.) The code-book constitutes a finite countable set of vectors of the adaptation parameters values. In the conventional AMC scheme, for instance, the code-book consists of pairs of parameters, namely the code rate and the constellation order that satisfy the required target BEP, and maximize the transmission bit rate. These schemes maximizing the bit rate for the constrained transmission power are called *rate-adaptive*. In other considered practical AMC schemes called *margin-adaptive*, the required bit rate and BER are constraints, and the total transmission power is the subject of minimization [66]. This strategy choice is often made for the energy-constrained wireless networks, and the motivations for this approach are as follows: to extend the battery life, to minimize the electromagnetic radiation in populated areas, to reduce the cost in infrastructure-based networks, and to reduce the interference [142].

It is up to the system designers to decide the adaptation strategy of a particular system. Note that in many standardized systems, the available flexibility is limited to the choice of one power level and one coding scheme for a group of subcarriers. The reason for choosing one coding scheme for a group of subcarriers is that the applied code should support robustness to subcarrier fading. Moreover, the code word should be long enough for the coding scheme to be useful within the channel coherence time.

The AMC schemes for multicarrier systems have been studied for various types of digital modulations, codes, pilot patterns, feedback channels, and so on. Moreover, these schemes have been considered for multiuser adaptive Orthogonal Frequency-Division Multiple Access (OFDMA) (e.g., in Ref. [143])

and extended to the multicarrier Multiple Input, Multiple Output (MIMO) transmission schemes (e.g., in Ref. [144]). The literature on multicarrier AMC is extremely rich; let us just mention an interesting paper on the trends in AMC [142] and a comprehensive book on this topic [100].

2.5 Reception Techniques and CFO Sensitivity

One of the advantages of OFDM systems is simple, yet efficient channel equalization in the frequency domain as visible in Figure 2.2. However, in order to allow for successful signal detection, a number of issues have to be solved, for example, efficient receiver front-end design in order to provide sufficient SINR, synchronization algorithms achieving sufficient alignment in time and in frequency and accurate channel estimation. Additionally, the OFDM receiver design should account for the effects of fixed-point arithmetic that provides another source of distortions and errors [145].

The reception starts in the analog front end where RF band of interest is observed, amplified, and down-converted to the BB. There are various receiver front-end architectures proposed in the literature [119]. Apart from the radio-channel distortions, two effects are expected in the OFDM received signal: (i) nonlinear distortion caused by high PAPR and nonlinear amplifier characteristic (as explained in the previous sections) and (ii) white noise power increase caused by nonzero Noise Figure. The mixers used for frequency conversion can introduce the so-called In-Phase and Quadrature (IQ) imbalance as well as the Local Oscillator (LO) leakage. As the LO leakage is observed at the 0-th subcarrier, it is a common approach to keep this carrier unmodulated at the transmitter and reject all the interference at this subcarrier at the receiver. However, there are some algorithms for removing these distortions in digital postprocessing [146]. A very important issue is the appropriate A/D selection and the input-signal level adjustment. High resolution of A/D is required in order to reliably represent the wanted signal of high dynamic range. This should prevent clipping of peak samples and maintain the quantization noise below the expected thermal noise floor. The A/D resolution should additionally digitalize the signal in the presence of other, high-power signals (e.g., NBI or adjacent channel signals not perfectly filtered out at the preceding reception stages). On the other hand, high resolution of A/D results in higher power consumption and higher cost. It is therefore essential to control the analog signal power level in order to use optimal A/D working point.

2.5.1 Synchronization

The synchronization algorithms have to detect an OFDM symbol/frame beginning sample (time synchronization) and CFO (frequency synchronization) in order to allow for reliable signal detection. In order to show the effect of time/frequency misalignment on received symbols, let us consider a single transmitted OFDM symbol consisting of samples s_m for $m \in \{-N_{CP}, \ldots, N-1\}$ (the OFDM-symbol index p from (2.5) is omitted for simplicity). It can be calculated using IDFT transform as

$$s_m = \frac{1}{\sqrt{N}} \sum_{n=0}^{N-1} d_n e^{j2\pi \frac{nm}{N}}, \tag{2.22}$$

where d_n is QAM/PSK symbol transmitted at subcarrier indexed n. Such a time-domain signal passes through the multipath channel of L taps with impulse response $h(l)$ for $l \in \{0, \ldots, L-1\}$. Assuming that Carrier Frequency Offset (CFO) equals v (normalized to subcarrier spacing), the received signal is

$$r_m = \sum_{l=0}^{L-1} h(l) s_{m-l} e^{j2\pi \frac{mv}{N}}. \tag{2.23}$$

For clarity, temporarily the white noise influence is neglected. Let us denote the time instant of the receiver DFT window beginning as $\hat{m}$. The *optimal timing point* $\hat{m} = 0$ is a time instant when the CP ends and N-samples of the main part of an OFDM symbol are aligned with the receiver DFT window (as shown in Figure 2.13 a)). Because of the use of CP of N_{CP} samples, the ISI is not expected to deteriorate the OFDM symbol reception for $\hat{m} \in \langle -N_{CP} + L - 1, 0 \rangle$. Otherwise, that is, when $\hat{m} \notin \langle -N_{CP} + L - 1, 0 \rangle$, the reception window spans partially the preceding or the following OFDM symbol samples. In this later case, interference is expected of the power proportional to the power of adjacent symbol samples observed within the reception window. This problem is shown in Figure 2.13 b.

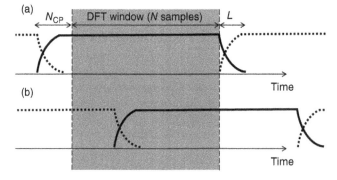

Figure 2.13 Illustration of time synchronization error effect: a) ideal time synchronization, b) ISI from previous OFDM symbol.

The DFT result at subcarrier n' can be formally presented as

$$R_{n'} = \frac{1}{\sqrt{N}} \sum_{m=0}^{N-1} r_{m+\hat{m}} e^{-j2\pi \frac{mn'}{N}}.$$ (2.24)

For the ISI-free region, that is, for $\hat{m} \in \langle -N_{\mathrm{CP}} + L - 1, 0 \rangle$, formula (2.22) and (2.23) can be substituted in (2.24) giving

$$R_{n'} = \frac{1}{\sqrt{N}} \sum_{m=0}^{N-1} \sum_{l=0}^{L-1} h(l) \frac{1}{\sqrt{N}} \sum_{n=0}^{N-1} d_n e^{j2\pi \frac{(m-l+\hat{m})n}{N}} e^{j2\pi \frac{(m+\hat{m})v}{N}} e^{-j2\pi \frac{mn'}{N}}$$

$$= \frac{1}{N} \sum_{n=0}^{N-1} d_n e^{j2\pi \frac{\hat{m}(n+v)}{N}} H_n e^{j\pi \left(1-\frac{1}{N}\right)(n+v-n')} \frac{\sin(\pi(n+v-n'))}{\sin\left(\pi \frac{n+v-n'}{N}\right)},$$ (2.25)

where

$$H_n = \sum_{l=0}^{L-1} h(l) e^{-j2\pi \frac{ln}{N}}$$ (2.26)

is DFT of the channel impulse response, that is, channel frequency response. Assuming no CFO, that is, for $v = 0$, the aforementioned formula simplifies to

$$R_{n'} = d_{n'} e^{j2\pi \frac{\hat{m}n'}{N}} H_{n'}.$$ (2.27)

It is visible that for $\hat{m} \in \langle -N_{\mathrm{CP}} + L - 1, 0 \rangle$, a phase change proportional to the subcarrier index occurs. From the receiver perspective, this phase change is estimated together with the channel frequency response, that is, $\hat{H}_n = e^{j2\pi \frac{\hat{m}n}{N}} H_n$, and reversed by the channel equalizer. Thus, no degradation of the reception performance should be observed. Interestingly, $\hat{m}$ is not limited to integer values, that is, time misalignment of a fraction of sampling period is also acceptable.

If we neglect the timing offset, that is, $\hat{m} = 0$, (2.25) simplifies to

$$R_{n'} = \frac{1}{N} \sum_{n=0}^{N-1} d_n H_n e^{j\pi \left(1-\frac{1}{N}\right)(n+v-n')} \frac{\sin(\pi(n+v-n'))}{\sin\left(\pi \frac{n+v-n'}{N}\right)}.$$ (2.28)

In this case, $R_{n'}$ is dependent not only on $d_{n'}$ but also on all other active subcarriers, that is, ICI occurs. Observe that $\sin\left(\pi \frac{n+v-n'}{N}\right)$ for small arguments (that is typical case) can be approximated by $\pi \frac{n+v-n'}{N}$ giving

$$R_{n'} \approx \sum_{n=0}^{N-1} d_n H_n e^{j\pi \left(1-\frac{1}{N}\right)(n+v-n')} \mathrm{Sinc}(\pi(n+v-n')).$$ (2.29)

For a relatively small CFO value v, for example, $|v| < 1$, a wanted symbol d_n is mostly interfered by adjacent symbols: d_{n-1}, d_{n+1}. Additionally, considering

$n = n'$, the wanted signals (on all subcarriers) are rotated by a common phase $\pi(1 - \frac{1}{N})v$.

The considerations presented earlier show that OFDM reception performance is dependent on precise synchronization both in time and in frequency. There is a number of synchronization algorithms proposed in the literature [147]. The two most commonly used approaches use a special OFDM symbol called preamble (e.g., the Scmidl&Cox algorithm presented in [148]) or CP in each OFDM symbol (e.g., the one presented in [149]).

The preamble-based synchronization can be exemplified by a well-established Schmidl&Cox (S&C) algorithm [148]. It uses a preamble generated by modulating only even-indexed subcarriers by nonzero symbols (other subcarriers are modulated by zeroes). In the time domain, it results in two equal sequences of samples to be transmitted, that is,

$$s_m = s_{m+\frac{N}{2}}, \tag{2.30}$$

for $m = \left\{ -N_{\text{CP}}, \dots, \frac{N}{2} - 1 \right\}$. Observe that because of the CP utilization, there are $N_{\text{CP}} + \frac{N}{2}$ pairs of identical samples distanced by $\frac{N}{2}$ sampling periods. Most importantly, the CP utilization preserves this repetition after passing multipath channel. To show this, let us temporarily assume no CFO ($v = 0$). In such a case, (2.23) simplifies to

$$r_m = \sum_{l=0}^{L-1} h(l)s_{m-l}. \tag{2.31}$$

Observation of samples $r_{m+\frac{N}{2}}$ with the assumption given in (2.30) results in

$$r_{m+\frac{N}{2}} = \sum_{l=0}^{L-1} h(l)s_{m+\frac{N}{2}-l} = \sum_{l=0}^{L-1} h(l)s_{m-l} = r_m, \tag{2.32}$$

for $m - l = \left\{ -N_{\text{CP}}, \dots, \frac{N}{2} - 1 \right\}$. Obviously, this range is dependent on the value of l. The range of m values for which this repetition is maintained, for all channel taps, is $m = \left\{ -N_{\text{CP}} + L - 1, \dots, \frac{N}{2} - 1 \right\}$. In case of the presence of CFO, and still considering the no-noise condition, we can write

$$r_{m+\frac{N}{2}} = r_m e^{j\pi v} \tag{2.33}$$

for $m = \left\{ -N_{\text{CP}} + L - 1, \dots, \frac{N}{2} - 1 \right\}$.

The S&C algorithm requires the autocorrelation to be calculated according to the following formula:

$$A_m = \sum_{m'=0}^{\frac{N}{2}-1} r^*_{m+m'} r_{m+\frac{N}{2}+m'}. \tag{2.34}$$

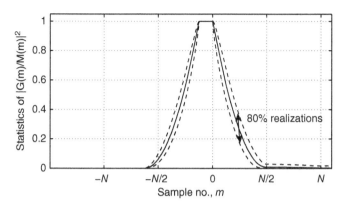

Figure 2.14 Timing synchronization metric of the S&C algorithm. Median (solid line) and the 10th and the 90th percentiles (dashed lines) are shown.

For $m = \left\{ -N_{\mathrm{CP}} + L - 1, \ldots, \frac{N}{2} - 1 \right\}$ all components add in-phase, giving

$$A_m = e^{j\pi v} \sum_{m'=0}^{\frac{N}{2}-1} |r_{m+m'}|^2. \tag{2.35}$$

Additionally, the received energy over the correlation window is

$$\mathcal{E}_m = \sum_{m'=0}^{\frac{N}{2}-1} |r_{m+m'}|^2. \tag{2.36}$$

Finally, the timing synchronization metric is

$$\hat{m} = \arg \max_{m} \left| \frac{A_m}{\mathcal{E}_m} \right|^2. \tag{2.37}$$

The nominator in (2.37) is maximized for $m = \left\{ -N_{\mathrm{CP}} + L - 1, \ldots, \frac{N}{2} - 1 \right\}$ when all summed components add in phase as shown in (2.35). In the absence of preamble within autocorrelation window, all components should add with random phases. An example of timing synchronization metric for a random CFO value and random symbols modulating preamble is shown in Figure 2.14.

The frequency synchronization is achieved by estimating the phase of $A_{\hat{m}}$:

$$\hat{v} = \frac{1}{\pi} \arg \{A_{\hat{m}}\}. \tag{2.38}$$

An important advantage of the S&C synchronization is its low computational complexity. Both functions in the nominator and denominator of the timing metric can be obtained iteratively as

$$A_m = A_{m-1} - r^*_{m-1}r_{m+\frac{N}{2}-1} + r^*_{m+\frac{N}{2}-1}r_{m+N-1}, \tag{2.39}$$

$$\mathcal{E}_m = \mathcal{E}_{m-1} - |r_{m-1}|^2 + |r_{m+\frac{N}{2}-1}|^2. \tag{2.40}$$

The CP-based synchronization algorithm [149] does not require a preamble to be present in the transmitted OFDM frame. It utilizes the fact that CP conveys repeated samples of the end of an OFDM symbol. The timing and frequency offsets estimates are obtained as follows [147]:

$$\hat{m} = \arg\max_m |A_{\text{CP}m}|, \tag{2.41}$$

$$\hat{v} = \frac{1}{2\pi}\arg\{A_{\text{CP}\hat{m}}\}, \tag{2.42}$$

where

$$A_{\text{CP}m} = \sum_{m'=0}^{N_{\text{CP}}-1} r^*_{m+m'-N_{\text{CP}}}r_{m+m'-N_{\text{CP}}+N}. \tag{2.43}$$

Performance of this metric is better, the longer is CP (the higher N_{CP}) in comparison to the multipath channel delay spread in samples. Additionally, it can be improved by averaging $A_{\text{CP}m}$ over a number of consecutive OFDM symbols, that is, delayed by $N + N_{\text{CP}}$ samples.

2.5.2 Channel Estimation and Equalization

One of the advantages of OFDM is relatively simple detection using N one-tap equalizers (one per subcarrier) in the properly designed system, in which application of CP at the transmitter and its omission at the receiver eliminate ISI. In a properly designed OFDM system, a single-subcarrier bandwidth should be a few times smaller than the channel coherence bandwidth, duration of the OFDM symbol should be shorter than the channel coherence time, and the CP duration should be at least as long as the channel delay spread spread minus one sample period. These conditions simplify the channel estimation and symbol detection. Similar requirements are stated when designing the reference symbols (e.g., pilots or preambles).

Let us assume transmission of data symbols d_n of variance σ^2 according to (2.22), flat fading in each subchannel defined around each subcarrier n with the frequency response H_n, and the presence of AWGN. In case of perfect time and frequency synchronization, (2.25) changes to

$$R_n = d_n H_n + W_n, \tag{2.44}$$

where W_n is the value of AWGN transformed to the frequency domain (by DFT applied at the receiver) at subcarrier n and of variance σ^2_W. The aforementioned equation can be written in the vector form as

$$\mathbf{R} = diag(\mathbf{d})\mathbf{H} + \mathbf{W}, \tag{2.45}$$

where $\mathbf{R} = [R_0, \dots, R_{N-1}]^T$, $\mathbf{H} = [H_0, \dots, H_{N-1}]^T$, $\mathbf{W} = [W_0, \dots, W_{N-1}]^T$, and $diag(\mathbf{d})$ denotes diagonal matrix with elements of vector $\mathbf{d} = [d_0, \dots, d_{N-1}]^T$ on its main diagonal.

The simplest approach to channel estimation is to transmit symbols d_n known at the receiver, that is, pilot symbols. Usually, a subset of subcarriers is devoted to transmit the pilots and their position subcarriers repeatedly change in each OFDM symbol. This way, the channel can be estimated after reception of a number of symbols. Because of the channel dynamics, this is a continuous process. For now, let us assume that at a certain point of time, all pilot symbols are known at the receiver. The Least-Squares approach [150] can be used to estimate the channel frequency response $\hat{\mathbf{H}}$ by the following optimization:

$$\hat{\mathbf{H}} = \arg \min_{\overline{\mathbf{H}}} \| \mathbf{R} - diag(\mathbf{d})\overline{\mathbf{H}} \|^2, \tag{2.46}$$

where $\overline{\mathbf{H}}$ is any considered channel frequency response estimation vector. The result of the aforementioned optimization is the following:

$$\hat{\mathbf{H}} = diag(\mathbf{d})^{-1}\mathbf{R}. \tag{2.47}$$

In the scalar form, it is

$$\hat{H}_n = \frac{R_n}{d_n}. \tag{2.48}$$

This estimate can be further improved by means of utilizing the knowledge on the maximal length of the channel impulse response $(L \leq N_{CP})$ [150]. Let us define a truncated DFT $N \times L$ matrix $\mathbf{F}_L$ as $F_{L n,l} = e^{-j2\pi \frac{nl}{N}}$ for $l = 0, \dots, L-1$ and $n = 0, \dots, N-1$. The channel impulse response $\mathbf{h} = [h_0, \dots, h_{L-1}]^T$ can be mapped to the frequency response as

$$\mathbf{H} = \mathbf{F}_L\mathbf{h}. \tag{2.49}$$

The Least-Squares estimate of $\mathbf{h}$ is

$$\hat{\mathbf{h}} = (diag(\mathbf{d})\mathbf{F}_L)^{\dagger}\mathbf{R} = (\mathbf{F}_L^{\mathcal{H}} diag(\mathbf{d})^{\mathcal{H}} diag(\mathbf{d})\mathbf{F}_L)^{-1} \mathbf{F}_L^{\mathcal{H}} diag(\mathbf{d})^{\mathcal{H}}\mathbf{R}, \tag{2.50}$$

where $\mathbf{X}^{\dagger}$ denotes pseudoinverse of matrix $\mathbf{X}$, and $\mathbf{X}^{\mathcal{H}}$ denotes Hermitian transposition of matrix $\mathbf{X}$. It can be mapped to frequency domain as

$$\hat{\mathbf{H}} = \mathbf{F}_L\hat{\mathbf{h}} = \mathbf{F}_L(\mathbf{F}_L^{\mathcal{H}} diag(\mathbf{d})^{\mathcal{H}} diag(\mathbf{d})\mathbf{F}_L)^{-1} \mathbf{F}_L^{\mathcal{H}} diag(\mathbf{d})^{\mathcal{H}}\mathbf{R}. \tag{2.51}$$

Observe that for $L = N$, this equation simplifies to (2.49). The considered Least-Squares approximation does not utilize information on the long-term channel statistics. For improved performance, the MMSE channel estimator can be used [150].

Equalization of the received signal is performed in the frequency domain, typically with a single-tap equalizer for each subcarrier defined by G_n coefficients. The estimated data symbol $\hat{d}_n$ is obtained as

$$\hat{d}_n = G_n R_n. \tag{2.52}$$

The simplest approach is to use Zero Forcing (ZF) equalizer defined as

$$G_n = \frac{1}{\hat{H}_n}. \tag{2.53}$$

By combining two aforementioned equations with (2.44), we obtain

$$\hat{d}_n = \frac{H_n}{\hat{H}_n} d_n + \frac{W_n}{\hat{H}_n}. \tag{2.54}$$

Even for perfect channel estimation, that is, $H_n = \hat{H}_n$, the noise power can be significantly increased for a given faded subcarrier, that is, when $|\hat{H}_n| \approx 0$. An approach to find G_n that is more suitable for low SNR values is the Minimum Mean Square Error (MMSE) approach optimizing

$$G_n = \arg \min_{\overline{G}_n} \mathbb{E}[|\overline{G}_n R_n - d_n|^2] = \arg \min_{\overline{G}_n} \mathbb{E}[|\overline{G}_n H_n - 1|^2 |d_n|^2 \tag{2.55}$$

$$+ 2\mathfrak{R}\{(\overline{G}_n H_n - 1) d_n \overline{G}_n^* W_n^*\} + |\overline{G}_n|^2 |W_n|^2], \tag{2.56}$$

where $\overline{G}_n$ is any equalizer coefficient value at subcarrier n considered in the optimization process, and $(\cdot)^*$ denotes complex conjugate. Assuming that $\mathbb{E}[|W_n|^2] = \sigma_W^2$, $\mathbb{E}[|d_n|^2] = \sigma^2$ and $\mathbb{E}[d_n W_n^*] = 0$ the optimization problem simplifies to

$$G_n = \min_{\overline{G}_n}[|\overline{G}_n R_n - 1|^2 \sigma^2 + |\overline{G}_n|^2 \sigma_W^2]. \tag{2.57}$$

The solution can be obtained by calculating the derivative of the function to be minimized and finding its root (for simplicity, separate derivatives over the real and imaginary parts of G_n can be calculated). The result is

$$G_n = \frac{H_n^*}{|H_n|^2 + \frac{\sigma_W^2}{\sigma^2}}. \tag{2.58}$$

For high SNR (SNR $= \frac{\sigma^2}{\sigma_W^2}$), the MMSE equalizer coefficients approach those of the ZF equalizer.

Apart from the most popular ZF and MMSE equalizers, there are other ideas for channel equalization in multicarrier systems, for example, such as *controlled equalization* [151], *partial equalization* [152], or *maximal ratio combining* [153] considered for multicarrier Code-Division Multiple

Access (CDMA). The literature on channel estimation and equalization for multicarrier systems is really vast. In the books dedicated to multicarrier systems [98–102], an interested reader can find chapters devoted to channel estimators and equalizers, including their practical implementation, complexity, and issues specific to multicarrier technologies.

3

Noncontiguous OFDM for Future Radio Communications

The Non-contiguous Orthogonal Frequency-Division Multiplexing (NC-OFDM) technique is considered as a strong candidate for air-interface definition in the future communication systems with cognitive capabilities. It possesses sufficient spectral agility in order to facilitate the transmission of data from unlicensed-system users (Secondary Users (SUs)) across several fragmented frequency bands simultaneously even in the presence of licensed-system users (Primary Users (PUs)) signals, thus, resulting in an increase of spectrum utilization [3, 4, 36–38]. In particular, SCs located in the frequency vicinity of unoccupied wireless spectrum can be used for transmitting data while those subcarriers that could potentially interfere with nearby PU signals can be deactivated or nulled. However, simply deactivating the subcarriers for the purposes of Out-of-Band (OOB) interference mitigation may not be sufficient for the neighboring PU interference tolerance level. In addition to achieving a required level of OOB emission power within a given spectrum mask, an SU transmitter must be capable of tailoring its spectral characteristics dynamically in order to avoid interference with the dynamically changing incumbent PU transmissions [4].

In order to the evaluate spectrum properties of the NC-OFDM waveform, let us consider the NC-OFDM transmitter diagram in Figure 3.1. The input data bits are mapped into α complex Quadrature Amplitude Modulation (QAM) symbols for each NC-OFDM symbol denoted as vector $\mathbf{d}_{DC}$. These symbols are Serial-to-Parallel (S/P) converted and fed to appropriate inputs of Inverse Fast Fourier Transform (IFFT) block of size N (within NC-OFDM subcarrier (SC) selection block). This block implements multicarrier modulation. Although the set of all possible SC indices is $\{-N/2, \ldots, N/2 - 1\}$, only SC of indices in $\mathbf{I}_{DC} = \{I_{DCj}\}$ (where $j = 1, \ldots, \alpha$) are modulated by nonzero symbols. After IFFT, N_{CP} samples of the Cyclic Prefix (CP) are inserted, and symbol samples s_m ($m = -N_{CP}, \ldots, N - 1$) can be described as

$$s_m = \sum_{n=-\frac{N}{2}}^{\frac{N}{2}-1} \tilde{d}_n e^{j2\pi \frac{nm}{N}},$$ (3.1)

Advanced Multicarrier Technologies for Future Radio Communication: 5G and Beyond, First Edition.
Hanna Bogucka, Adrian Kliks, and Paweł Kryszkiewicz.

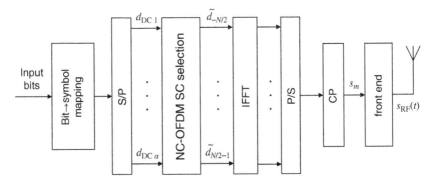

Figure 3.1 The diagram of basic NC-OFDM modulator.

where $\tilde{d}_n$ is the value of a data symbol for $n \in \mathbf{I}_{\mathrm{DC}}$ and otherwise equals zero. These samples are subject to the Digital-to-Analog (D/A) conversion (with D/A converter modeled as low-pass filter of impulse response $h_{\mathrm{DAC}}(t)$) providing continuous-time representation

$$s(t) = \sum_{m=-N_{\mathrm{CP}}}^{N-1} s_m h_{\mathrm{DAC}}(t - mT_{\mathrm{s}}), \tag{3.2}$$

where T_{s} is the sampling period. This signal is up-converted to carrier frequency f_{c}, and only its real part can be transmitted via antenna port giving

$$s_{\mathrm{RF}}(t) = \Re\{s(t)e^{j2\pi f_{\mathrm{c}}t}\}, \tag{3.3}$$

where $\Re$ denotes real part of a complex number. The spectrum of the continuous-time-domain signal $x(t)$ can be obtained by means of Fourier transform, that is,

$$S(f) = \mathcal{F}\{s(t)\} = \mathcal{F}\left\{\left[\sum_{m=-N_{\mathrm{CP}}}^{N-1} s_m \delta(t - mT_{\mathrm{s}})\right] * h_{\mathrm{DAC}}(t)\right\} =$$

$$\mathcal{F}\left\{\sum_{m=-N_{\mathrm{CP}}}^{N-1} s_m \delta(t - mT_{\mathrm{s}})\right\} H_{\mathrm{DAC}}(f) =$$

$$= H_{\mathrm{DAC}}(f) \int_{-\infty}^{\infty} \sum_{m=-N_{\mathrm{CP}}}^{N-1} \sum_{n=-\frac{N}{2}}^{\frac{N}{2}-1} \tilde{d}_n e^{j2\pi\frac{nm}{N}} \delta(t - mT_{\mathrm{s}})e^{-j2\pi ft}dt =$$

$$= H_{\mathrm{DAC}}(f) \sum_{n=-\frac{N}{2}}^{\frac{N}{2}-1} \tilde{d}_n \sum_{m=-N_{\mathrm{CP}}}^{N-1} e^{j2\pi\left(\frac{nm}{N}-fmT_{\mathrm{s}}\right)}, \tag{3.4}$$

where $\mathcal{F}$ denotes Fourier transform, $*$ denotes convolution, $H_{\mathrm{DAC}}(f)$ is the frequency response of the DAC-modeling low-pass filter, and $\delta(t)$ denotes the Dirac delta function. Essentially, the second factor is the digital-domain NC-OFDM symbol spectrum, infinitive in frequency and periodic with period $1/T_s$. The spectrum of a single nth subcarrier $S(n, v)$ can be calculated at normalized frequency v by substitution of $f = \frac{v}{NT_s}$, giving [43]

$$S(n, v) = \sum_{m=-N_{CP}}^{N-1} e^{j2\pi \frac{n-v}{N}m} = \frac{e^{j2\pi \frac{n-v}{N}(-N_{CP})} - e^{j2\pi(n-v)}}{1 - e^{j2\pi \frac{n-v}{N}}} =$$

$$e^{j\pi(n-v)\left(1 - \frac{1+N_{CP}}{N}\right)} \cdot \frac{\sin\left(\pi(n-v)\left(1 + \frac{N_{CP}}{N}\right)\right)}{\sin\left(\pi \frac{n-v}{N}\right)} \tag{3.5}$$

which was simplified by the means of summing the geometric sequence and trigonometric identities. Now formula (3.4) can be rewritten as

$$S(f) = H_{DAC}(f)S(v) = H_{DAC}(f) \sum_{n=-\frac{N}{2}}^{\frac{N}{2}-1} \tilde{d}_n S(n, v), \tag{3.6}$$

where $S(v) = S(fNT_s)$ is the spectrum of all modulated subcarriers in a single NC-OFDM symbol observed at frequency v in the digital domain. Although function $S(n, v)$ is infinitive in frequency, the D/A converter should attenuate the components outside the frequency range from $-1/(2T_s)$ to $-1/(2T_s)$ and cause negligible distortion to in-band components. From this perspective, further investigations can be limited to the in-band frequencies, and H_{DAC} can be omitted.

Essentially, the single-subcarrier spectrum model given in (3.5) assumes digital realization of the OFDM modulation (by means of IFFT) and as such is different than commonly considered *Sinc*-like spectrum, for example, in [154–157]. The *Sinc*-like spectrum is an approximation of the one provided in (3.5) for large N and $|n - v| \approx 0$. In [158], this simplification was used on purpose to simplify optimization. As shown in [159], the Power Spectral Density (PSD) of a signal can be estimated by its periodogram. Assuming that d_n has zero mean and unitary power, the PSD of a single subcarrier can be estimated as

$$|S(n, v)|^2 = \left| \frac{\sin\left(\pi(n-v)\left(1 + \frac{N_{CP}}{N}\right)\right)}{\sin\left(\pi \frac{n-v}{N}\right)} \right|^2$$

$$\approx \left| \frac{\sin\left(\pi(n-v)\left(1 + \frac{N_{CP}}{N}\right)\right)}{\pi \frac{n-v}{N}} \right|^2, \tag{3.7}$$

while the approximation is valid for $|\pi \frac{n-v}{N}| \approx 0$. The maximum is observed at $v = n$, and the region $(n - v) \in (-0.5; 0.5)$ is called the *main lobe*, while the components at other frequencies are called *sidelobes*. It is visible that the envelope of PSD equals approximately $\frac{1}{(\pi(n-v)/N)^2}$, so the sidelobes decrease by 20 dB per decade.

To counteract the potential for significant OOB interference resulting from the NC-OFDM transmission, which can negatively affect the neighboring wireless signals, several techniques have been proposed in the literature that are designed to significantly suppress these sidelobes in order to make coexistence between PUs and SUs feasible. On the other hand, the OOB power reduction process can potentially increase the computational complexity and energy (power) utilization. Given the possible constraints of limited computational and energy resources available at a user equipment, a practical approach to this problem that achieves a balance between the OOB interference mitigation efficiency and its associated costs is needed. Next, a short overview of such practical approaches is presented. (Fragments of this overview have been published by us in [4].)

Digital Filtering (DF)

The most straightforward approach is to filter out the OOB components. However, this approach does not use the properties of an NC-OFDM modulator and causes major practical issues. The filter has to be adjusted to the occupied band and has to be redesigned as the allocated frequency resources change. Moreover, low-order filters can distort the useful signal, without providing sufficient OOB radiation reduction, while high-order filters have high complexity and can contribute to the time-domain signal dispersion, so that the joint channel-filter impulse response spans beyond the CP [160]. As exemplified in [39], to obtain the spectrum notch of 24 dB, the power consumption of a Finite Impulse Response (FIR) filter is 300 mW. Such suppression achieved by Cancellation Carrier (CC) method (described as follows) needs only 2 mW of power.

Guard Subcarriers (GS)

A simple OOB interference reduction method is used to reserve a number of SU subcarriers lying closest to the PU frequencies to serve as a spectral buffer [161], that is, deactivation of those subcarriers decreases OOB power as described by formula (3.5) because $|n - v|$ increases assuming that n represents utilized SU NC-OFDM subcarrier and v is within the PU band. This is called the GS method. Although simple to implement, this method significantly decreases the spectral efficiency and does not provide sufficient OOB power reduction in most scenarios [4].

Windowing (WIN)

WIN is a method usually applied to the Orthogonal Frequency-Division Multiplexing (OFDM) symbol time-domain samples containing CP [161, 162]. The

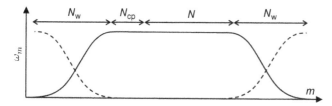

Figure 3.2 Windowed OFDM symbols diagram; dashed lines denote the preceding and the following OFDM symbol. (Based on [4].)

concept of windowing is shown in Figure 3.2 [4]. There, one can see that the time-domain OFDM symbol of duration $N + N_{CP}$ samples is extended cyclically with N_W samples at each end of the considered symbol. Let us denote the time-domain signal of a single OFDM symbol as vector $\mathbf{s} = \{s_{-N_W-N_{CP}}, \dots, s_{N-1+N_W}\}$. In such a case, the OFDM-symbol time-domain samples after windowing are defined by vector $\mathbf{y}_{WIN} = \{y_{WINm}\}$, which results from the multiplication of the vector $\mathbf{s} = \{s_m\}$ by the window shape $\mathbf{w} = \{w_m\}$ [4], namely

$$y_m = w_m s_m \,, \tag{3.8}$$

where $m \in \{-N_W - N_{CP}, \dots, N - 1 + N_W\}$. In [162], it has been shown that the largest sidelobe suppression is achieved when the Hanning window is applied. In such a case, $\mathbf{w} = \{w_n\}$ has the following form [4]:

$$w_m = \begin{cases} 0.5 + 0.5 \cos\left(\pi \frac{m + N_{CP}}{N_W}\right), & m \in \{-N_W - N_{CP}, \dots, -1 - N_{CP}\} \\ 1, & m \in \{-N_{CP}, \dots, N-1\} \\ 0.5 + 0.5 \cos\left(\pi \frac{m - N}{N_W}\right), & m \in \{N, \dots, N - 1 + N_W\} \end{cases} \cdot \tag{3.9}$$

As visible in Figure 3.2, consecutive symbols overlap each other by N_W samples, which results in an effective OFDM symbol duration of $N + N_{CP} + N_W$ samples. Although, it was shown in [163] that windowing can be applied without prolonging an NC-OFDM symbol, such an approach causes an increase of Inter-Carrier Interference (ICI) and Bit Error Rate (BER).

The advantages of windowing as the OOB reduction method are as follows: relatively low computational complexity, independence of the modulated data, and suitability for noncontiguous multicarrier transmission, for example NC-OFDM. In [164], the performance of WIN in OOB power reduction has been confirmed in Software-Defined Radio (SDR) implementation. When employed by a Cognitive Radio (CR) communication system attempting to access the available spectrum in a dynamically varying radio environment, it is also important that the length and shape of the applied window can also be altered dynamically. This method is the most suitable for minimizing the interference in the PU transmission that is relatively distant in frequency from SU transmission band [4, 161, 162]. The main drawback of this method is the decrease of throughput caused by increased duration of a single OFDM symbol.

Adaptive Symbol Transition (AST)

In the AST method [4, 165], similarly to WIN, the time-domain samples in the transition region between consecutive symbols are chosen in order to minimize the OOB power by *smoothing* the transition between consecutive symbols. For the AST algorithm, the information about symbols mapped to each subcarrier is needed in order to assess the amount of OOB interference in the neighboring frequency bands. The Mean Squared Error (MSE) minimization method is used to determine dynamically, that is, independently for each OFDM symbol, the values of the time-domain samples in the transition region. The primary drawbacks of this method are as follows: high computational complexity and reduced throughput (by prolonging the NC-OFDM symbol) [4].

Constellation Expansion (CE)

Another method, called CE [4, 166], adjusts the modulated data symbols transmitted per subcarrier so that the OOB power can be reduced while simultaneously not losing any data information or causing distortion. This is achieved by enlarging the modulation constellation and by allowing data symbols to be represented by any one of the two constellation points. As a result, the minimum distance between the constellation points is reduced, and the BER performance decreases [4]. Moreover, the optimization is computationally complex, and the resultant OOB power reduction is limited.

Subcarrier Weighting (SW)

Another method, called SW [4, 167, 168], minimizes the signal OOB power level by multiplying the data subcarriers by optimized real, positive weighting coefficients [4]. At the receiver, data symbols transmitted using the weighted subcarriers can be viewed as distorted, particularly for the high or low values of the weighting coefficients. Consequently, the authors in [167, 168] suggest to impose a constraint on the weighting coefficients' values. Simulation results exhibit moderate OOB power suppression and the BER increase while optimization is quite computationally complex.

Multiple-Choice Sequences (MCS)

In the MCS [4, 169] method, for each sequence of data symbols to be transmitted in an OFDM symbol, a set of corresponding sequences representing it is calculated. The sequence yielding the lowest interference to adjacent bands is then chosen from this set and transmitted. To retrieve the initial data sequence at the receiver, the identification number of the selected sequence has to be provided, which requires an additional control channel for this side information. The authors presented three variants of MCS generation: by changing constellation symbols, by symbol interleaving, and symbol phase rotation. Variants of the MCS method with reduced computational complexity are the Constellation Adjustment method [170] and the Phase Shift method [156]. As the OFDM edge subcarriers have the strongest influence on the OOB radiation, only constellation points or symbol phase of those subcarriers are altered. This limits the number of possible sets of sequences to be examined in order to choose the optimum one [4]. Another variant of the MCS method involves its merging with other spectrum shaping algorithms, that is, SW or Cancellation Carriers (CCs) as shown in [171].

Polynomial Cancellation Coding (PCC)

The PCC has been proposed in [4, 172] and revisited in [173, 174]. This method not only reduces the OOB power but also lowers the OFDM signal sensitivity to phase and frequency errors [4]. As neighboring subcarriers have firmly aligned spectra, the adjacent subcarriers are modulated by the same, appropriately scaled data symbol in order to reduce the sidelobe power. This is usually done for the groups of two or three subcarriers. Although this method reduces the system throughput, the coding redundancy can be used to increase the Signal-to-Noise Ratio (SNR). Although PCC has very low computational complexity requirements, the OOB power is mostly reduced when CP is not used.

Active Set (AS)

In [175], the AS method has been proposed. It allows each data symbol to be modified by an additive complex symbol calculated in order to decrease the OOB radiation power. As this modification, from the receiver point of view, results in noise-like distortion, its power has to be limited. All the QAM symbols after spectrum shaping belong to the same decision region as that before shaping. The optimization process is computationally complex, and strong OOB power attenuation requires serious degradation of BER in the receiver.

Spectrum Precoding (SP)

Recently, many authors have proposed to use different kinds of SP. While all these methods provide strong OOB power reduction, typically they all involve relatively high computational complexity [176]. In this method, the correlation between the data symbols transmitted on subcarriers is introduced by complex-field block coding. The approach in [177] not only reduces the OOB power but also decreases BER. However, it requires some redundancy tones to be used; thus, lower bit rate is obtained. Additionally, computationally complex reception is required. All precoding matrixes derived in [157, 158, 178–181] are based on orthogonal projection operation. While [158] can have quite low computational complexity, it works best in zero-padded OFDM. While the reception performance in [178] is the same as in a typical OFDM system, the OOB power reduction performance is steered by the number of redundancy tones (for the code rate equal to 1, no OOB radiation reduction is observed). In the case presented in [157], the code rate is equal to 1, and the stronger the OOB power reduction, the more spectrum-sampling points in the OOB region are defined. However, the BER floor is observed for high SNR values. The improved orthogonal projection operation proposed in [180] was used for multiuser system in [179]. Moreover, the authors propose iterative approach to precoding matrix designed in order to make the resultant spectrum obey the given SEM. While [179] requires computationally complex reception, the precoder in [181] is suitable for a standard NC-OFDM receiver. It is done by limiting Error Vector Magnitude (EVM) introduced by precoding. In this approach, the OOB power is most reduced at the frequencies mostly distanced (in frequency) from the occupied band. It might be therefore not suitable for the creation of deep and narrowband spectrum notch.

N-continuous OFDM (N-OFDM)

Another approach similar to SP but very distinctive and commonly cited is N-OFDM method [182]. It has been observed that the OOB power components are the result of the time-domain noncontinuity between subsequent OFDM symbols. The proposed method aims at forcing consecutive symbols to be continuous in time. The continuity of the 0th to the N_{N-OFDM}th-order derivatives at the ends of the OFDM symbols is achieved by adding low power, complex-valued quantities to each active data subcarrier at the input of an IFFT block . The stronger the OOB radiation attenuation, the higher derivative order is maintained continuous. The main difficulty here is the memory-based precoder that makes precoding computationally complex and reliable signal reception difficult. Moreover, the strongest OOB power attenuation is observed at the frequencies more distanced from the occupied frequencies band. It makes this method not suitable when narrow and deep PSD notches are required. Recently, in [183], an improved version of the N-OFDM precoder and the adequate reception method have been proposed, which improve BER to the level of a typical OFDM signal reception, but at the cost of reduced bit rate, that is, the coding rate is below 1. The N-OFDM technique also has the potential to be implemented with reduced computational complexity by precoding in time domain [184]. While computational complexity of the standard N-OFDM signal generation is $\mathcal{O}(\alpha^2)$, the time-domain approach typically requires $\mathcal{O}(N \cdot N_{N-OFDM})$. As typically N_{N-OFDM} is small, the proposed method allows its implementation in real-time transmitters. The computational complexity of signal reception is the same as in the standard approach. Moreover, N-OFDM has been designed based on continuity of continuous-time OFDM representation. The implementation utilizing the Field Programmable Gate Array (FPGA) reported in [185] suggests that the N-OFDM transmitter requires high oversampling factor of the signal to obtain low OOB power.

Cancellation Carriers (CCs)

The CC method [4, 41] takes advantage of the spectrum shape of each subcarrier and applies some additional, specifically modulated subcarriers (CCs) in order to reduce the resulting OOB power level [4]. This method has particularly promising capabilities in OOB power reduction and is discussed in the next section in greater detail.

Extended Active Interference Cancellation (EAIC)

Another method for reducing OOB power is called the EAIC [186, 187]. Similarly to Active Interference Cancellation (AIC), it is based on the insertion of special carriers that are designed to negatively combine with high-power sidelobes caused by the data subcarriers. While the AIC method utilizes only subcarriers orthogonal to the DC, EAIC uses nonorthogonal frequencies as well. The main drawback of this method results from this lack of orthogonality, that is, data-symbol distortion occurs. A variant of the EAIC method has been presented in [188], where the sidelobe-power suppression approach is improved by using a long time-domain cancellation signal spanning over a number of consecutive OFDM symbols [4]. indexOFDM This method results in an increase of BER due to increased interference relative to the method presented in [186]. In [187] improvement of the EAIC method is presented, which bases on the constraint optimization when fixing the allowable self-interference power. Although the obtained OOB power attenuation is significant, it requires a high number of EAIC subcarriers, which is the reason for high computational complexity.

Filter-Bank Multi-Carrier (FBMC)

In contrast to the previously mentioned algorithms, the Filter-Bank Multicarrier (FBMC) technique is a multicarrier modulation scheme that is a competitor to the NC-OFDM. It is commonly considered a successor to OFDM for the future wireless communication systems [3, 37, 42]. Its transmitter can be efficiently implemented by means of an IFFT block that is appended with filtering. Because of the precise design of filters, each subcarrier introduces a very low level of interference into its OOB region. Additionally, CP is not needed in this scheme that rapidly increases the throughput in comparison to

NC-OFDM. The drawback of this scheme is high computational complexity of signal generation and reception, because of filtering, as well as a multitap equalizer needed for each subcarrier. Additionally, PAPR in a FBMC system may be higher than for OFDM systems, depending on the applied filters. (In the best case, Peak-to-Average Power Ratio (PAPR) can be the same as for OFDM.)

3.1 Enhanced NC-OFDM with Cancellation Carriers

An OFDM symbol has limited time duration, and therefore, this can be interpreted as cutting out a part of an infinitely long OFDM symbol by a rectangular window [4]. Thus, each subcarrier spectrum is convolved with a *Sinc* function, which results in widening the spectral overlapping regions with the other subcarrier spectra as shown in (3.5). Although this is generally the primary reason for the existence of high OOB power, this phenomenon can be used positively by the CC method, where a subset of active subcarriers are selected for the sole purpose of canceling the OOB interference of the adjacent subcarriers [4]. It is typically assumed that the subcarriers closest to the spectrum edge have the strongest influence on the OOB power; therefore, they are usually chosen to be modulated by the canceling signal values, which are not independent data symbols to be transmitted. The sidelobes of these subcarriers are intended to negatively combine with the sum of the sidelobes resulting from the active data-bearing subcarriers, thus, potentially reducing the overall OOB power levels as shown in Figure 3.3 [4].

The values of the cancellation subcarriers have to be calculated for each OFDM symbol separately since the independent modulated data symbols cause different OOB power levels. Thus, several frequency-sampling points are defined in order to determine the values of the OOB signal spectrum at the corresponding frequencies. These γ frequency-sampling points describe the optimization region in which the estimates of the spectrum values resulting from the spectral superposition of the Data Carriers (DCs) and the CC have to be calculated. The optimization problem to be solved for each OFDM symbol can be defined as follows [4]:

$$\mathbf{d}_{CC}^{\star} = \arg\min_{\mathbf{d}_{CC}} \|\mathbf{P}_{CC}\mathbf{d}_{CC} + \mathbf{P}_{DC}\mathbf{d}_{DC}\|_2^2, \tag{3.10}$$

where $\mathbf{P}_{CC}$ is the matrix of dimensions ($\gamma \times \beta$) transforming the vector of the CC values $\mathbf{d}_{CC}$ of length β to the spectrum estimates. For DCs, the matrix $\mathbf{P}_{DC}$ of dimensions ($\gamma \times \alpha$) and vector $\mathbf{d}_{DC}$ of length α have the same role. However, as the inventors of this method have found, such an optimization approach may result in a higher power level for the CCs relative to mean DC power. Consequently, the additional constraint has been introduced to limit this effect [4]:

$$\|\mathbf{d}_{CC}\|_2^2 \leq \Pi_{CC}, \tag{3.11}$$

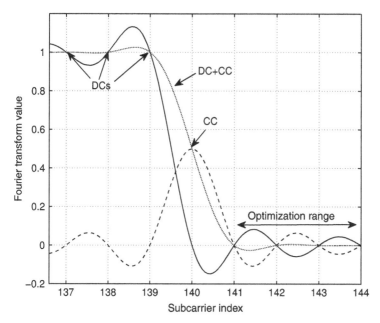

Figure 3.3 The simplified diagram of spectrum resulting from superposition of data subcarriers spectrum and the cancellation subcarrier spectrum for a single OFDM symbol; no CP used. (Based on [4].)

where Π_{CC} is the maximum allowable power for CCs. Although the solution of (3.10) is widely known, and has been presented in [4, 39, 40], that is,

$$\mathbf{d}^{\star}_{CC} = -\mathbf{P}^{\dagger}_{CC}\mathbf{P}_{DC}\mathbf{d}_{DC}, \tag{3.12}$$

where $\mathbf{A}^{\dagger}$ denotes pseudoinverse of matrix $\mathbf{A}$, the constraint (3.11) increases the computational complexity of the optimization problem significantly, requiring to solve the Lagrange inequality for each OFDM symbol, which might become infeasible for wideband transmissions possessing a large number of subcarriers [4]. In [189], an algorithm for solving this problem is provided in the chapter *LS Minimization over a Sphere.*

Another drawback of the CC method, apart from the computational complexity, is the link-performance deterioration, that is, an increase of BER. This is due to the fact that an OFDM system usually operates under the total power constraint. If part of an OFDM symbol energy is sacrificed to the cancellation carriers, the remaining energy that can be used for data transmission is reduced, and this naturally results in an SNR loss and corresponding BER degradation [4]. It is important to mention that the increased number of subcarriers used by the OFDM system causes increase of PAPR. However, in practice, this is rather acceptable, for example, about 0.2 dB for 10^{-2} of PAPR Complementary Cumulative Distribution Function (CCDF) [154].

The CC algorithm has been extensively investigated, and a number of modifications and combinations of the CC algorithm with other methods has been presented in the literature [4]. For example, AIC [39] is a method similar to CC, where in addition to the OFDM edge subcarriers, several other subcarriers inside the PU transmission band are also used to minimize the OOB radiation. However, as shown in the aforementioned paper, the AIC subcarriers inside the PU transmission band have a negligible influence on the OOB interference. Moreover, they can significantly increase the computational complexity of the resulting implementation [4].

Another approach to optimization, based on Genetic Algorithm, was presented in [190]. This method is highly computationally complex, that is, the execution time measured during simulation is more than 100 times longer than that with standard solution and enables less than 1 dB improvement in the OOB power level.

Far less computationally complex is the method proposed in [191]. The complexity is reduced over the standard approach [154] via simplifying the most laborious steps of the CC method. First of all, spectrum value caused at a given frequency-sampling point v by DCs is established by taking into account only the most contributing subcarriers. To further reduce the system complexity, each CC is supposed to nullify the interference in just one frequency-sampling point. Thus, optimization is reduced to seeking for a CC coefficient that guarantees that its sidelobe has the same value as the DC-based spectrum value but with opposite sign for the given frequency-sampling point. It is expected that the OOB power is much less reduced than in the case of a standard optimization approach. A number of different approaches to AIC calculation are provided in [192]. The authors utilize statistical dependency between subcarriers in the OOB region to reduce complexity. Additionally, $min–max$ optimization problem is solved, which aims at reducing the maximum OOB power over all defined frequencies in the optimization range rather than summarized OOB power value. Moreover, the AIC algorithm has been extended with limitation of the CC power as in (3.11). Although advanced math tools were used, the improvement in the system performance in comparison to [39, 154] is rather modest.

In case of very deep spectrum notches requirement, it is possible to combine the CC method with WIN as presented in [40, 154]. The basic OFDM system is equipped additionally with a CC insertion unit placed after mapping symbols to subcarriers and windowing applied to an extended OFDM symbol in time domain. While there is no change needed in the case of the windowing method, the CC insertion block has to take into account the shape of single-subcarrier spectrum changed by the window. Dependent on the used window type and its length, the formula (3.5) has to be changed appropriately. Simulation results show that the combination of CC and WIN methods leads to a significant decrease of the OOB radiation power compared to windowing or CC method used separately.

Another method, proposed in [193], is low computationally complex combination of the CC algorithm and the CE method. The obtained OOB power attenuation is quite weak, that is, about 5 dB in the simulated cases. Strong OOB power attenuation, especially in narrow notches, can be achieved by combination of the CC and AST methods [194]. The combination of the CC method with digital filtering has been considered in [155]. It uses the fact that CC can very effectively reduce the OOB power components lying close in frequency to the occupied subcarriers band. The components lying further away in frequency can be effectively reduced by relatively low-order raised-cosine filtering. Although strong OOB power attenuation is achieved, this approach is still quite computationally complex, and BER degradation caused by filtering is observed.

3.1.1 Reception Quality Improvement for Cancellation Carrier Method

As mentioned in the previous section, one of the drawbacks of the CC method is the BER increase in the NC-OFDM receiver. If considering a given transmit power, introduction of CC requires lower amount of power to be utilized for DC. However, it can be observed that complex symbols transmitted on CC are calculated based on the DC values in a given NC-OFDM symbol. From this perspective, complex values on CC can be treated as redundancy introduced by complex-field block coding [195]. In order to utilize this fact, we start with derivation of the coding matrix used in the transmitter that allows for calculation of decoding matrix to be used in the receiver [196].

The transmitter considered here consists of a modified NC-OFDM modulator as shown in Figure 3.4. The "CCs calculation" unit calculates the CC values for each NC-OFDM symbol separately (for each data symbol vector $\mathbf{d}_{DC}$) and

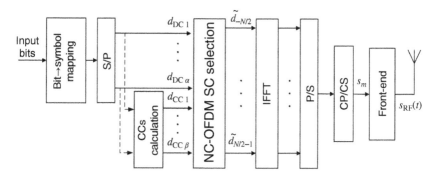

Figure 3.4 NC-OFDM transmitter applying the CC method.

is employed prior to the N-size IFFT block. The output of an IFFT block is extended with N_{CP} samples of CP in the time domain. Essentially, now the total number of occupied subcarriers is $\alpha + \beta$. As defined previously, DC occupy subcarriers indexed by the elements of vector $\mathbf{I}_{DC} = \{I_{DCj}\}$ for $j \in \{1, \ldots, \alpha\}$, and CCs occupy subcarriers indexed by the elements of vector $\mathbf{I}_{CC} = \{I_{CCj}\}$ for $j \in \{1, \ldots, \beta\}$. Both I_{DCj} and I_{CCj} belong to the set $\{ -N/2, \ldots, N/2 - 1\}$.

It is required to estimate the influence of each DC and CC on the resultant spectrum in a number of spectrum-sampling points. The CC complex symbols are to be chosen in order to make their sidelobes add negatively to the sidelobes caused by DC at these frequencies. For a set of frequency-sampling points $\mathbf{V} = \{V_l\}$ for $l \in \{1, \ldots, \gamma\}$ defined in the optimization region, and for $j \in \{1, \ldots, \alpha\}$, the elements of the $\gamma \times \alpha$ matrix $\mathbf{P}_{DC}$ can be defined as $P_{DCl,j} = S(I_{DCj}, V_l)$, where $S(n, v)$ is defined in (3.5). Similarly, for the cancellation carriers, when $j \in \{1, \ldots, \beta\}$, the elements of the $\gamma \times \beta$ matrix $\mathbf{P}_{CC}$ can be defined as $P_{CCl,j} = S(I_{CCj}, V_l)$ and can be calculated off-line.

Recall that the aim of CCs is to minimize the OOB power level, which implies solving the following optimization problem [4]:

$$\mathbf{d}_{CC}^\star = \arg\min_{\mathbf{d}_{CC}} \|\mathbf{P}_{CC}\mathbf{d}_{CC} + \mathbf{P}_{DC}\mathbf{d}_{DC}\|_2^2. \tag{3.13}$$

The solution of this problem yields the values of cancellation carriers $\mathbf{d}_{CC}$, namely

$$\mathbf{d}_{CC}^\star = -\mathbf{P}_{CC}^\dagger \mathbf{P}_{DC}\mathbf{d}_{DC}, \tag{3.14}$$

which is the same as in (3.12). Although such a solution is relatively fast with respect to computational complexity, since the multiplication of vector $\mathbf{d}_{DC}$ by a precalculated matrix is performed for each NC-OFDM symbol, the power of CCs is not limited, which can cause serious BER increase in the NC-OFDM receiver. As shown in [41], the decrease of the SNR value can be defined as

$$\text{SNR}_{loss} = 10 \log_{10} \left(\frac{\|\mathbf{d}_{CC}\|_2^2 + \|\mathbf{d}_{DC}\|_2^2}{\|\mathbf{d}_{DC}\|_2^2} \right). \tag{3.15}$$

The reference system in this case is the one that employs the guard subcarriers on the subcarriers used by the CC method in the proposed system.

As the CC are correlated with data symbols, these additional subcarriers can be used in the signal reception that might not only regain the power devoted to these subcarriers in the first place but also make use of the frequency diversity for achieving a higher degree of robustness with respect to the frequency-selective fading. It is possible, thanks to maintaining the matrix $\mathbf{W}$ constant for a given set of system parameters [4]. Let us consider (3.14) as a process of generating redundancy symbols $\mathbf{d}_{CC}$ transmitted in parallel to data symbols $\mathbf{d}_{DC}$. This operation is conducted on the complex symbols, thus allowing us to employ the theory of complex-field block codes [195] for this problem formulation. To do so, let us rewrite (3.14) in order to determine the systematic code generation matrix $\mathbf{G}$ of size $(\beta + \alpha) \times \alpha$:

$$\begin{pmatrix} \mathbf{d}_{DC} \\ \mathbf{d}_{CC} \end{pmatrix} = \begin{pmatrix} \mathbf{I} \\ \mathbf{W} \end{pmatrix} \mathbf{d}_{DC} = \mathbf{G}\mathbf{d}_{DC}. \tag{3.16}$$

By changing the row order of the presented matrix, the order of data and cancellation symbols can be unchanged, but for simplicity's sake, this operation will be omitted. A simple reception mechanism designed for such codes is based on the ZF criterion [195], for which the reception matrix is defined as [4]

$$\mathbf{R}_{ZF} = (\mathbf{HG})^{\dagger}, \tag{3.17}$$

where $\mathbf{H}$ is $(\beta + \alpha) \times (\beta + \alpha)$ diagonal matrix with channel coefficients for each of the used subcarriers on its diagonal. Although more advanced reception schemes are proposed in [195], they introduce additional computational complexity with no significant improvement in BER in comparison to reception using $\mathbf{R}_{ZF}$ matrix. This matrix should be used in the receiver after the Fast Fourier Transform (FFT) processing instead of an equalizer used in standard reception chain [4]. The estimate of the data symbols $\hat{\mathbf{d}}_{DC}$ is achieved by the following operation:

$$\hat{\mathbf{d}}_{DC} = \mathbf{R}_{ZF} \begin{pmatrix} \mathbf{r}_{DC} \\ \mathbf{r}_{CC} \end{pmatrix}, \tag{3.18}$$

where $\mathbf{r}_{DC}$ and $\mathbf{r}_{CC}$ are the received vertical vectors at the output of FFT block containing distorted and noisy values of data and cancellation subcarriers, respectively. Although the calculation of matrix $\mathbf{R}_{ZF}$ can be quite complex, it needs to be performed only once for each channel instance and subcarrier pattern. Moreover, with a systematic code implementation, this method may be treated as optional, reserved only for high-performance, high-quality reception [4].

Example Results

The evaluation of the system performance improvement caused by the reception technique utilizing CC discussed earlier has been carried out via computer simulations. The system that has been considered is based on FFT/IFFT of size $N = 256$. It operates in the multipath Rayleigh fading channel with maximal path delay equal to 62 samples, so CP of length $N/4$ is used. The channel consists of four paths whose relative gains are as follows: 0, −3, −6 and −9 dB. If not stated differently, the system spans $\alpha = 11$ DC modulated by Quadrature Phase-Shift Keying (QPSK) symbols and 2 CC at each side of the used band, that is, $\beta = 4$. The CC optimization is based on $\gamma = 76$ sampling points (38 for each side of occupied band) spanned evenly in the band of 20 subcarriers next to the used spectrum (10 for each band side).

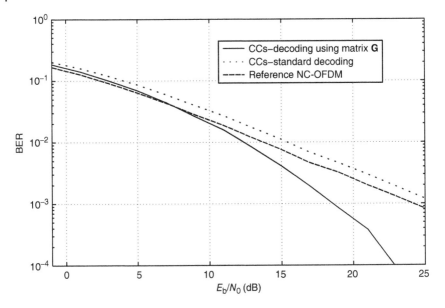

Figure 3.5 BER vs E_b/N_0 obtained in multipath Rayleigh fading channel without CCs (reference), with CCs and standard detection, and with CCs utilizing coding matrix for symbol detection.

The BER curves are shown in Figure 3.5, for three different systems based on the one described earlier.

The simulation has been carried out for 1 100 000 random input bits for each of 2000 randomly generated channel instances. The reference curve relates to the system performance that uses all 11 DCs, but the CCs always equal to zero, so that a system with guard subcarriers is obtained. Although such a system does not provide sufficient OOB power reduction, it has the whole OFDM symbol energy devoted to DCs, so it is a kind of a lower limit for the BER performance of a system with CCs. The figure presents the results of the standard ZF equalization technique, that is, when the complex symbols received on CCs are omitted in the receiver, for the system applying complex-field code decoding according to (3.18).

The results show that, while standard decoding of CCs loses for BER $= 10^{-3}$ about 1.8 dB in SNR values in comparison to the reference system, the new decoding provides 7.2 dB SNR improvement in comparison to standard CC decoding leading to 5.4 dB enhancement of SNR in reference to the system with guard subcarriers. The frequency diversity provided by CCs used as complex-field code improves the performance for high SNR values that is similar to the results shown in [195]. In the low-SNR region (SNR < 7 dB), the

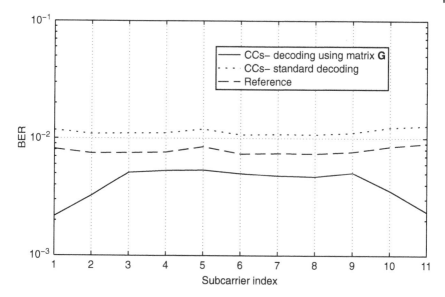

Figure 3.6 BER vs. subcarrier index obtained in multipath Rayleigh fading channel without CCs (reference), with CCs and standard detection, and with CCs and proposed detection. $E_b/N_0 = 15$ dB.

performance of the system with proposed decoding, as expected, is a bit worse than for the reference one.

Some information on the mechanism of the observed quality improvement is given by Figure 3.6, where separate BERs for each of used DCs are drawn. The simulation parameters are as follows: 550 000 input bits for each of 6000 randomly generated channels. The E_b/N_0 value chosen for this figure is 15 dB. There are two additional plots for standard decoding of CC and the reference system that are aimed at the comparison. Although the standard CC decoding and the reference system have both flat BER characteristics, the new CC reception technique improves the quality of data tones placed near the occupied spectrum edges, that is, closest to CC location, more than those located in the middle. This is caused by the highest influence of these tones on the OOB power so that most power of CC is correlated with them and then recovered in the receiver.

Next, the system has been evaluated for varying number of DC. Computer simulations have been carried out for a system similar to the one previously used. The only difference is in the number of DC that changes from 1 to 30 for $E_b/N_0 = 15$ dB. The results are shown for the transmission of 5000 OFDM symbols over 1000 random instances of Rayleigh fading channel. The results for the obtained BER are shown in Figure 3.7a. Although the curve for the reference system is flat in the whole observed range (as expected), the standard

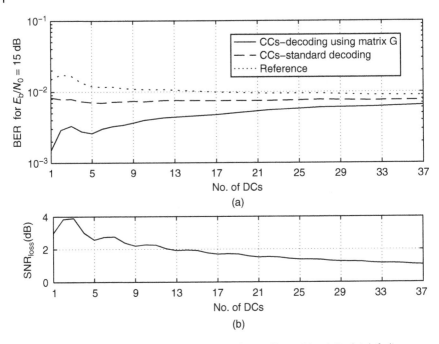

Figure 3.7 (a) BER versus the number of DCs (α) obtained in multipath Rayleigh fading channel without CCs (reference), with CCs and standard detection, and with CCs utilizing coding matrix for detection. (b) SNR loss caused by CCs according to (3.15). $E_b/N_0 = 15$ dB.

decoding of OFDM symbol carrying CCs has high values when there are just a few DCs. It is the situation when most of the symbol's energy is devoted to CCs and none of it is used for the reception. It is confirmed by Figure 3.7b showing the loss in SNR caused by the introduction of CCs. For the same reason, the decoding used in data detection for the CC method achieves the highest BER improvements for a small number of used data carriers as it utilizes the information from CC. For an increasing number of DCs, the curves for both reception methods in CC-employing systems approach the quality of the reference system. This effect is caused by a relatively small amount of power devoted to CC in comparison to the power of DC when there are many DCs employed. This can be also explained observing Figure 3.7b. When the number of DCs increases, the SNR_{loss} decreases.

It can be concluded that the proposed method of reception quality improvement can boost reception performance in an NC-OFDM system where spectrum shaping is obtained by using CCs. In the considered scenario, E_b/N_0 can be reduced by 7.2 dB for BER = 10^{-3} in comparison to the system utilizing standard decoding. The requirement for the method to be feasible is constant coding matrix known at the receiver. The cost of this method is in terms of some extra processing power at the receiver.

3.1.2 Reduced-Complexity Reduced-Power Combination of CCs and Windowing

The combination of CC with windowing (CC&WIN) seems to be a promising spectrum shaping mechanism. While the windowing method provides higher OOB power reduction for spectrum components more distant from the occupied NC-OFDM band on the frequency axis, the CC method reduces mostly the PSD components located closer to the NC-OFDM nominal band as shown in [40]. Thus, the combination of both methods provides additional degrees of freedom as the number of cancellation carriers and window shapes can be altered to fulfill the transmission requirements [4]. It has been presented in [40, 154]. In [40], the power sacrificed for CC is not upper-bounded. While in [154], the CCs power is limited, and the optimization process is computationally complex. Let us discuss several enhancements to this combined approach, yielding reduction of the computational complexity, reduction of the energy loss, and an improvement of the BER performance that has been presented in [4].

The transmitter now consists of a modified NC-OFDM modulator. The difference between its structure and the one presented in Figure 3.4 is the introduction of windowing. Windowing is applied to the time-domain signal after extending it with the cyclic prefix and suffix (N_W samples, additionally to N_{CP} samples of CP). The size of IFFT and sets of DCs and CCs are the same as described in the previous section. However, introduction of windowing operation changes the single-subcarrier spectrum so that "CC calculation" block has to be redesigned. The CC optimization formula must be designed for the time-domain windowed signal $w_m s_m$, where w_m is the window shape, and $m \in \{ -N_{CP} - N_W, \ldots, N - 1 + N_W \}$. In this case, the single-subcarrier spectrum is different than the one defined in (3.5). The frequency response of subcarrier $n \in \{ -N/2, \ldots, N/2 - 1 \}$ at normalized frequency v can now be calculated as

$$S(n, v) = \sum_{m=-N_{CP}-N_W}^{N-1+N_W} w_m e^{j2\pi \frac{n-v}{N} m}. \tag{3.19}$$

Now, matrices $\mathbf{P}_{DC}$ and $\mathbf{P}_{CC}$ defined in the previous section can be obtained using the aforementioned equation. The simplest approach to find CC values is to solve the following optimization problem:

$$\mathbf{d}_{CC}^{\star} = \arg \min_{\mathbf{d}_{CC}} \| \mathbf{P}_{CC} \mathbf{d}_{CC} + \mathbf{P}_{DC} \mathbf{d}_{DC} \|_2^2 \tag{3.20}$$

which is similar to (3.13) (although matrixes $\mathbf{P}_{DC}$ and $\mathbf{P}_{CC}$ are different). A computationally simple solution is

$$\mathbf{d}_{CC}^{\star} = -\mathbf{P}_{CC}^{\dagger} \mathbf{P}_{DC} \mathbf{d}_{DC}. \tag{3.21}$$

3.1.3 Rate and Power Issues with the CC Method

As mentioned previously, CC can use a significant part of the NC-OFDM symbol power causing the SNR loss defined by (3.15). Although the reception method proposed in the previous section can be utilized to mitigate this effect, it sets some additional requirements on the NC-OFDM receiver. Thus, an important phenomenon when applying the CC method for the OOB power reduction is the occurrence of frequency-domain power peaks for frequencies assigned to the CCs. This is due to the fact that the CCs have to compensate for a number of DC sidelobes. A large power increase at the edges of the NC-OFDM frequency spectrum (where CCs are located) may be unacceptable according to the existing regulations that impose constraints on the Spectrum Emission Masks (SEMs). In order to provide some metric reflecting this problem, let us define the Spectrum Overshooting Ratio (SOR) for a given probability p_ϱ of exceeding level ϱ of the spectrum mask by the cancellation carrier power [4]:

$$
\text{SOR} = 10 \log_{10} \left(\frac{\arg_\varrho [\Pr(S(f_{\text{CC}}) > \varrho) = p_\varrho]}{\frac{1}{B_{-\text{CC}}} \int_{B_{-\text{CC}}} S(f)\, df} \right),
\tag{3.22}
$$

where $S(f)$ is the continuous PSD function of the considered NC-OFDM signal, $B_{-\text{CC}}$ is the useful transmission bandwidth (the bandwidth used by the data sub-carriers excluding cancellation subcarriers frequency bands), and f_{CC} is any one of the frequencies belonging to CCs bands. This definition of SOR can be interpreted as the logarithm of the PSD peaks of cancellation carriers with respect to the mean power level in data-carrier band. The occurrence of these peaks is measured with the probability p_ϱ [4].

To overcome the aforementioned problem of an unacceptable power increase, it has been proposed in [4] to supplement the optimization problem described by (3.20) with an additional, indirect constraint whose aim is to limit the CC power. The optimization problem is now defined as to find the optimal vector $\mathbf{d}_{\text{CC}}^\star$:

$$
\mathbf{d}_{\text{CC}}^\star = \arg \min_{\mathbf{d}_{\text{CC}}} \{ \|\mathbf{P}_{\text{CC}}\mathbf{d}_{\text{CC}} + \mathbf{P}_{\text{DC}}\mathbf{d}_{\text{DC}}\|_2^2 + \mu \|\mathbf{d}_{\text{CC}}\|_2^2 \},
\tag{3.23}
$$

where μ factor is used to balance between the CC power and the resulting OOB power reduction. Essentially, the value of μ is assumed to be constant for a given system configuration, that is, constant over a number of consecutive NC-OFDM symbols when system parameters do not change. The solution of this problem can be derived by merging both conditions into block-matrix optimization problem, that is, to find

$$
\mathbf{d}_{\text{CC}}^\star = \arg \min_{\mathbf{d}_{\text{CC}}} \left\| \begin{pmatrix} \mathbf{P}_{\text{CC}} \\ \sqrt{\mu}\mathbf{I} \end{pmatrix} \mathbf{d}_{\text{CC}} + \begin{pmatrix} \mathbf{P}_{\text{DC}} \\ \mathbf{0} \end{pmatrix} \mathbf{d}_{\text{DC}} \right\|_2^2,
\tag{3.24}
$$

where $\mathbf{I}$ is an identity matrix, here of size $\beta \times \beta$. It results in the following optimal vector CC:

$$\mathbf{d}_{CC}^{\star} = -\left(\frac{\mathbf{P}_{CC}}{\sqrt{\mu}\mathbf{I}} \right)^{\dagger} \left(\begin{matrix} \mathbf{P}_{DC} \\ 0 \end{matrix} \right) \mathbf{d}_{DC} = \mathbf{W}\mathbf{d}_{DC}, \tag{3.25}$$

where $\mathbf{W}$ results from multiplication of the first two matrices in the aforementioned equation. Such an optimization has similar computational complexity to the optimization problem of (3.20), because the optimization (calculation of matrix $\mathbf{W}$) is implemented only once for a given spectrum mask and after the indices of DC and CC are determined. Then, for each NC-OFDM symbol, matrix-by-vector multiplication is carried out with precalculated matrix $\mathbf{W}$.

The optimization procedure described earlier significantly reduces the SNR loss typically encountered when a CC method is applied. This is obtained by imposing a constraint on the power assigned to CCs, which results in decreased SOR and higher power reserved for the DCs. Nevertheless, the reduced power available for data subcarriers still causes some deterioration of the reception quality [4]. Therefore, the reception technique defined by formula (3.18) can be applied to this system as well.

Finally, following the idea discussed in [4], let us derive a metric that indicates the potential throughput loss caused by the introduction of CC, windowing, or the combination of both. This throughput loss can be assessed in comparison to a system not employing any OOB power reduction method, in which all subcarriers, that is, $\alpha + \beta$, are occupied by DC. Note that the actual system throughput depends not only on the number of data subcarriers but also on the power assigned to these subcarriers and the channel characteristic observed. Therefore, this proposed metric indicates only the potential throughput loss that results from the information-signal bandwidth reduction due to introduction of the cancellation carriers and window duration extension, thus assuming the same transmit power and channel quality at each subcarrier. It is defined by the following expression [4]:

$$R_{\text{loss}} = \left(1 - \frac{1 - \frac{\beta}{\beta+\alpha}}{1 + \frac{N_{W}}{N+N_{CP}}} \right) \cdot 100\%. \tag{3.26}$$

Note that the reference system for this definition, employing all subcarriers for data transmission, is prohibited from operating in the considered scenario, where the PU transmission may be in place in the frequency proximity and its protection is required by the SU signal OOB power reduction.

Finally, a drawback of standard CC is a substantial increase of PAPR that is caused by the high-power values modulating CCs and being correlated with the DC. Apart from the PAPR value, usually the probability of peaks occurrence is also taken into account since it is conceivable that the time-domain peaks having moderate instantaneous power can cause nonlinear distortions and performance deterioration that can prove to be much worse than the high power (strong) peaks occurring relatively infrequently.

Example Results [4]

Some example results of the reduced-complexity reduced-power combination of Cancellation Carriers and Windowing methods are presented as follows. (Some of these results and their discussion have been published by us in [4].)The results show that the introduced modifications of the combined CC and WIN methods improve the overall performance of the NC-OFDM system in several ways. In the experiments, $N = 256$ subcarriers have been adopted, where the subcarriers of indices in $\mathbf{I}_{DC} = \{-100, \ldots, -62\} \cup \{-41, \ldots, -11\} \cup \{10, \ldots, 40\}$ $\cup \{61, \ldots, 101\}$ are occupied by the QPSK data symbols ($\alpha = 142$), and there are three cancellation carriers placed on each side of the data carrier blocks, that is, $\mathbf{I}_{CC} = \{-103, -102, -101\} \cup \{-61, -60, -59\} \cup \{-44, -43, -42\} \cup$ $\{-10, -9, -8\} \cup \{7, 8, 9\} \cup \{41, 42, 43\} \cup \{58, 59, 60\} \cup \{102, 103, 104\}$. The subcarrier pattern of four data subcarrier blocks is separated by narrowband PU spectrum. The duration of CP equals $N_{CP} = 16$ samples, but $\beta = 16$ samples of the Hanning window extension (equal to the cyclic suffix) are also used on each side of an OFDM symbol. The number of CCs and the shaping window duration have been chosen in such a way that the mean OOB interference power level is achieved at least 40 dB below the mean in-band power level for the reasonable value of μ, that is, $\mu = 0.01$. This OOB power attenuation is sufficient in order to respect several regulatory SEMs, e.g., defined for IEEE802.11g [197] or Long Term Evolution (LTE) user equipment [198].

First, Figure 3.8 shows the results of the OOB power reduction obtained for the following four methods under consideration: GS method, CC method, windowing, and combined CC and WIN scheme. The comparison has been performed for the schemes that present the same potential throughput loss metric, which for the evaluated system equals $R_{loss} = 19.2\%$. Such a potential throughput loss is obtained either from the GS or CC method with four guard or cancellation subcarriers per edge of the DC band, from the WIN method with Hanning window extension of $\beta = 65$ samples, or from the combined CCs and WIN method with three CCs per edge and $N_W = 16$, that is, the scenario described earlier. The PSD plots have been obtained for the signal before High Power Amplifier (HPA) using Welch's method for 10 000 random NC-OFDM transmitted symbols. The spectrum has been estimated in $4N$ frequency-sampling points using $3N$-length Hanning windows.

It can be observed that the GS method provides the highest OOB power levels among other compared methods. Note that the windowing method achieves a high OOB power attenuation, but it requires several frequency guard bands for the OOB attenuation slope. Thus, it is potentially unsuitable

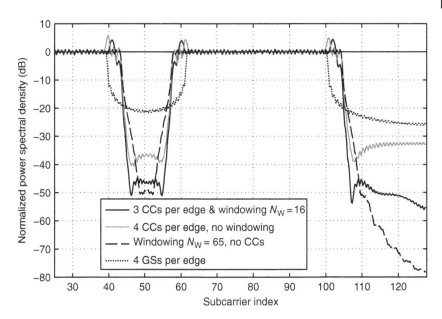

Figure 3.8 A fragment of the normalized PSD of the NC-OFDM transmission signal in the experimental scenario in the case of the application of GS, CC, WIN, and combined CC and WIN methods. (Based on [4].)

for protecting narrowband PU signals from unintentional secondary OOB interference. Conversely, the CC method alone results in a relatively steep OOB power reduction, but the resulting OOB attenuation is not very high for the assumed number of applied CCs. The combination of both methods achieves decent performance in terms of high and steep OOB attenuation, thus confirming that such a combination of methods possesses the potential for protecting both wideband and narrowband PU signals employing strict requirements with respect to the Signal-to-Interference Ratio (SIR). According to the other experiments for QAM schemes and to their results not presented here, the normalized PSD plots are very similar.

In Figure 3.9, several of the system performance metrics, such as the SOR for $p_o = 10^{-1}$, PAPR increase for $\Pr(\text{PAPR} > \text{PAPR}_0) = 10^{-3}$, SNR loss for BER $= 10^{-4}$, and mean OOB power level, are presented in relation to the optimization constraint parameter $\mu \in \langle 10^{-6}, 10^0 \rangle$. Thus, the optimization procedure is considered in the range of μ, defining scenarios from a weak constraint on the CCs power, close to no power constraint, to a constraint on the strictly limited CC power maintaining the result close to the GS approach. For these performance metrics under consideration, measurements have been obtained after the transmission of $2 \cdot 10^5$ NC-OFDM symbols. The SNR loss has been calculated at the receiver, for example, four-path

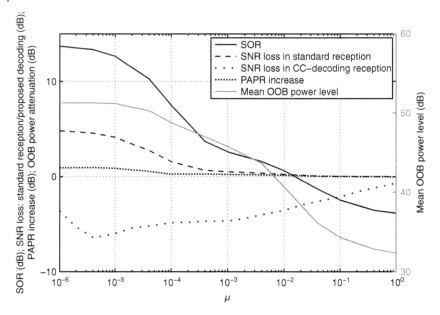

Figure 3.9 Simulation results for the NC-OFDM system versus the optimization factor μ; SNR_{loss} calculated for BER = 10^{-4}. (Kryszkiewicz *et al.*, 2012 [4]. Reproduced with the permission of Springer.)

Rayleigh fading channel defined in Case 3 test scenario for Universal Mobile Telecommunications System (UMTS) user equipment [199]. Averaging of the results has been performed using 10^4 channel realizations.

It can be observed in Figure 3.9 that the OOB power attenuation decreases slowly with an increase of μ for small values of μ. Thus, when μ is low, there is no use in spending additional power on CC since the spurious OOB emissions remain the same. On the other hand, low-power CC (for high μ values) do not achieve improvement in OOB power over the results obtained for windowing method without the application of CC. However, the other metrics improve when μ increases. For example, the fluctuation of SOR ranges from 13.7 to −3.9 dB. It is worth mentioning that the rest of the system performance metrics are calculated with respect to the reference system, which does not use windowing and CCs for OOB power reduction. Instead, the CCs are replaced with zeros. A significant improvement is observed in the PAPR-increase value that approaches zero, when μ becomes high. Both the new optimization goal defined by (3.23) and the proposed reception algorithm have influence on the values of an SNR loss with standard detection and with the proposed detection making use of the CC redundancy. The stronger limit on the CC power (the higher μ), the more power is assigned to DC. Thus, the SNR loss changes from 4.8 dB for $\mu = 10^{-6}$ to nearly 0 dB for $\mu = 10^0$. The results after employing the proposed CC-decoding detection method show that not only is the DC power recovered, but also an additional improvement is achieved thanks (in part) to the frequency diversity introduced by cancellation carriers treated as parity symbols of the block code. It can be observed that the coding gain for BER = 10^{-4} (with respect to system without CCs) varies from 6.42 dB for $\mu = 4 \cdot 10^{-6}$ to 0.7 dB for $\mu = 1$. For very low values of μ ($\mu < 4 \cdot 10^{-6}$), the SNR loss caused by the introduction of CCs becomes higher than can be compensated even by using high-power CCs, which yields a decrease in coding gain. The results presented in Figure 3.9 show that the detection algorithm making use of the CC redundancy yields decent performance even in the assumed case of large fragmentation of the available (not occupied by the PUs) frequency bands.

The proposed combination of CC and WIN methods allows to significantly reduce the OOB power, while maintaining high NC-OFDM throughput. It outperforms the CC or WIN method when used separately. The used optimization method decreases computational complexity at the transmitter in comparison to the standard CC method and allows to control the OOB power attenuation, the power assigned to CC, and the PAPR increase. Additionally, the CC method can be used for reception quality improvement at the receiver.

3.2 Reduction of Subcarrier Spectrum Sidelobes by Flexible Quasi-Systematic Precoding[1]

The combination of CC and windowing methods allows for relatively strong attenuation of the OOB power while the computational complexity of the NC-OFDM transmitter and receiver is maintained at an acceptable level. However, if the transmitter and receiver have some extra processing power, some more sophisticated methods can be used. Spectral precoding methods achieve the strongest OOB power reduction among other techniques. However, it has been shown that some of these methods have a number of flaws, for example, BER degradation, throughput decrease, and computational complexity of detection at the receiver.

A flexible Quasi-Systematic Precoding (QSP) method for OOB power reduction in the NC-OFDM transmission has been proposed in [200]. It is called *quasi-systematic* because each information data symbol appears at the output of the precoder almost the same as in its input, only slightly distorted in a noise-like manner. Contrary to the methods presented in [157] and [181], the coding rate is allowed to be lower than 1, that is, it allows for the *redundancy* values modulating edge SC that are used exclusively for the OOB power suppression (similarly to the CC method). This in turn reduces the interference introduced to data and pilot SC and decreases BER. Moreover,

---

1 © 2012 IEEE. Fragments of this section reprinted, with permission, from Ref. [200].

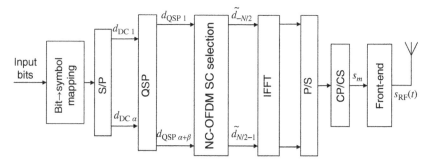

Figure 3.10 Diagram of an NC-OFDM transmitter with QSP. (Based on [200].)

seamless detection of the data symbols (i.e., without decoding) is possible at the receiver, significantly decreasing its computational complexity. This mechanism also has some flexibility embedded in a parametrically defined number, position, and weight of the sampling points in the OOB optimization region and in the degree of protection of the pilot SC. The method supports flexible NC-OFDM spectrum shaping of the SU transmission, meeting dynamically varying spectrum-mask requirements in a CR network.

The NC-OFDM transmitter employing QSP is depicted in Figure 3.10. The data bits are mapped to symbols that, after the S/P conversion, are fed to the QSP block, which operates on α complex data symbols in the vertical vector $\mathbf{d}_{DC}$ (as previously defined). The QSP coding matrix $\mathbf{W}_{QSP}$ of size $(\alpha + \beta) \times \alpha$ is used to obtain the output $\alpha + \beta$-length vertical vector of coded symbol values $\mathbf{d}_{QSP} = [d_{QSP_1}, d_{QSP_2}, \dots, d_{QSP_{\alpha+\beta}}]$:

$$\mathbf{d}_{QSP} = \mathbf{W}_{QSP}\mathbf{d}_{DC}. \tag{3.27}$$

As mentioned in the previous section, the total number of modulated subcarriers equals $\alpha + \beta$. Note that the elements of vector $\mathbf{d}_{QSP}$ are from the finite set of complex values. That is why they are called *symbols* although they are not complex integers and not from the same set of digital symbols as the elements of $\mathbf{d}_{DC}$. Vector $\mathbf{d}_{DC}$ is used to modulate SC indexed $\mathbf{I}_{QSP} = \{I_{QSP_j}\}$ for $j \in \{1, \dots, \alpha + \beta\}$ and $I_{QSP_j} \in \{-N/2, \dots, N/2 - 1\}$.

3.2.1 Precoder Design

Matrix $\mathbf{W}_{QSP}$ should be designed in such a way that the spectrum of a resultant NC-OFDM symbol is properly shaped with the required power level of the OOB radiation attaining sufficient data and pilot SC quality. For this purpose, let us consider sampling the spectrum of each occupied subcarrier in the OOB frequency range. For the set of spectrum-sampling points $\mathbf{V} = \{V_l\}$ ($l \in \{1, \dots, \gamma\}$) and indices of occupied SC $\mathbf{I}_{QSP} = \{I_{QSP_j}\}$, the γ-by-$(\alpha + \beta)$ matrix $\mathbf{P}_{QSP} = \{P_{QSP_{l,j}}\}$ can be defined. It can be calculated using formula (3.5) for

$P_{QSP_{l,j}} = S(I_{QSP_j}, V_l)$. Consequently, multiplication of matrix $\mathbf{P}_{QSP}$ by vector $\mathbf{d}_{QSP}$ results in a vector of the transmit signal spectrum values at the sampling points $\mathbf{V}$. The goal is to minimize the signal energy in this range.

Let us consider vector $\mathbf{s}_{res}$ defined in the following way:

$$\mathbf{s}_{res}(\mathbf{d}_{QSP}) = \begin{pmatrix} \mathbf{A_1 P}_{QSP} \\ \mathbf{A_2} \end{pmatrix} \mathbf{d}_{QSP} - \begin{pmatrix} \mathbf{0} \\ \mathbf{A_3} \end{pmatrix} \mathbf{d}_{DC} , \qquad (3.28)$$

which will be *the residual* of the optimization problem, that is, the first γ elements of this vector are the weighted OOB spectrum samples, while the following $\alpha + \beta$ elements are differences between the precoder weighted quasi-systematic output and related input symbols, with the assumption that the redundant QSP output symbols are related to a fictitious input zero-valued symbol. The idea behind such a definition of the residual is to minimize the OOB power, control the power of the redundancy symbols, and minimize the difference between the original data symbols and their quasi-systematic representation.

In the aforementioned formula, $\mathbf{A_1}$ is a $\gamma \times \gamma$ diagonal matrix that contains the weighting factors on the diagonal that value the OOB power suppression level at the sampling points, that is, element $A_{1_{l,l}}$ ($l \in \{1, \ldots, \gamma\}$) having high value indicates that the OOB spectrum at pulsation V_l should be suppressed to a higher degree. The null matrix $\mathbf{0}$ of size $\gamma \times \alpha$ is used to drive the target value of zero for the first γ elements of $\mathbf{s}_{res}$ (the OOB spectrum samples). The real nonnegative diagonal matrix $\mathbf{A_2}$ of size $(\alpha + \beta) \times (\alpha + \beta)$ and real nonnegative matrix $\mathbf{A_3}$ are used to control the power of redundancy tones and the self-interference generated by QSP to the data and pilot symbols at the precoder output. The higher the coefficient in the same row of both matrices corresponding to the quasi-systematic symbols, the lower the self-interference. The rows of these matrices corresponding to the redundancy SC are to limit the power of these tones. These rows in $\mathbf{A_3}$ consist of zeros. In case of $\mathbf{A_2}$, the respective diagonal entries are nonnegative real coefficients. The higher these coefficients, the lower the power of the redundancy tones.

The optimization goal is now defined as finding the optimum vector $\mathbf{d}_{QSP}$ that minimizes the norm of vector $\mathbf{s}_{res}(\mathbf{d}_{QSP}$ defined by (3.28):

$$\mathbf{d}^\star_{QSP} = \arg\min_{\mathbf{d}_{QSP}} \|\mathbf{s}_{res}(\mathbf{d}_{QSP})\|^2_2. \qquad (3.29)$$

Note that (3.29) defines unconstrained optimization problem, which is easier to solve than the constraint optimization. The constraints (limits) on the redundancy tone power and self-interference power are introduced indirectly by the design of matrices $\mathbf{A_2}$ and $\mathbf{A_3}$. The problem (3.29) can be solved in the least squares sense by pseudoinverse operation, that is,

$$\mathbf{d}^\star_{QSP} = \begin{pmatrix} \mathbf{A_1 P}_{QSP} \\ \mathbf{A_2} \end{pmatrix}^\dagger \begin{pmatrix} \mathbf{0} \\ \mathbf{A_3} \end{pmatrix} \mathbf{d}_{DC} , \qquad (3.30)$$

and thus, the optimal coding matrix is specified as

$$\mathbf{W}_{QSP} = \begin{pmatrix} \mathbf{A}_1 \mathbf{P}_{QSP} \\ \mathbf{A}_2 \end{pmatrix}^{\dagger} \begin{pmatrix} \mathbf{0} \\ \mathbf{A}_3 \end{pmatrix}. \tag{3.31}$$

The matrices $\mathbf{A}_1$, $\mathbf{A}_2$, $\mathbf{A}_3$ can be adjusted to fit different requirements. They can be summarized as follows:

- **Nonunitary coding rate.** As mentioned before, in order to decrease the power of in-band self-interference, β SC on the edges (or in the middle) of the occupied SC block can be used exclusively for the OOB power attenuation. For the j-th SC in vector $\mathbf{d}_{QSP}$ to act as CC, the j-th row of matrix $\mathbf{A}_3$ must be all zeros. Moreover, these tones usually have high power [41], which may be considered as *wasted* on the redundant transmission of the CCs, and may exceed the power level allowable in the spectrum mask. Therefore, the jth diagonal entry of matrix $\mathbf{A}_2$ should be determined to control the power of these SC; that is, if it is equal to zero, no power constraint is placed on redundant SC, and when it has a higher positive value, the power of the redundancy tone is limited.

- **Protection of data and pilot carriers.** Definition of the $\mathbf{s}_{res}$ and optimization (3.29) implies that the proposed precoding is quasi-systematic thanks to minimization of the square error between the weighted precoder input and related weighted output symbol. Definition of constraint matrices $\mathbf{A}_2$ and $\mathbf{A}_3$ gives us another degree of freedom to control the relative quality of DC (their power to self-interference ratio) (see [200] for more details).

- **Adjustment of OOB power reduction level.** The degree of the OOB power reduction is achieved by defining the number of sampled OOB spectrum pulsation points γ. The higher the γ value, the more *flat* OOB spectrum of the precoded transmit signal. Note that the OOB spectrum-sampling points do not have to be equally spaced but can be freely chosen to define vector $\mathbf{V}$. Furthermore, the $\mathbf{A}_1$ matrix is used to balance the achieved sidelobe attenuation against the self-interference power. For a simple case, when the same attenuation is required at all pulsation sampling points, $\mathbf{A}_1$ has a single value at the diagonal, that is, $A_{1_{l,l}} = \mu_{QSP}$, where $l \in \{1, \ldots, \gamma\}$ and $\mu_{QSP} \geq 0$. The higher the μ_{QSP} value, the higher the OOB power attenuation. For a more complicated scenario, when different PUs operating in adjacent bands have different allowed interference levels, the $\mathbf{A}_1$ matrix can have different elements on its diagonal, weighting each row of the $\mathbf{P}_{QSP}$ matrix.

3.2.2 Reception Quality Improvement for NC-OFDM with Quasi-Systematic Precoding

Let us first consider *seamless reception* of the quasi-systematically precoded NC-OFDM signal that suffers from self-interference added to data symbols due to applied QSP. This self-interference acts as an additional noise, and in seemless-reception scheme, the receiver does not try to remove it in any decoding algorithm. As this reception method is also the least computationally complex, it can be used by low-power devices at the expense of deteriorated reception quality.

In the considered scenario, the transmitter has to control the self-interference introduced by QSP to data symbols; therefore, it has to estimate the lower bound of BER based on the Signal-to-Self Interference Ratio (SSIR). It can be shown that the central limit theorem allows for approximation of this self-interference by the AWGN. Thus, the lower-bound BER performance of the proposed QSP NC-OFDM scheme with QAM mapping for the n-th DC can be approximated as [201]:

$$\mathrm{BER}_n \approx \frac{\sqrt{M_{\mathrm{QAM}}} - 1}{\sqrt{M_{\mathrm{QAM}}} \log_2 \sqrt{M_{\mathrm{QAM}}}} \mathrm{erfc} \left(\sqrt{\frac{3 \cdot \mathrm{SSIR}_n}{2(M_{\mathrm{QAM}} - 1)}} \right), \tag{3.32}$$

where $M_{\mathrm{QAM}} = \mathcal{M}$ is the number of QAM constellation points (symbols), and SSIR_n is the SSIR at data subcarrier n, which can be calculated knowing the matrix $\mathbf{W}_{\mathrm{QSP}}$ [200]. The actual BER for seamless reception in a noisy fading channel is always be higher than the overall BER lower bound averaging BER_n for all DCs.

The reception performance can be improved if the precoding matrix is known at the receiver (similarly as for the CCs in Section 3.1.1). Decoding of a complex-field code defined by $\mathbf{W}_{\mathrm{QSP}}$ can follow general reception rules defined in [195]. The wireless multipath fading channel characteristic can be defined as in Section 3.1.1, that is, as a diagonal matrix $\mathbf{H}$ containing frequency-domain channel coefficients for $\alpha + \beta$ occupied SC on its diagonal. The QSP Minimum Mean Square Error (MMSE) reception matrix has the following form:

$$\mathbf{R}_{\mathrm{MMSE}} = (\mathbf{HW}_{\mathrm{QSP}})^{\mathcal{H}} \cdot (\sigma_n^2 \mathbf{I} + \mathbf{HW}_{\mathrm{QSP}}(\mathbf{HW}_{\mathrm{QSP}})^{\mathcal{H}})^{-1}, \tag{3.33}$$

where $\mathbf{X}^{\mathcal{H}}$ denotes the Hermitian transposition of matrix $\mathbf{X}$, σ_n^2 is the AWGN power observed at each subcarrier, and $\mathbf{I}$ is the $(\alpha + \beta)$-by-$(\alpha + \beta)$ identity matrix. For the vertical vector $\mathbf{r}_{\mathrm{QSP}} = [r_{\mathrm{QSP}\,1}, \dots, r_{\mathrm{QSP}\,\alpha+\beta}]^{\mathrm{T}}$ of the received symbol values at the output of FFT, the estimate of $\mathbf{d}_{\mathrm{DC}}$ is vector $\hat{\mathbf{d}}_{\mathrm{DC}}$:

$$\hat{\mathbf{d}}_{\mathrm{DC}} = \mathbf{R}_{\mathrm{MMSE}} \mathbf{r}_{\mathrm{QSP}}. \tag{3.34}$$

Reception quality can be improved by a Decision Directed (DD) receiver as it performs nonlinear decision operation using the knowledge of limited symbols alphabet with relatively low computational complexity increase [195].

Moreover, an iterative receiver can be considered, as in [182] and [157]. Because the code used therein does not utilize redundancy symbols, for the proposed QSP NC-OFDM signal detection, let us use truncated $\alpha \times \alpha$ coding matrix $\tilde{\mathbf{W}}_{\mathrm{QSP}}$ resulting from the selection of rows of matrix $\mathbf{W}_{\mathrm{QSP}}$ corresponding to data and pilot SC (redundancy tones are neglected). In the same way, by discarding β rows and β columns, $\tilde{\mathbf{H}}$ is obtained from $\mathbf{H}$.

In the ith iteration, the estimate of the data symbol vector is calculated in the following way:

$$\hat{\mathbf{d}}_{DC}^{(i)} = \tilde{\mathbf{W}}_{QSP}(\tilde{\mathbf{H}}_{DC})^{-1}\mathbf{r}_{QSP} + (\mathbf{I} - \tilde{\mathbf{W}}_{QSP})\check{\mathbf{d}}_{DC}^{(i-1)}, \qquad (3.35)$$

where $\check{\mathbf{d}}_{DC}^{(i-1)}$ is the decision vector after demapping and mapping in the $(i-1)$th iteration.

Example Results

Let us discuss some example results of the QSP method for the NC-OFDM spectrum shaping and signal reception. An LTE-like system has been considered with the maximum of $N = 512$ SC spaced by 15 kHz. The channel model is the Extended Vehicular A channel with nine taps and the tap delay spread of 2510 ns [198]. The duration of the cyclic prefix is one-sixteenth of the orthogonality period in the OFDM symbol. The indices of the occupied SC are $\mathbf{I}_{QSP} = \{ -100, -99, \ldots, -1, \ 1, \ldots, 16, 33, \ldots, 100 \}$. The notch from the 17-th to the 32-nd subcarrier can accommodate a PU transmission, whose bandwidth is 200 kHz (e.g., wireless microphone). There are 17 pilots spaced by about 10 SC. The SSIR for the pilot subcarriers equals 12.04 dB. Subcarriers $\{ -100, -99, 14, 15, 16, 33, 34, 35, 99, 100 \}$ are devoted only to sidelobe suppression with their power-limiting coefficient (on the diagonal of matrix $\mathbf{A}_2$) equal to 0.2.

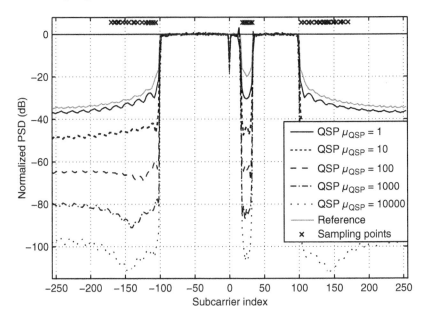

Figure 3.11 Normalized PSD for different suppression scalars μ_{QSP} of QSP. (© 2013 IEEE. Reprinted, with permission, from [200].)

In Figure 3.11, the normalized PSDs are shown for various suppression parameter values μ_{QSP}, which is same as the value of all diagonal entries of A_1 when the same attenuation is required at all sampling points in the OOB region. There are $\gamma = 249$ sampling points defined, whose location is shown above the PSD plots. This number allows for proper spectrum *flattening*, and though it results in some computational complexity for the calculation of the precoding matrix W_{QSP}, it does not influence the runtime precoding complexity at all. An important observation is that alteration of QAM constellation order has no visible influence on the PSD plots. The reference curve in Figure 3.11 is obtained for the standard NC-OFDM transmission without QSP, in which all SCs indexed by I_{QSP} are used by data symbols. Let us note that QSP results in the steep OOB power suppression close to the occupied band. Moreover, the level of spurious emission can be flexibly adjusted by the choice of μ_{QSP}. The level of sidelobe suppression has a direct impact on self-interference that can be evaluated by determining the BER lower limit in case of seamless reception, which is presented in Table 3.1.

For further simulations, $\mu_{QSP} = 8$ has been used, which provides more than -40 dB PSD level in OOB region. In this case, $\gamma = 220$ sampling points have been defined. In Figure 3.12, complex plain plots for QPSK constellation after coding for this example are shown for two subcarriers with $SSIR_n = 11.2$ dB and $SSIR_n = 32.7$ dB. It is shown to confirm the AWGN-like character of self-interference.

In Figure 3.13, BER performance of a system employing QPSK modulation is presented. Now, the reference system is the NC-OFDM system without precoding, where α SC are occupied by the QPSK data symbols, and with MMSE reception. The other β subcarriers are those used as redundancy symbols in QSP, but for the reference system, they are modulated by zeros, that is, Guard Subcarriers (GS) spectrum shaping method is employed. In both systems, the same QPSK symbol rate is obtained. There have been 10 000 OFDM symbols transmitted in each of 1000 independent Rayleigh channel instances to obtain the presented BER results.

It can be observed that BER for seamless reception approaches its lower limit of $4 \cdot 10^{-5}$. Although the simulated BER lower bound is slightly below the theoretical lower bound, the difference is quite small. It is caused by the usage of QPSK modulation for which the self-interference does not deteriorate the

Table 3.1 Lower bound of BER for seamless reception after QSP in NC-OFDM.

μ_{QSP}	1	10	100	1000	10 000
BPSK	$1.4 \cdot 10^{-263}$	$4.3 \cdot 10^{-11}$	$5.4 \cdot 10^{-6}$	$1.1 \cdot 10^{-4}$	$8.7 \cdot 10^{-4}$
QPSK	$4.3 \cdot 10^{-134}$	$5.7 \cdot 10^{-7}$	$3.4 \cdot 10^{-4}$	$2.3 \cdot 10^{-3}$	$6.9 \cdot 10^{-3}$
16-QAM	$2.5 \cdot 10^{-30}$	$2.6 \cdot 10^{-3}$	$2.3 \cdot 10^{-2}$	$4.9 \cdot 10^{-2}$	$7.4 \cdot 10^{-2}$

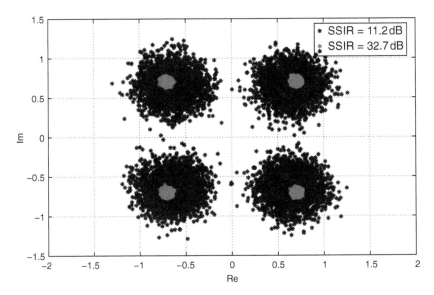

Figure 3.12 Constellation plot for 10 000 transmitted symbols over two data subcarriers and $\text{SSIR}_j = 11.2$ dB and $\text{SSIR}_j = 32.7$ dB. (© 2013 IEEE. Reprinted, with permission, from [200].)

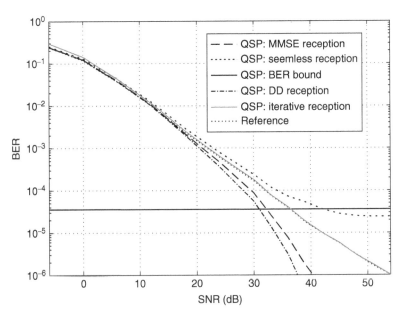

Figure 3.13 BER performance in Rayleigh channel using QPSK constellation. (© 2013 IEEE. Reprinted, with permission, from [200].)

signal exactly as if it was AWGN. The reference system's performance is similar to the QSP signal reception with iterative detector. MMSE detection defined by formula (3.34) or DD significantly detection improves the QSP signal reception performance for SNR > 20 dB. For the BER of 10^{-5}, MMSE and DD reception gain by about 5.7 and 7.3 dB, respectively, in comparison to the reference system. This result is quite impressive as the coding rate is about 0.95.

To summarize, the QSP scheme allows for significant OOB power reduction in the secondary NC-OFDM transmission. Contrary to the other precoding methods, the QSP is nearly systematic with adjustable coding rate (lower than or equal to 1). Increasing the number of the redundancy SCs used exclusively for the OOB power suppression results in the reduction of the self-interference. Analytical and simulation results show that the QSP method is very efficient in the NC-OFDM sidelobe reduction also in narrowband spectrum notches. By the proper choice of the parameters defining the coding matrix, a number of features can be controlled: the OOB power, reception quality, and balance between the computational complexity of the method and the throughput loss due to the use of the redundant tones.

3.3 Optimized Cancellation Carriers Selection[2]

Let us now discuss a method of the CC calculation based on stochastic approach, which dynamically adjusts to the subcarrier pattern. The mean CC power is now directly constrained (in contrast to indirect approach proposed in Section 3.1.2). Moreover, the location of CC is also revised. Typically, an equal number of CCs are placed on each side of the data-occupied subcarrier blocks [39, 41]. In [43], the Optimized Cancellation Carriers Selection (OCCS) is proposed, which is the heuristic approach to choose CC locations iteratively. This method significantly outperforms the traditional approach in terms of the OOB power reduction for a given number of CCs, which is obtained at the only cost of the iterative off-line low computationally complex design of the CC calculation matrix.

The NC-OFDM system model considered in this section is the same as that shown in Figure 3.4.

The optimization problem for the calculation of CC with direct power constraint [41] can be rewritten using notation from Section 3.1.1. The aim is to find vector $\mathbf{d}_{CC}^{\star}$ resulting in the minimization of the OOB power with limited CC power:

$$\mathbf{d}_{CC}^{\star} = \arg\min_{\mathbf{d}_{CC}} \|\mathbf{P}_{CC}\mathbf{d}_{CC} + \mathbf{P}_{DC}\mathbf{d}_{DC}\|_2^2$$

$$\text{s.t. } \|\mathbf{d}_{CC}\|_2^2 \leq \beta \,, \tag{3.36}$$

---

2 © 2013 IEEE. Fragments of this section reprinted, with permission, from Ref. [43].

where $\|\cdot\|_2$ is the Euclidean vector norm. Here, it is assumed that the power of CC should be equal to or lower than β, making the mean CC power equal to or lower than the normalized data symbol power, which is assumed to be 1.

Let us study the Lagrange function for problem (3.36):

$$
\begin{aligned}
f(\mathbf{d}_{CC}, \theta) &= \| \mathbf{P}_{CC}\mathbf{d}_{CC} + \mathbf{P}_{DC}\mathbf{d}_{DC} \|_2^2 + \theta(\| \mathbf{d}_{CC} \|_2^2 - \beta) = \\
&= \mathbf{d}_{CC}^H \mathbf{P}_{CC}^H \mathbf{P}_{CC}\mathbf{d}_{CC} + \mathbf{d}_{CC}^H \mathbf{P}_{CC}^H \mathbf{P}_{DC}\mathbf{d}_{DC} + \mathbf{d}_{DC}^H \mathbf{P}_{DC}^H \mathbf{P}_{CC}\mathbf{d}_{CC} \\
&\quad + \mathbf{d}_{DC}^H \mathbf{P}_{DC}^H \mathbf{P}_{DC}\mathbf{d}_{DC} + \theta(\mathbf{d}_{CC}^H \mathbf{d}_{CC} - \beta),
\end{aligned}
\tag{3.37}
$$

where θ is the active Lagrange multiplier, that is, taking a value higher than zero, when the CC power is to be limited and 0 otherwise, according to the Karush–Kuhn–Tucker conditions. The solution of (3.36) can be obtained by finding $\mathbf{d}_{CC}$ and θ that meet

$$
\begin{cases}
\frac{\partial f(\mathbf{d}_{CC}, \theta)}{\partial \mathbf{d}_{CC}} = 0 \\
\frac{\partial f(\mathbf{d}_{CC}, \theta)}{\partial \theta} = 0.
\end{cases}
\tag{3.38}
$$

After calculation of the partial derivative of (3.37), this can be rewritten as

$$
\begin{cases}
2\mathbf{P}_{CC}^H \mathbf{P}_{CC}\mathbf{d}_{CC} + 2\mathbf{P}_{CC}^H \mathbf{P}_{DC}\mathbf{d}_{DC} + 2\theta\mathbf{d}_{CC} = 0 \\
\mathbf{d}_{CC}^H \mathbf{d}_{CC} - \beta = 0.
\end{cases}
\tag{3.39}
$$

Solution of the first equation in (3.39) gives

$$
\mathbf{d}_{CC}^\star = -(\mathbf{P}_{CC}^H \mathbf{P}_{CC} + \theta\mathbf{I})^{-1}\mathbf{P}_{CC}^H \mathbf{P}_{DC}\mathbf{d}_{DC} = \mathbf{W}\mathbf{d}_{DC},
\tag{3.40}
$$

where $\mathbf{W}$ is the new CC calculation matrix. If the problem (3.36) is to be solved directly, θ has to be found to satisfy the second equation from (3.39) for each NC-OFDM symbol separately. As shown in [189] (in section *LS Minimization over a Sphere*), the algorithm for solving this problem exists although running it for every NC-OFDM symbol will be computationally unfeasible.

Here, the focus is on satisfying the condition of the mean CC power assuming that random symbols in $\mathbf{d}_{DC}$ are independent with zero mean and unit variance. This mean CC power equals

$$
\begin{aligned}
\mathbb{E}[\|\mathbf{d}_{CC}\|_2^2] &= \operatorname{trace}\left(\mathbb{E}\left[\mathbf{d}_{DC}^H \mathbf{W}^H \mathbf{W}\mathbf{d}_{DC}\right]\right) = \operatorname{trace}\left(\mathbb{E}\left[\mathbf{d}_{DC}\mathbf{d}_{DC}^H\right]\mathbf{W}^H \mathbf{W}\right) \\
&= \operatorname{trace}\left(\mathbf{W}^H \mathbf{W}\right) = \operatorname{trace}\left(\mathbf{P}_{DC}^H \mathbf{P}_{CC}(\mathbf{P}_{CC}^H \mathbf{P}_{CC} + \theta\mathbf{I})^{-2}\mathbf{P}_{CC}^H \mathbf{P}_{DC}\right).
\end{aligned}
\tag{3.41}
$$

Expression (3.41) is obtained due to the linearity of trace and expectation operators and cyclic property of trace. As this equation is nonlinear, finding the value of θ, for which $\mathbb{E}[\mathbf{d}_{CC}^H \mathbf{d}_{CC}] \le \beta$ requires the use of the Newton method, where calculation of matrix inverse and a number of matrix-by-matrix multiplications is performed in each iteration.

Decreased computational complexity of the CC value calculation (without accuracy deterioration) can be obtained by replacing $\mathbf{P}_{CC}$ by its Singular Value

Decomposition (SVD), that is, $\mathbf{P}_{CC} = \mathbf{U}_1 \mathbf{\Sigma} \mathbf{U}_2^H$, where $\mathbf{U}_1$ and $\mathbf{U}_2$ are unitary matrices, and $\mathbf{\Sigma}$ is $\gamma \times \beta$ diagonal matrix with δ singular values on its diagonal (δ being the minimum of β and γ). We can also assume full-rank $\mathbf{P}_{CC}$. By using the properties of trace, unitary matrix, and matrix inversion, we can obtain the following result [43]:

$$\mathbb{E}[\|\mathbf{d}_{CC}\|_2^2] = \sum_{i=1}^{\delta} \frac{A_{i,i}|\Sigma_{i,i}|^2}{(\theta + |\Sigma_{i,i}|^2)^2} \le \beta \,, \tag{3.42}$$

where $A_{i,i}$ is the element of the positive semidefinite Hermitian matrix $\mathbf{A}$ defined as $\mathbf{A} = \mathbf{U}_1^H \mathbf{P}_{DC} \mathbf{P}_{DC}^H \mathbf{U}_1$. Note that for a given set of CC and DC, only three matrix operations (one SVD and two matrix-by-matrix multiplications) and a few scalar-based Newton algorithm iterations have to be performed to find θ.

Additionally, matrices $\mathbf{U}_1$, $\mathbf{\Sigma}$, $\mathbf{U}_2$, and $\mathbf{A}$ can be used for calculation of the mean OOB power and for the final CC calculation matrix. The OOB power minimization problem (3.36) can be reformulated by using (3.40) as follows:

$$\mathbf{d}_{CC}^{\star} = \arg \min_{\mathbf{d}_{CC}} \|\mathbf{G}\mathbf{d}_{DC}\|_2^2 \,, \tag{3.43}$$

where

$$\mathbf{G} = \mathbf{P}_{CC}\mathbf{W} + \mathbf{P}_{DC} = \left(\mathbf{I} - \mathbf{P}_{CC} \left(\mathbf{P}_{CC}^H \mathbf{P}_{CC} + \theta \mathbf{I} \right)^{-1} \mathbf{P}_{CC}^H \right) \mathbf{P}_{DC}. \tag{3.44}$$

The mean OOB radiation power

$$\text{OOB} = \frac{1}{\gamma (N + N_{CP})^2} \mathbb{E}[\|\mathbf{G}\mathbf{d}_{DC}\|_2^2] \tag{3.45}$$

can be found by averaging the norm over all possible $\mathbf{d}_{DC}$ vectors (by repeating operations presented in (3.41)) for the number of spectrum-sampling points γ. Again, using the properties of diagonal matrices and trace as in [43], the final formula can be obtained:

$$\text{OOB} = \frac{1}{\gamma (N + N_{CP})^2} \left(\sum_{i=\delta+1}^{\gamma} A_{i,i} + \sum_{i=1}^{\delta} A_{i,i} \left(\frac{\theta}{\theta + |\Sigma_{i,i}|^2} \right)^2 \right). \tag{3.46}$$

Note that for a basic OFDM system not using CC, the OOB power equals $\text{OOB} = \frac{1}{\gamma(N+N_{CP})^2} \|\mathbf{P}_{DC}\|_2^2$. Finally, CC calculation matrix can be obtained as

$$\mathbf{W} = -\mathbf{U}_2 (\mathbf{\Sigma}^H \mathbf{\Sigma} + \theta \mathbf{I})^{-1} \mathbf{\Sigma}^H \mathbf{U}_1^H \mathbf{P}_{DC} =$$

$$-\mathbf{U}_{2\delta} \text{diag} \left(\frac{\Sigma_{1,1}^H}{\theta + |\Sigma_{1,1}|^2}, \ldots, \frac{\Sigma_{\delta,\delta}^H}{\theta + |\Sigma_{\delta,\delta}|^2} \right) \mathbf{U}_{1\delta}^H \mathbf{P}_{DC} \,, \tag{3.47}$$

where $\mathbf{U}_{2\delta}$ and $\mathbf{U}_{1\delta}$ are submatrices containing only first δ columns of matrices $\mathbf{U}_2$ and $\mathbf{U}_1$, respectively. The optimum $\mathbf{d}_{CC}$ vector can be now easily calculated using this matrix $\mathbf{W}$ as in formula (3.40).

3.3.1 Computational Complexity

The computational complexity of the original CC algorithm is relatively high. According to [189], for each NC-OFDM symbol, about $\alpha\gamma + \delta\beta + 2\delta M_{\mathrm{N}}$ operations are needed for the calculation of the CC symbols, where M_{N} is the number of steps in the Newton method. In the OCCS method, after calculation of matrix $\mathbf{W}$, the number of operations needed for each NC-OFDM symbol is relatively low. Matrix $\mathbf{W}$ can be used directly or with precalculation of $\mathbf{P}_{\mathrm{DC}}\mathbf{d}_{\mathrm{DC}}$ ($\mathbf{P}_{\mathrm{DC}}$ decomposed from $\mathbf{W}$ for $\gamma \ll \beta$), which requires $\alpha\beta$ or $\alpha\gamma + \beta\gamma$ operations, respectively. The computational complexity of the proposed method in any case is significantly lower than the original one from [41].

The computational speed-up on the stage of multiplier θ determination should also be estimated. For this purpose, a number of operations needed for the calculation of $\mathbb{E}[\|\mathbf{d}_{\mathrm{CC}}\|_2^2]$ can be compared when using (3.41) or (3.42). In the first case, the order of the number of computations is $\beta^3 + \alpha\beta + \alpha\beta\gamma$ (assuming matrix inversion using Gauss–Jordan elimination), while in the second, it is only δ. In both cases, precomputation of matrices independent from θ is assumed.

3.3.2 Heuristic Approach to OCCS

Let us note that matrix $\mathbf{G}$ defined in (3.44) projects data symbols $\mathbf{d}_{\mathrm{DC}}$ onto spectrum samples in the normalized frequencies $\mathbf{V}$ with the CC insertion. As data symbols are independent random variables with zero mean and unit mean power, coefficient $|G_{l,j}|^2$ is the mean normalized OOB radiation power caused at frequency-sampling point V_l by data subcarrier indexed by $I_{\mathrm{DC}j}$. Matrix $\mathbf{G}$ can be partitioned into vertical vectors $\mathbf{g}_j$, that is, $\mathbf{G} = [\mathbf{g}_1, \mathbf{g}_2, \ldots, \mathbf{g}_\alpha]$, the mean OOB power over all spectrum-sampling points caused by $I_{\mathrm{DC}j}$-th data subcarrier equals:

$$\|\mathbf{g}_j\|_2^2 = \sum_{l=1}^{\gamma} |G_{l,j}|^2. \tag{3.48}$$

It is obvious that the data subcarrier causing the highest OOB power has the biggest influence on the OOB region. As such, it has the highest potential to be used as a CC to be added in the counterphase to strong OOB components caused by other subcarriers, that is, to cancel them. If it is used as CC, it should reduce some $|G_{l,j}|^2$ components, that is, decrease the OOB radiation power. The OCCS criterion for finding index $\hat{j}$ of the DC to be used as CC is the following:

$$\hat{j} = \arg\max_j \|\mathbf{g}_j\|_2^2. \tag{3.49}$$

The CC selection has to be performed iteratively, that is, after the introduction of each CC, matrix $\mathbf{G}$ has to be recalculated. Single CC selection changes matrix $\mathbf{W}$, also changing the correlation properties between subcarriers. Typically, selection of a single CC is such that the other DCs in its neighborhood are

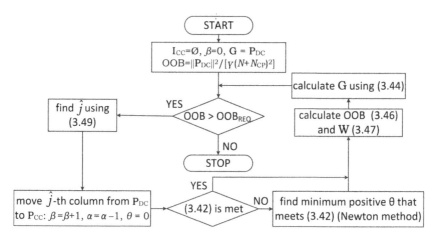

Figure 3.14 Flow diagram of the heuristic OCCS algorithm. (Based on [43].)

not chosen to be the CCs in the next step because their influence on the OOB radiation is similar (highly correlated). The algorithm stops when the OOB radiation power achieves the required level OOB_{req} (see the flow diagram of the OCCS in Figure 3.14).

Finally, note that the OCCS can be combined with the time-domain windowing similarly to the standard CC method [4].

Example Results

For the heuristic OCCS algorithm evaluation, an example NC-OFDM system has been considered with $N = 256$ subcarriers spaced by 15 kHz (as in the LTE system [198]) and a set of occupied subcarriers indexed by $I_{DC} = \{ -80, \dots, 16\} \cup \{49, \dots, 80\}$. Notch-spanning subcarriers from 17 to 48 (480 kHz) can be occupied by a narrowband licensed system. The subcarrier indexed 0 is unoccupied. The algorithm uses $\gamma = 485$ spectrum-sampling points (in vector **V**) distributed equally in the normalized frequency region $\langle-125.75; -81\rangle \cup \langle17; 48\rangle \cup \langle81; 125.75\rangle$. Matrix **W** is calculated for a number of CCs varying from 0 to 40, for various CP durations, for both the standard CC selection method and heuristic OCCS.For OCCS, the heuristic algorithm has been used. For the standard selection of CC, index $\hat{\jmath}$ indicates a data subcarrier lying closest to the NC-OFDM band edges (not (3.49)).

The resulting mean OOB power versus CP duration and β is presented in Figure 3.15. Note that while the standard CC selection needs different number of CCs to obtain a given mean OOB power, the OCCS method is nearly independent from N_{CP}. Moreover, for each number of CCs, the OCCS method is not worse (in terms of the mean OOB power) than the standard method while significantly outperforming it as the required OOB power level decreases.

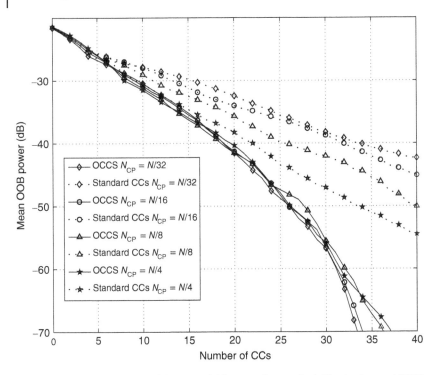

Figure 3.15 The comparison of the mean OOB power for standard CC selection and OCCS scheme for various CP durations at the PA input. (© 2013 IEEE. Reprinted, with permission, from [43].)

Additionally, the shorter the CP, the higher the saving in the number of required CCs. For example, if the required mean OOB power is -40 dB, for $N_{CP} = N/32$, the saving is about 44% of CCs (the decrease from 34 to 19 in the number of CCs). These saved subcarriers can be used as DCs, which increase the bit rate.

The remaining results are presented for the system with a fixed number of CC $\beta = 19$, $N_{CP} = N/16$, and Gray-mapped QPSK symbols. The comparison is performed among the NC-OFDM *reference system I*, i.e., without any spectrum shaping mechanism where all $\alpha + \beta$ subcarriers are DC, the systems applying standard CC, the heuristic OCCS method and OCCS combined with time-domain windowing with $\beta = 15$, and window cyclic suffix $N_W = 10$, that is, the parameters resulting in the same bit rate as systems using CC only.

In Figure 3.16, the PSD of the NC-OFDM signal can be observed. The locations of the selected CCs are also marked. It is visible that subcarriers used as CCs are distributed over a wider band in the OCCS method. While the standard selection of CC decreases the OOB power by 12 dB in comparison to reference system I, the proposed OCCS method provides additional 6 dB of the OOB power reduction. Moreover, the OCCS method lowers the peaks rising in the

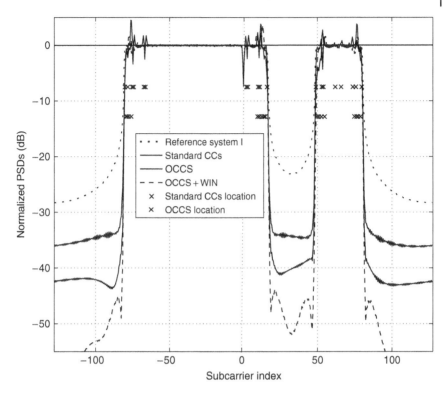

Figure 3.16 PSDs of signals in reference system I, the system with standard CCs, OCCS ($\beta = 19, N_{CP} = N/16$), and OCCS combined with windowing ($\beta = 15, N_W = 10$).

in-band region, which helps the issue of satisfying the SEM. Additionally, it can be observed that the OCCS can be effectively combined with the windowing scheme. Considering the constant bit rate, it can be observed that combination of OCCS and windowing lowers the mean OOB radiation power by about 8 dB in comparison to the OCCS method. Interestingly, the OCCS method also results in lower values of the Complementary Cumulative Distribution Function (CCDF) of PAPR compared to the standard CC selection scheme, for example, by 0.4 dB for $\Pr(\text{PAPR} > \text{PAPR}_0) = 10^{-4}$.

In Figure 3.17, the BER curves are shown for both CC selection schemes considered with the option of matrix $\mathbf{W}$ being available and not being available at the receiver. Both schemes are compared with *reference system II*, in which subcarriers that could be used as standard CCs (at the edges of the used band) are instead modulated by zeros and become simply the so-called guard subcarriers. This is assumed for the sake of fair comparison of systems with equal bit rate. Matrix $\mathbf{W}$ can be used at the receiver to improve the performance

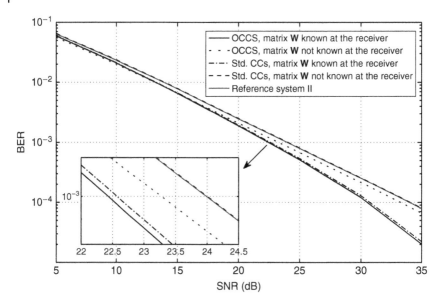

Figure 3.17 BER versus SNR for reference system II, the system with standard CCs allocation and OCCS; $\beta = 19$, $N_{CP} = N/16$. (© 2013 IEEE. Reprinted, with permission, from [43].)

making use of the redundancy symbols modulating CC as shown in [4]. Here, the nine-path Rayleigh channel model recommended for LTE [198] has been considered. There have been 50 000 channel instances simulated with 1000 random NC-OFDM symbols in each of them.

Note that the SNR loss with respect to the reference system II is the same for both CC selection methods when the **W** matrix is not known in the receiver. This is due to sacrificing some transmission power to the CC. This SNR loss equals $10\log_{10}(1 + \beta/\alpha)$, that is, about 0.7 dB in the considered scenario. The knowledge of matrix **W** allows the receiver to decrease the BER and obtain the SNR gain even over reference system II (of 0.8 dB for BER $= 10^{-3}$). The OCCS scheme outperforms the standard approach to CC selection, although the improvement of BER performance is relatively small (0.15 dB for BER $= 10^{-3}$). As the CCs improve mostly the reception quality of data subcarriers in their frequency neighborhood, the OCCS providing sparse CC pattern causes more DCs to be positively influenced.

The OCCS algorithm presented in this section allows lower computational complexity, lower OOB power, and lower PAPR in comparison with the standard CC algorithm. Moreover, the OCCS approach increases the SNR gain at the receiver if the decoder makes use of the known CC calculation (coding) matrix.

3.4 Reduction of Nonlinear Effects in NC-OFDM

In the previous chapter, flexible PSD shaping methods of the NC-OFDM signal have been discussed and evaluated (by the means of computer simulations) assuming ideal radio front end. Implementation of the spectrum shaping methods on a practical platform reveals the OOB power attenuation floor independent of the subcarrier spectrum sidelobe suppression parameters. This floor is caused by nonlinear effects of the radio front end, mostly nonlinear input–output characteristic of the power amplifier [119]. Although some distortions, for example, IQ imbalance, harmonics, or LO leakage, can be introduced by other front-end elements, the nonlinearity of HPA is seen to be the most severe problem.

Example PSDs are shown in Figure 3.18 for the same simulation setup as assumed for Figure 3.16 with additional curves for the NC-OFDM PSD at the output of HPA [43]. The HPA model used to obtain these curves is the Rapp model with nonlinearity hardness parameter $p = 10$. First, it can be seen that OOB power is always lower while observing the signal at the input of HPA. Unfortunately, nonlinear HPA causes the rise of the observed PSD floor in the OOB region. Interestingly, for the relatively low values of Input Back-Off (IBO), that is, 6 dB, the OOB floor in PSD is nearly the same for both OCCS and OCCS combined with WIN spectrum shaping methods, even though at the

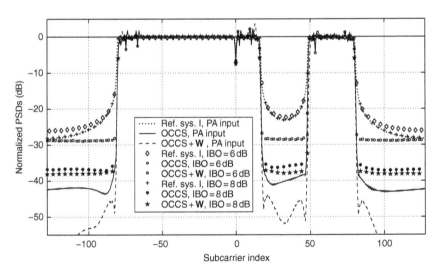

Figure 3.18 Comparison of NC-OFDM waveform PSD at the input and at the output of the Rapp-modeled HPA.

HPA input, the difference is quite substantial. It can be observed, at both ends of the observed NC-OFDM band, that spectrum shaping provides hardly any improvement in PSD over the reference system when relatively low IBO values are used. The conclusion is that reduction of OOB power in practical systems requires both methods for subcarriers-spectrum-sidelobe power reduction in the OOB region and measures to combat nonlinear effects in the analog RF front end. In the case of an NC-OFDM signal passing through highly nonlinear HPA, application of spectrum sidelobe reduction techniques may not bring the required effects in limiting the generated interference in the adjacent frequency bands.

The nature of nonlinear distortion in a multicarrier system could be revealed by removing the subcarrier spectrum sidelobe effects from the PSD plot, so that only nonlinear effects would result in the OOB power radiation. In order to do so, the NC-OFDM symbols can be prolonged cyclically in time. From the frequency-domain point of view, this is equivalent to changing the spectrum of each subcarrier from the *Sinc*-like shape to the Dirac delta. In case of a lack of nonlinear distortion, the comb of Dirac deltas should be observed at the frequencies of the occupied NC-OFDM subcarriers. An example of PSD when such a comb passes through a practical radio front end (Universal Software Radio Peripheral (USRP) N210 with WBX daughterboard [202]) is shown in Figure 3.19 [94]. On the left, the comb signal is observed over the band of 10 MHz in gray. The black curve is obtained by placing only zeros at the USRP input; that is, it depicts the spectrum analyzer noise floor and the USRP local oscillator leakage. On the right plot, the spectrum notch is zoomed. The nonlinear distortion floor is nothing else but a low-power comb of complex sinusoids separated by the same subcarrier spacing as used in NC-OFDM. The OOB radiation caused by nonlinear distortion of multicarrier waveform can be therefore explained by the Intermodulation Distortion (IMD) phenomenon.

An HPA can be modeled in a more suitable way in order to obtain the described comb. The memoryless polynomial model can be applied to complex baseband input signal $\tilde{s}_{BB}(t)$ giving the HPA output signal:

$$s_{BB}(t) = \sum_{k=1}^{K} b_k |\tilde{s}_{BB}(t)|^{2(k-1)} \tilde{s}_{BB}(t), \tag{3.50}$$

where b_k are the model coefficients, real-valued for a memoryless model or complex-valued for a *quasi-memoryless* model, that is, when the output phase is changed according to the input amplitude. The Rapp model can be approximated with the aforementioned memoryless polynomial. In the aforementioned equation, only uneven powers of $\tilde{s}_{BB}(t)$ are used, which is the effect of complex baseband representation [119]. A single NC-OFDM symbol as in (3.1) can be analyzed using the memoryless polynomial model. For simplicity, $K = 2$, and (3.50), as the memoryless model, is considered over the discrete

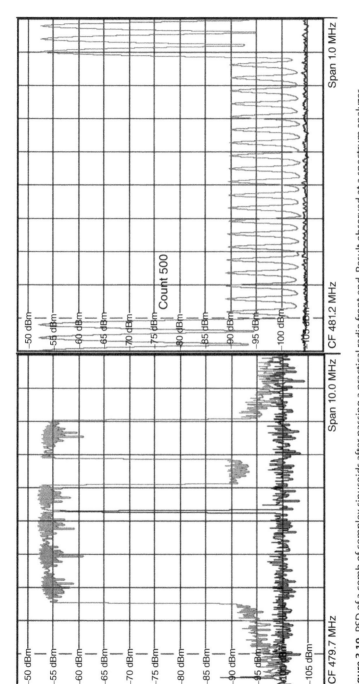

Figure 3.19 PSD of a comb of complex sinusoids after passing a practical radio front end. Result observed at a spectrum analyzer.

time domain. For the input signal sample $\tilde{s}_{\mathrm{BB}\ n}$, the output sample $s_{\mathrm{BB}\ n}$ is

$$s_{\mathrm{BB}\ n} = b_1 \tilde{s}_{\mathrm{BB}\ n} + b_2 |\tilde{s}_{\mathrm{BB}\ n}|^2 \tilde{s}_{\mathrm{BB}\ n} \tag{3.51}$$

for $n \in \{ -N_{\mathrm{CP}}, \dots, N - 1 \}$ and

$$\tilde{s}_{\mathrm{BB}\ n} = \sum_{j=1}^{\alpha} d_{\mathrm{DC}\ j} \exp \left(j 2\pi \frac{I_{\mathrm{DC}\ j} n}{N} \right). \tag{3.52}$$

The OOB components are complex sinusoids at frequencies being linear combinations of the occupied subcarrier frequencies. For higher order of nonlinearity considered, that is, $K > 2$, the linear combination of frequencies has higher order as well.

3.4.1 Sequential PAPR and OOB Power Reduction

The IMD reduction can be achieved by utilization of highly linear HPA or linearization of existing HPA characteristic (by means of predistortion). However, in practical HPA, there is always the saturation region in the input–output characteristic, that is, the signal clipping occurs for the input signal amplitude exceeding the saturation voltage V_{sat}. Even if the IBO value is higher than 0 dB, that is, the mean signal power is in the linear part of the HPA characteristic, occasionally, a sample amplitude can exceed V_{sat}, resulting in signal clipping. It is therefore essential to reduce the probability of high-power signal peaks, that is, to reduce the PAPR.

A number of various PAPR reduction methods can be found in the literature [121]. Some of these techniques can be considered as preferable ones for a number of practical reasons. For example, the PAPR methods that do not require transmission of the side information necessary to restore the original data signal (distorted intentionally at the transmitter by the PAPR reducing algorithm) may be considered for their basically costless development. Another preference for a PAPR reduction method could be to choose a scheme that does not require any modifications in the reception algorithm, ensuring backward compatibility with the standard receivers. In the context of preferable methods, the Active Constellation Extension (ACE) method [108] is considered to be very practical in the OFDM-based wireless transmission [203] and has been proposed for the new Digital Video Broadcasting-Terrestrial (DVB-T)2 standard, released in 2009 [204]. The principle of the ACE method is the amplitude distortion (increase or *expansion*) of some selected data symbols at the input of IFFT. The symbols to be distorted are selected so as to decrease PAPR at the IFFT output. As discussed in Section 2.3, in this method, only the outer constellation points can be predistorted (expanded) in order to maintain a fixed minimum distance between the constellation points. The optimization problem defined for ACE is to find a set of data symbols to be distorted to minimize the infinite norm of the IFFT-output signal samples [203]. A detailed analysis of this optimization problem for the ACE method can be found in [131]. Finally note that, when PAPR is low in comparison to IBO, the main source of the OOB power leakage is the shape of the subcarrier spectrum, while when PAPR is relatively high, reduction of the spectrum sidelobes is not sufficient, and the PAPR reduction is of higher importance.

In the literature, one can find various approaches to the problem of the baseband signal OOB power and the PAPR reduction. For example, [205, 206] present the combination of PAPR and OOB reduction algorithms. In most cases, however, there do not exist optimal solutions that jointly optimize all cost functions targeting optimal bit and power loading, OOB power reduction and PAPR reduction [205]. Some works propose the joint PAPR-OOB optimization problem, where the weighted impact of the specific criterion (i.e., PAPR or OOB power) on the goal function is considered, for example, paper [205]. Other works promote the sequential application of the OOB and PAPR reduction algorithms, that is, to implement respective PAPR and OOB power reduction procedures in a consecutive manner.

Example Results

Let us first evaluate the effectiveness of the ACE method (used for PAPR reduction), implemented as in [108], combined with the CC algorithm described before in Section 3.1.2. Let us consider the following example system setup (as in [203]): the IFFT order is set to $N = 256$, the cyclic prefix of $N_{CP} = N/16$ samples is used, and the utilized data subcarriers, modulated by the QPSK symbols, have indices $\mathbf{I}_{DC} = \{-100, \ldots, -1, 1, 2, 3, 46, \ldots, 100\}$, that is, the frequency notch of center frequency 24.5 is placed inside the NC-OFDM system band. Moreover, the considered ACE algorithm has the following parameters: the scaling parameter of the method (introduced in Section 2.3) is flexible and can be modified in the range from 1 to 3, the number of predistorted symbols is set to 25% of all data subcarriers, the samples to be attenuated must have their amplitude at least 1.4 times higher than the mean amplitude of all time-domain samples in one NC-OFDM symbol.

The number of CCs on each edge of the occupied subcarrier block is 4, that is, $\mathbf{I}_{CC} = \{-104, \ldots, -101, 4, \ldots, 7, 42, \ldots, 45\}$, and the factor used for limitation of CC power in (3.25) equals $\sqrt{\mu} = 0.04$. There are $\gamma = 152$ spectrum-sampling points defined in the OOB region. It was also assumed that the ACE method cannot modify the values carried on the CC. The results have been obtained for the transmission of 10^5 randomly modulated NC-OFDM symbols.

Similarly, as in [203], the CCDFs of PAPR for five cases have been investigated. The first case does not encompass any PAPR nor OOB power reduction method, that is, the original unmodified NC-OFDM signal has been considered. For the next two cases, the application of CC without ACE and the application of ACE without the CC method have been considered, respectively. For the last two cases, both methods have been implemented in the following

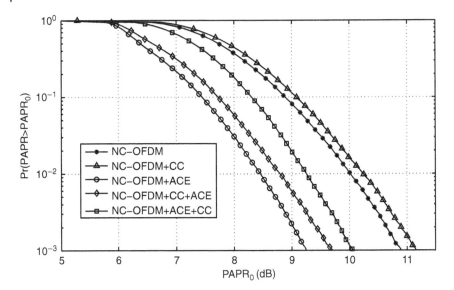

Figure 3.20 PAPR distribution (CCDF of PAPR) for the considered scenarios of joint approach to PAPR and OOB power reduction. (Based on [203].)

configurations: the ACE method performed before the CC method and the CC algorithm preceding the ACE. The results in terms of CCDF of PAPR are presented in Figure 3.20 for these five cases.

Note that an increase of CCDF of PAPR can be observed with respect to the reference signal when the CC method is applied alone. This means that the power back-off of HPA should be increased if the aim is to maintain the OOB level unchanged. However, this leads to lower energy efficiency of the power amplifier. Meaningful PAPR decrease (by more than 1 dB at the level of 10^{-3}) can be observed when both OOB power reduction and ACE methods are applied; however, it seems that the better choice of the joint PAPR and OOB-power reduction methods combination is for the case when the OOB power reduction is applied first. In such a case, PAPR is only minimally higher than that for the reference signal with the ACE method applied.

The impact of both the sidelobes of the subcarrier spectrum and nonlinearities of HPA on the signal spectrum shape, particularly in the adjacent bands at the output of HPA, can be observed in Figure 3.21 for the IBO parameter equal to 7 dB [203]. The HPA is Rapp-modeled with parameter $p = 4$. For this IBO parameter value, the most advantageous, in terms of the OOB radiation level at the output of HPA, is the combination of CC method, followed by PAPR reduction using ACE method.Use of only PAPR reduction or sidelobe suppression provides worse results. However, it is visible that the lowest PSD OOB radiation level before HPA is obtained for CC only or ACE followed by the CC

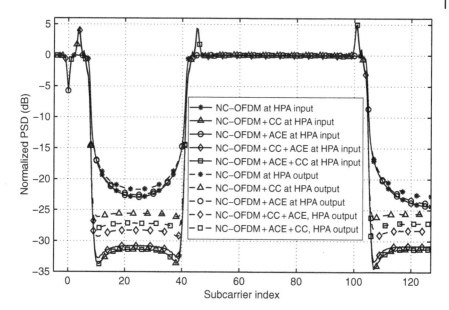

Figure 3.21 PSDs observed at the input and at the output of HPA (p=4) for IBO=7 dB.

methods. This means that for a sufficiently linear HPA, these setups will achieve the lowest OOB radiation level at the HPA output.

3.4.2 Joint Non-linear Effects Reduction with Extra Carriers[3]

As shown earlier in this chapter, the OOB radiation phenomenon can be seen as a sum of high-power Subcarrier Spectrum Sidelobes (SSSs) and IMD at HPA. Both problems are usually solved separately. The most promising solutions assume that some dedicated subcarriers can be sacrificed in the NC-OFDM signal to reduce either IMD or SSS power. In the context of the PAPR reduction, such an approach is known as the TR technique [208], while for IMD power reduction, these tones are called CC [41]. It is therefore reasonable to combine both approaches to minimize the resultant OOB power regardless of its source [207]. Although a combination of both methods has already been proposed in [209], different subcarriers were used for CC and Tone Reservation (TR) algorithms, resulting in a relatively weak OOB power attenuation. Furthermore, both algorithms have been used separately, that is, the TR method followed by the CC algorithm.

In [4], it has been shown that unwanted PAPR increase caused by the CC insertion can be observed. An advanced solution for this problem has been proposed in [205] where the same subcarriers are used for both PAPR and

---

3 © 2013 IEEE. Fragments of this subsection reprinted, with permission, from Ref. [207].

SSS mitigation. The optimization formula showed therein is based on a convex function being a weighted sum of infinitive norm for PAPR and second norm for subcarriers sidelobe power. In this case, the resultant OOB power attenuation depends on the weighting factor and the HPA characteristic. A closed-form solution of this problem is not provided, and the solution is obtained using computationally complex method from CVX package [210], impractical in a real hardware implementation.

A low-complexity algorithm for reduction of the OOB power observed at the output of the power amplifier regardless of its origin has been presented in [207]. There, the dedicated tones reserved to reduce the OOB radiation are called Extra Carriers (ECs).The proposed algorithm is based on computationally efficient implementations of CC [4, 43] and the gradient TR method described in [211]. The advantage of this solution comparing to the optimal one presented in [205] is the possibility of relatively simple hardware implementation of the proposed algorithm at the expense of slightly lower OOB power reduction at the output of HPA. Moreover, in this solution, the gradient TR method is used to reduce the power of IMD, not PAPR. The set of EC is used for the purpose of joint IMD and SSS power reduction at the same time. The main concept is to partially cancel the *Sinc*-shaped sidelobes at each iteration of the gradient TR method not allowing the TR algorithm to increase the SSS power. The side effect of the proposed method is a decreased value of the PAPR metric that can be used to increase the energy efficiency of the power amplifier [207].

In the considered NC-OFDM system, the block of α complex data symbols (e.g., QAM symbols) is delivered to the IFFT block of the size $N > \alpha$. These data symbols constitute vector $\mathbf{d}_{\mathrm{DC}}$ while the indexes of α useful DC are represented by vector $\mathbf{I}_{\mathrm{DC}}$. The entries of $\mathbf{I}_{\mathrm{DC}}$ are taken from the set $\{-N/2, \ldots, -1, 1, \ldots, N/2 - 1\}$. The $\mathbf{d}_{\mathrm{DC}}$ vector is also delivered to the dedicated module used for the EC generation.The complex symbols modulating EC constitute vector $\mathbf{d}_{\mathrm{EC}}$ of β elements whose indexes are in vector $\mathbf{I}_{\mathrm{EC}}$. The DC and EC sets are disjoint –the subcarriers used to accommodate EC are different from those used for DC. After the transformation from the frequency domain to the time domain, the CP of N_{CP} samples is added to the vector at the output of the IFFT block. Next, the digital time-domain signal is converted to the analog form in the D/A converter and then amplified and modulated to a certain radio-frequency band by means of dedicated RF front end.

The m-th sample of the time-domain NC-OFDM symbol **s** after CP insertion can be described as

$$s_m = \sum_{j=1}^{\alpha} d_{\mathrm{DC}j} e^{j2\pi \frac{m I_{\mathrm{DC}j}}{N}} + \sum_{j=1}^{\beta} d_{\mathrm{EC}j} e^{j2\pi \frac{m I_{\mathrm{EC}j}}{N}} \tag{3.53}$$

for $m \in \{ -N_{CP}, \ldots, N-1 \}$. The aforementioned formula can be presented in the matrix form as

$$\mathbf{s} = \mathbf{F}_{DC}\mathbf{d}_{DC} + \mathbf{F}_{EC}\mathbf{d}_{EC}, \tag{3.54}$$

where $\mathbf{F}_{DC}$ and $\mathbf{F}_{DC}$ denote the Fourier transform matrices with each column corresponding to one entry of $\mathbf{I}_{DC}$ and $\mathbf{I}_{EC}$, respectively. In order to minimize the SSS power, the spectrum values outside the nominal band (i.e., in the OOB region) have to be evaluated, for a given NC-OFDM symbol. Vector $\mathbf{V}$ of length γ containing the normalized OOB spectrum-sampling points taken from the normalized frequency range $\langle -N/2, N/2 \rangle$ can be defined as in the previous section. Utilizing the formula for a single subcarrier spectrum derived in (3.5), the vector $\mathbf{S}$ of Fourier spectrum samples at frequencies $\mathbf{V}$ can be defined as

$$\mathbf{S} = \mathbf{P}_{DC}\mathbf{d}_{DC} + \mathbf{P}_{EC}\mathbf{d}_{EC}, \tag{3.55}$$

where the (l,j)-th element of matrices $\mathbf{P}_{DC}$ and $\mathbf{P}_{EC}$ represents $S(I_{DCj}, V_l)$ and $S(I_{ECj}, V_l)$, respectively. For simplicity, in all the described algorithms (e.g., TR or CC) used for comparison, the same vector notation will be used.

Tone Reservation (TR)

The basic approach to PAPR reduction that utilizes additional dedicated tones is to minimize the peak power while maintaining the mean power of each EC not higher than the limit (equal to 1 when power normalization is assumed). Thus, the MinMax optimization problem can be presented as [207]:

$$\mathbf{d}_{EC}^{\star} = \arg \min_{\mathbf{d}_{EC}} \max |\mathbf{F}_{DC}\mathbf{d}_{DC} + \mathbf{F}_{EC}\mathbf{d}_{EC}|$$

$$s.t. \ \|\mathbf{d}_{EC}\|_2^2 \leq \beta. \tag{3.56}$$

Details on such an approach including mathematical analysis of the problem have been presented in [211]. One can observe that the inner maximum can also be written as an infinitive norm, that is,

$$\mathbf{d}_{EC}^{\star} = \arg \min_{\mathbf{d}_{EC}} \|\mathbf{F}_{DC}\mathbf{d}_{DC} + \mathbf{F}_{EC}\mathbf{d}_{EC}\|_{\infty}$$

$$s.t. \ \|\mathbf{d}_{EC}\|_2^2 \leq \beta, \tag{3.57}$$

where $\| \cdot \|_{\infty}$ denotes the infinity norm.

Cancellation Carriers (CC) Method

Similarly, as for the TR method, the CC values are calculated to minimize the mean power of the SSS in the frequency-sampling points $\mathbf{V}$ while maintaining the mean CC power equal to or below 1. Thus, the optimization problem is formulated analogously to the TR method , as shown earlier in Section 3.1 by (3.10), that is,

$$\mathbf{d}_{EC}^{\star} = \arg \min_{\mathbf{d}_{EC}} \|\mathbf{P}_{DC}\mathbf{d}_{DC} + \mathbf{P}_{EC}\mathbf{d}_{EC}\|_2^2$$

$$s.t. \ \|\mathbf{d}_{EC}\|_2^2 \leq \beta. \tag{3.58}$$

The computationally efficient solution of this problem, limiting mean ensemble power of CC, instead of instantaneous, has been shown earlier. Thus, following the derivation presented in Section 3.1, the exact solution can be obtained by simple matrix-vector multiplication

$$\mathbf{d}^\star_{EC} = -\left(\mathbf{P}^{\mathcal{H}}_{EC}\mathbf{P}_{EC} + \theta\mathbf{I}\right)^{-1}\mathbf{P}^{\mathcal{H}}_{EC}\mathbf{P}_{DC}\mathbf{d}_{DC} = \mathbf{W}\mathbf{d}_{DC}. \tag{3.59}$$

The aforementioned scalar θ, representing the constant value for a given sub-carrier pattern, is chosen in order to hold the EC power constraint. Matrix $\mathbf{W}$ can be directly used to calculate $\mathbf{d}^\star_{EC}$ from vector $\mathbf{d}_{DC}$.

Combination of CC and TR

In order to combine both PAPR and SSS reduction, a simple weighting of (3.57) and (3.58) can be used as proposed in [205]:

$$\mathbf{d}^\star_{EC} = \arg\min_{\mathbf{d}_{EC}} \lambda\xi\|\mathbf{P}_{DC}\mathbf{d}_{DC} + \mathbf{P}_{EC}\mathbf{d}_{EC}\|^2_2$$

$$+ (1 - \lambda)\|\mathbf{F}_{DC}\mathbf{d}_{DC} + \mathbf{F}_{EC}\mathbf{d}_{EC}\|_\infty$$

$$\text{s.t. } \|\mathbf{d}_{EC}\|^2_2 \leq \beta, \tag{3.60}$$

where $\lambda \in \langle 0; 1 \rangle$ is a factor for finding balance between PAPR and SSS reduction and factor $\xi = \|\mathbf{F}_{DC}\mathbf{d}_{DC}\|_\infty / \|\mathbf{P}_{DC}\mathbf{d}_{DC}\|^2_2$ is used to equalize the impact of both norms on minimized function for $\lambda = 0.5$. Although the optimization problem is convex, its solution (e.g., using the CVX package [210]) is highly computationally complex. Additionally, there is no method provided in [205] for adjusting λ for a given HPA characteristic to obtain the minimum OOB radiation power coming from both IMD and SSS [207].

Reduced-Complexity EC Method

The EC method makes use of computationally efficient implementation of TR method for PAPR reduction (the gradient method presented in [211]) and of the OCCS algorithm for SSS minimization described in the previous section and in [43]. The gradient TR method is iterative, so is the EC algorithm presented in Figure 3.22. It is worth mentioning that the solution of the optimization problem is not a trivial serial combination of both algorithms. Let us discuss it step by step.

The first approximation of the EC symbols, represented hereafter by vector $\mathbf{d}_{EC0}$, is nothing else but CC calculated according to (3.59). Afterward, IFFT of both DC and EC approximation is calculated. As in the original gradient TR algorithm [211], the time-domain signal is clipped with a soft limiter of clipping threshold $PAPR_0$. Here, $PAPR_0$ is the ratio of the maximum allowed peak power value to the mean signal power that is close to the IBO value characterizing the HPA operating point for a given mean power of the input signal. In consequence, $PAPR_0$ value can be more easily adjusted to the given HPA parameters than the value of λ parameter used in (3.60). In the next step, the original signal

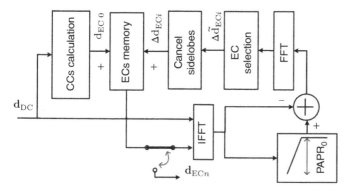

Figure 3.22 The block diagram of the EC calculation algorithm. (© 2013 IEEE. Reprinted, with permission, from [207].)

is subtracted from the signal generated by the clipper giving the time-domain representation (with the minus sign) of the original signal above the clipping level. After transformation of this signal to the frequency domain (FFT block shown in Figure 3.22), the auxiliary vector of N values is created from which the set of β EC of indices $\mathbf{I}_{EC}$ is selected in the *EC Selection* module. These identified ECs are appropriately scaled by a positive real value, which is the parameter of the TR gradient method used therein for its convergence tuning [207].

Let us denote the vector of β EC correction factors after the *EC selection* block in the i-th iteration as $\Delta \hat{\mathbf{d}}_{ECi}$. These correction factors are subject of SSS canceling in the *Cancel sidelobe* block. Before describing its functionality, let us observe that vector $\mathbf{d}_{ECn}$ obtained after n iterations, and defined as [207]

$$\mathbf{d}_{ECn} = \mathbf{d}_{EC0} + \sum_{i=1}^{n} \Delta \tilde{\mathbf{d}}_{ECi} \tag{3.61}$$

is the combination of CC (the first component of the sum) and $\Delta \tilde{\mathbf{d}}_{ECi}$ for $i = 1, \ldots, n$ obtained in the TRgradient method (the other elements of the sum) occupying the same subcarriers. The problem is that although $\mathbf{d}_{EC0}$ cancels the DC sidelobes at the first stage, the added component $\sum_{i=1}^{n} \Delta \tilde{\mathbf{d}}_{ECi}$ causes subcarrier sidelobe power regrowth (to the relatively high level $\|\mathbf{P}_{EC} \sum_{i=1}^{n} \Delta \tilde{\mathbf{d}}_{EC\,i}\|_2^2$), which is caused by the EC sidelobes. It can be stressed that this regrowth is not caused by a data carrier whose sidelobes have been substantially reduced by $\mathbf{d}_{EC0}$. Thus, only the sidelobes of $\sum_{i=1}^{n} \Delta \tilde{\mathbf{d}}_{ECi}$ are to be reduced. Moreover, the optimization problem (3.58) can be modified in order to minimize the spectrum sidelobes of EC(with the same power constraint):

$$\Delta \hat{\mathbf{d}}_{ECi} = \arg \min_{\Delta \hat{\mathbf{d}}_{ECi}} \|\mathbf{P}_{EC} \Delta \hat{\mathbf{d}}_{ECi} + \mathbf{P}_{EC} \Delta \tilde{\mathbf{d}}_{ECi}\|_2^2, \tag{3.62}$$

where $\Delta \hat{\mathbf{d}}_{\mathrm{EC}i}$ is an update of EC to reduce their spectrum sidelobes. The solution of this problem is obtained similarly as in (3.59), that is,

$$\Delta \hat{\mathbf{d}}_{\mathrm{EC}\,i} = -\left(\mathbf{P}_{\mathrm{EC}}^{H} \mathbf{P}_{\mathrm{EC}} + \theta \mathbf{I}\right)^{-1} \mathbf{P}_{\mathrm{EC}}^{H} \mathbf{P}_{\mathrm{EC}} \Delta \tilde{\mathbf{d}}_{\mathrm{EC}i} = \mathbf{W}_{2} \Delta \tilde{\mathbf{d}}_{\mathrm{EC}i}, \tag{3.63}$$

where, again, θ is chosen to maintain the power constraint. Finally, the values modulating EC are obtained at the nth iteration [207]:

$$\mathbf{d}_{\mathrm{EC}} = \mathbf{d}_{\mathrm{EC}n} = \mathbf{d}_{\mathrm{EC}0} + \sum_{i=1}^{n} \Delta \mathbf{d}_{\mathrm{EC}i}, \tag{3.64}$$

where

$$\Delta \mathbf{d}_{\mathrm{EC}i} = \Delta \tilde{\mathbf{d}}_{\mathrm{EC}i} + \Delta \hat{\mathbf{d}}_{\mathrm{EC}i} = (\mathbf{I} + \mathbf{W}_{2}) \Delta \tilde{\mathbf{d}}_{\mathrm{EC}i}. \tag{3.65}$$

Although high improvement in the OOB power attenuation is achieved in comparison to the straightforward combination of CC followed by the TR method, the computational complexity increase is relatively small. In each iteration, one additional matrix-vector multiplication needs to be performed. After n iterations, the EC method needs about $n\beta^2$ operations. Although the performance of the EC method could be expected to be worse in the PAPR reduction compared to the optimal solution proposed in [205], the iterative suboptimal approach has tractable computational complexity, making the solution more practical. Moreover, the implemented gradient method takes into account all samples exceeding PAPR_0, not just the highest one. From this perspective, the iterative EC algorithm should perform better in IMD reduction compared to the one presented in [205].

Example Results

Similarly, as in [207], let us evaluate the proposed methods in the following system setup. The simulated system uses $N = 256$ subcarriers with $\alpha + \beta = 144$ occupied subcarriers (data and EC) of indexes $\mathbf{I}_{\mathrm{DC}} \cup \mathbf{I}_{\mathrm{EC}} = \{ -80, -1, 1, 16, 33, 80 \}$. The system is noncontiguous as it uses two sets of subcarriers for data transmission, whose indexes vary from -80 to 16, without zero subcarrier, and from 33 to 80. If ECs are used, 5 subcarriers per subcarrier block edge are reserved (thus, 20 in total), that is, the vector of indexes is defined as $\mathbf{I}_{\mathrm{EC}} = \{ -80, -76, 12, 16, 33, 37, 76, 80 \}$. The duration of the cyclic prefix is of $N_{\mathrm{CP}} = N/4$ samples. The data carriers are modulated with random QPSK symbols, and 10^4 of NC-OFDM symbols in each simulation are transmitted. The following simulation cases are considered for comparison (as in Ref. [207]):

- Case A: the reference system where all 144 subcarriers carry data symbols;
- Case B: only the CC method is applied, so the EC are calculated using (3.59);
- Case C: only TR method is applied, and the ECs are calculated using TR gradient method;

- Case D: the TR method is implemented before the CC method; 10 ECs are determined using the gradient TR method, while another 10 ECs (closest to the subcarrier block edges) are calculated using (3.59);
- Case E: the CC method is implemented before the TR method; 10 ECs (closest to the subcarrier block edge) are calculated using (3.59) followed by 10 ECs generated by TR gradient algorithm;
- Case F: the reduced-complexity iterative OOB power reduction EC scheme described in the previous section, in which all ECs are modulated according to (3.64);
- Case G: optimal CC&TR combination scheme presented in [205], in which the EC symbols are calculated according to (3.60).

The results obtained for both SSS and PAPR reduction are apparent on a PSD plot of a signal at the output of HPA. For a realistic OOB power calculation, the Welch method has been used preceded with four times of oversampling and a low-pass filtering. The selected filters have the same characteristics as in the commonly used radio platform USRP [202]. The power amplifier model is a simple soft limiter. For the sake of clarity, an example PSD plot (for positive subcarrier indexes), for a subset of cases described earlier, is shown in Figure 3.23.

The following values of the parameters used in the gradient-based methods were defined: the number of iterations was set to 40, and $PAPR_0$ was assumed to be equal to 5 dB. For case G, the parameter $\lambda = 0.5$ was used. Moreover, the

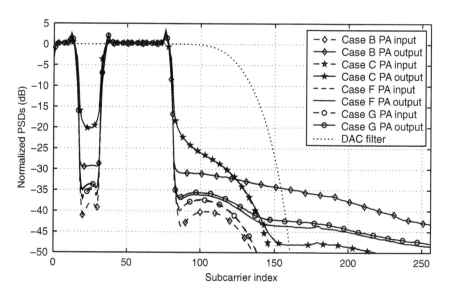

Figure 3.23 PSDs for a subset of all systems at the input and at the output of HPA (PAPR0=5 dB), for iterative methods 40 iterations per symbol were used, for Case G $\lambda = 0.5$. (© 2013 IEEE. Reprinted, with permission, from [207].)

IBO of the HPA was set to 6 dB. The dashed lines, that is, power spectrum of signals at the input of HPA, show that the lowest OOB power is obtained in case B. Nearly equal values are achieved for case G and for the iterative EC method (case F).The result for case C is the same as for case A (not shown on plot), that is, the OOB power is higher than in all other considered cases.

The PSDs of signals at the output of HPA, illustrated in Figure 3.23 by solid lines, have the most advantageous shape for cases F and G. Slightly better performance is achieved with the iterative EC algorithm (case F). The reason for such a result is that the approach to PAPR reduction used in that case reduces all signal values above some threshold, not only the highest peak as in the algorithm suggested in [205]. The OOB power attenuation is much lower in cases B and C.

Moreover, the influence of D/A filtering on the shapes of signals PSD can also be observed. At the input of HPA, all spectrum sidelobe components being out of the filter passband, that is, above subcarrier index 128, are practically removed. However, at the output of HPA, intermodulations are introduced in the whole band. A good example, in which the intermodulations can be distinguished from subcarrier spectrum sidelobes can be observed when analyzing the results for case C. The OOB power within the /A passband region is quite high as spectrum sidelobes are not canceled here, but it is the lowest, among other compared systems, outside the D/A passband, where the intermodulations are the dominant OOB power component. As the most significant PAPR reduction is obtained in the TR method alone (case C), the intermodulation power is the lowest, as expected.

Let us now focus on the D/A passband. The mean OOB powers within the D/A passband at the output of HPA as the function of IBO are shown in Figures 3.24–3.26. Moreover, for case G, optimal λ value has been found numerically. For cases C, D, E, and F, maximum of 40 iterations of an algorithm have been applied. The scaling factor used in the EC selection block equal to 5 was experimentally chosen.

In Figure 3.24, the mean OOB power is compared for all systems presented earlier. One can see that the reference system (case A) and the system using only a TR algorithm (case C) do not provide significant OOB power attenuation. For high IBO values, that is, when hardly anything is clipped in HPA and intermodulations are negligible, case C has some advantages over case A as ECs have much lower power than data carriers, resulting in lower spectrum sidelobe power. The introduction of CC achieves strong OOB power attenuation improvement. While combinations of the CC and TR methods, that is, cases D and E, outperform the CC method alone (case B) for low IBO (when intermodulations play a dominant role), for high IBO values using all ECs for sidelobe suppression, CC is a beneficial strategy. Note that case G and EC algorithm with reduced complexity (case F)outperform all other methods in terms of OOB power attenuation. It should be noted that for any IBO value,

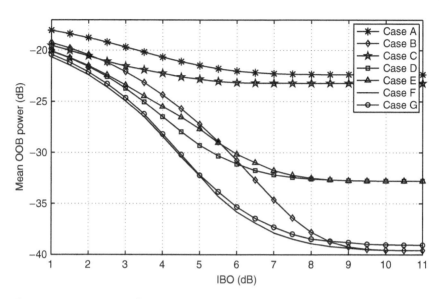

Figure 3.24 Comparison of mean OOB power at the output of HPA of all considered systems. About 40 iterations are used per NC-OFDM symbol in the iterative methods. (© 2013 IEEE. Reprinted, with permission, from [207].)

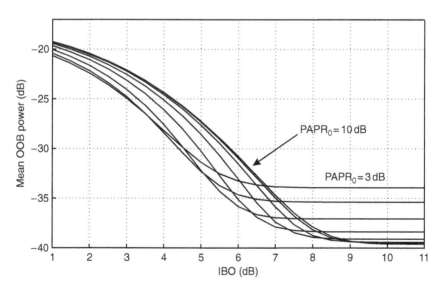

Figure 3.25 Mean OOB power at the output HPA of the NC-OFDM signal employing EC method with 40 iterations per symbol while varying $PAPR_0$ from 3 to 10 dB. (© 2013 IEEE. Reprinted, with permission, from [207].)

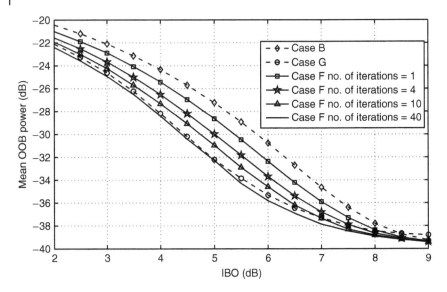

Figure 3.26 Comparison of the mean OOB radiation power of the proposed EC algorithm while changing the number of iterations. (© 2013 IEEE. Reprinted, with permission, from [207].)

the reduced-complexity EC algorithm is even slightly better than optimal one defined in case G. The explanation for this effect is the same as for the results observed in Figure 3.23.

In Figure 3.25, the mean OOB power for 40 iterations of EC algorithm is plotted while changing $PAPR_0$ parameter from 3 to 10 dB. It can be noticed that low $PAPR_0$ values should be chosen for low IBO values; however, they cause high OOB power floor for high IBO values. The $PAPR_0$ value should be therefore adjusted according to the HPA configuration used. The value of $PAPR_0$ should be similar to the HPA IBO that, in practice, is much easier to be established than the correct value of λ parameter used in [205].

An interesting issue is how to balance the number of iterations for each NC-OFDM symbol of the reduced-complexity EC algorithm in order to obtain strong OOB power attenuation while maintaining computational complexity low. The value used so far, that is, 40 iterations, gives a result close to an ideal one, that is, case G, but will be impractical in real hardware implementation. In Figure 3.26, the mean OOB power values for the EC algorithm for 1, 4, 10, and 40 iterations are presented. For the mean OOB power of −35 dB, this increase in the number of iterations of the mentioned algorithm allows for an increase in the HPA efficiency, that is, a decrease in IBO value, by 0.4, 0.7, 1, 1.4 dB, respectively, in comparison to the CC method alone (case B).

To summarize, the reduced-complexity EC method aims at lowering the out-of-band power in NC-OFDM systems, which are highly prone to intermodulation effects caused by nonlinear front-end components, in particular, by a high-power amplifier. The idea is to use the same set of carriers for joint IMD and SSS reduction. Based on the obtained results, it is apparent that it is possible to merge the two approaches, used traditionally separately for PAPR and SSS reduction, and minimize the OOB radiation in a computationally efficient manner. Utilization of such an algorithm in the NC-OFDM transmitter reduces the interference caused to the systems adjacent on the frequency axis and decreases the *electromagnetic pollution.* Additionally, the decreased PAPR value of an NC-OFDM signal allows for the use of highly efficient power amplifiers reducing the energy consumption.

3.5   NC-OFDM Receiver Design

The receiver of an NC-OFDM system has similar functionalities as those in the OFDM receiver. The key properties of OFDM, such as the use of CP, orthogonality of subcarriers, subcarrier spacing, are maintained for NC-OFDM that allow for a simple adoption of the existing OFDM reception algorithms for noncontiguous subcarrier blocks. On the other hand, the presence of possibly strong interference in the band of the NC-OFDM receiver, that is, in the NC-OFDM spectrum notch, significantly deteriorates the reception performance. Moreover, spectral agility of the NC-OFDM waveform at the transmitter must also be tracked at the receiver. The main issues of NC-OFDM reception are discussed as follows.

The key issue while considering NC-OFDM reception is that strong interference may be present in the spectrum notches of the NC-OFDM receiver band. This interference may originate from other systems using the band unused by the considered NC-OFDM system. This is not the case for OFDM systems, where signals of systems operating in adjacent bands are typically outside the bandwidth of an OFDM radio front end, and rejection of this interference signal takes place already at the receiver input. In case of NC-OFDM with flexible choice of bands, fixed filters are applied at the RF front end, and the received signal with interference is conveyed up to the digital domain where decision on the utilized subcarriers is made. All the receiver front-end analog components (amplifiers, mixers, A/D converters) have to have a high dynamic range in order not to distort the wanted signals [212].

Theoretically, in the digital domain, the unwanted signal can be rejected at the output of DFT block by nulling the appropriate subcarriers. However, first, time- and frequency-domain synchronization has to be achieved in order not to observe ISI and ICI (as explained in the case of OFDM in the previous chapter). On the other hand, most of the OFDM-based synchronization algorithms operate in the time domain, for example, [148, 213], being unable to reject the band-limited interference. This issue is covered in the next section focusing on synchronization.

In case of the NC-OFDM receiver perfectly synchronized in time and in frequency with the wanted signal, the same channel estimation and equalization algorithms can adopted as the ones typically used for OFDM. However, let us note that interfering signals utilizing bands nonoverlapping with the considered NC-OFDM system occupied band can still cause interference to NC-OFDM subcarriers. Let us consider just an interference signal $i(t)$ present at the input of the NC-OFDM receiver. Its Fourier transform representation $I(f)$ is band-limited, so that inverse Fourier transform can be defined as

$$i(t) = \int_{-\frac{f_s}{2}}^{\frac{f_s}{2}} I(f)e^{j2\pi ft}df, \tag{3.66}$$

where $f_s = N\Delta f$ is NC-OFDM sampling frequency. The signal received at the n-th output of DFT is

$$\hat{d}_n = \frac{1}{\sqrt{N}} \sum_{m=0}^{N-1} i(m\Delta t)e^{-j2\pi\frac{nm}{N}}, \tag{3.67}$$

that after substitution of (3.66) gives

$$\begin{aligned}
\hat{d}_n &= \frac{1}{\sqrt{N}} \sum_{m=0}^{N-1} \int_{-\frac{f_s}{2}}^{\frac{f_s}{2}} I(f)e^{j2\pi fm\Delta t}df\, e^{-j2\pi\frac{nm}{N}} \\
&= \frac{1}{\sqrt{N}} \int_{-\frac{f_s}{2}}^{\frac{f_s}{2}} I(f) \sum_{m=0}^{N-1} e^{j2\pi\frac{m}{N}\left(\frac{f}{\Delta f}-n\right)}df.
\end{aligned} \tag{3.68}$$

Using formula for the sum of geometric progression, we get

$$\hat{d}_n = \frac{1}{\sqrt{N}} \int_{-\frac{f_s}{2}}^{\frac{f_s}{2}} I(f)e^{j\pi\left(1-\frac{1}{N}\right)\left(\frac{f}{\Delta f}-n\right)} \frac{\sin\left(\pi\left(\frac{f}{\Delta f}-n\right)\right)}{\sin\left(\frac{\pi}{N}\left(\frac{f}{\Delta f}-n\right)\right)}df, \tag{3.69}$$

which can be interpreted as convolution:

$$\hat{d}_n = \int_{-\frac{f_s}{2}}^{\frac{f_s}{2}} I(f)\Gamma(n\Delta f - f)df, \tag{3.70}$$

where

$$\Gamma(x) = \frac{1}{\sqrt{N}} e^{-j\frac{\pi}{\Delta f}\left(1-\frac{1}{N}\right)x} \frac{\sin\left(\frac{\pi}{\Delta f}x\right)}{\sin\left(\frac{\pi}{N\Delta f}x\right)} \tag{3.71}$$

is the function convolved with $I(f)$ in the frequency domain, which results in the interference signal components occurring at all DFT subcarriers.

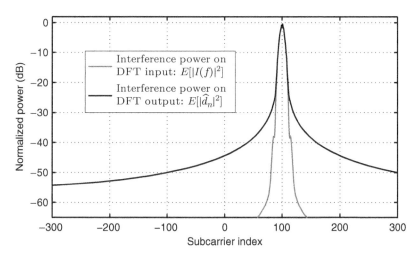

Figure 3.27 Interference power caused by GSM carrier on the input/output of DFT.

An example of this phenomenon is shown in Figure 3.27 for the considered NC-OFDM receiver using subcarrier spacing of 15 kHz (as in LTE) and receiving coexisting GSM signal. There, the interference power of a single Global System for Mobile Communications (GSM) carrier at the input and at the output of the NC-OFDM receiver DFT is presented. Most importantly, the interference components have the highest power at subcarriers closest to the band occupied by the interfering signal. If the interference is too strong, it is possible to apply some advanced signal processing methods, such as windowing at the receiver [214]. In such a case, rectangular window applied at DFT is replaced by a window having stronger sidelobe attenuation, for example, raised-cosine window.

Another issue that has to be considered is the design of reference signals, for example, pilots and preambles, in order to be used for efficient reception of the signals transmitted in systems with frequency agility and noncontiguous bands. However, it seems to be an open topic. For instance, the Primary Synchronization Signal in LTE utilizes Zadoff–Chu sequences due to their constant amplitude (minimal PAPR), and different sequences can be designed to be orthogonal to each other [215]. In the case of noncontiguous subcarrier allocation, the Zadoff–Chu sequence will lose these properties.

3.5.1 NC-OFDM Receiver Synchronization[4]

In the future network scenario of spectrum sharing (or CR), , a secondary system using spectrum opportunities is required to protect the incumbent systems, that is, to maintain the interference level below the set limit. Thus, in the

---

4 © 2016 IEEE. Fragments of this subsection reprinted, with permission, from Ref. [216].

considered secondary NC-OFDM system, the OOB power must be low enough to maintain the created interference power observed at a LU receiver below the required level. This can be achieved by applying the GS on the edges of the NC-OFDM signal useful bands (neighboring the Licensed User (LU)'s band) or by another method as described in the previous sections. However, an LU system is not required to take any measures to protect the secondary transmission. The NC-OFDM signal detection algorithm can take advantage of the fact that an LU system uses frequencies unused by the secondary NC-OFDM system inside its band [161] and suppresses the interference by means of the frequency-domain data symbol detection using an FFT block. However, it requires precise prior synchronization of the NC-OFDM frame in the time- and frequency domain, which is recognized as a challenge when the in-band interference is present [217].

In the literature, synchronization algorithms developed for OFDM have also been considered for NC-OFDM. The preamble-based Schmidl&Cox (S&C) algorithm [148] is the most known. There, the preamble preceding the data symbols consists of two identical series of time samples. It is referred to as the *S&C preamble* as follows. The receiver determines the autocorrelation in the time domain to find the optimal timing point, that is, the first effective sample of the first OFDM symbol in a frame (after omission of the CP). The result of this autocorrelation is also used for the estimation of fractional Carrier Frequency Offset (CFO), which is the fractional part of the CFO lying in interval $(-1, 1)$ of the frequency normalized to the OFDM subcarrier spacing. One additional OFDM symbol following the preamble is used to estimate the integer part of the CFO. The S&C algorithm has been improved in [213] and [218], where the MSE of time and frequency synchronization are reduced, respectively. The overview of the synchronization techniques proposed for OFDM is presented in [147].

The aforementioned works consider multipath fading channels with AWGN. In our considered case of a CR system, performance deterioration can be caused by the presence of high-power LU-originating interference. Recently, the S&C algorithm has been evaluated for an OFDM system operating in the presence of various interference types: narrowband modeled as a complex sinusoid [219], narrowband digitally modulated signal [220], and wideband interference signal occupying all SCs adjacent to the block of SCs used by an OFDM CR system [221]. These works consider OFDM with contiguous SC, but conclusions are valid for the NC-OFDM as well. While wideband interference is modeled in [221] as AWGN modulating OFDM-system-adjacent SC, the synchronization performance is degraded as if the preamble was distorted only by AWGN of the same power as the interference. This is because the wideband interfering signal does not have high autocorrelation peaks apart from the main one.

In [221], the authors make the assumption that interference occupies SC orthogonal to a secondary OFDM system SC. However, the first stage of the considered S&C synchronization algorithm is implemented in the time domain where separation of symbols modulating SC is not possible. Even more difficult is NC-OFDM synchronization in case of the Narrowband Interference (NBI) (modeled as complex sinusoid). As shown in [219], in the absence of a useful signal, the synchronization algorithm can result in the false detection of a frame (interference can be detected as a useful signal). This is caused by the autocorrelation properties of the complex sinusoid. As the bandwidth of the interfering signal increases, this effect is suppressed [220]. Because of this *false synchronization* effect, NBI is a much worse case than the Wideband Interference (WBI) for the NC-OFDM synchronization. In [222], the S&C synchronization algorithm performance in the presence of ideal and practical NBI is evaluated. In the ideal case, NBI is modeled as a complex sinusoid while practical NBI is a narrowband Frequency Modulation (FM) signal. The probability of synchronization error is slightly higher in the case of ideal interference as the false synchronization is more frequent.

It has been shown that extension of the range of the autocorrelation calculation does not improve the S&C algorithm [223] in the presence of NBI. One approach to the NBI rejection at the NC-OFDM receiver is to suppress it by filtering. As shown in [219], the a priori knowledge or correct detection of interference center frequency is required for this purpose, which is impractical. Similarly, subtraction of an estimated NBI signal proposed in [217] requires correct detection of the frequency, amplitude, and phase of an interfering signal. Low probabilities of frame false-and-miss detection can be obtained by the frequency-domain filtering and cross-correlation. However, this approach proposed in [224] assumes an idealistic channel with no fading and no CFO. Finally, rejection of interference by filtering has major practical issues. The filter has to be adjusted to the interference characteristic and redesigned as the CFO or an interfering signal frequency changes. Moreover, low-order filters can distort the useful signal [217] while high-order filters have high complexity and contribute to the time-domain signal dispersion [160].

As discussed earlier, standard autocorrelation-based synchronization algorithms are not suitable for a CRsecondary NC-OFDM system in the presence of an in-band LU-generated NBI. Utilization of cross-correlation between the received preamble and the reference (original) preamble can be more advantageous [94, 225]. Combination of cross-correlation-based synchronization with spectrum sensing has been proposed in [226]. There, however, no measures to fight the multipath effects were applied, which causes detection of the strongest channel path as the optimal timing point. One can expect that nonzero CFO will prevent correct detection of a preamble in this algorithm. In order to increase the probability of frame detection in the presence of interference and noise, it was proposed to use pseudonoise sequence as a preamble [227]. However, such a preamble generated in the time domain is not filtered, and therefore, it does not protect the LU transmission.

A multistage synchronization proposed in [228] improves the performance of the S&C algorithm for an OFDM system. The S&C algorithm is used therein to obtain coarse time and fractional frequency synchronization. Afterward, cross-correlation-based synchronization is used to localize the first channel-path component and to estimate the integer CFO. However, this framework seems to be not useful in the presence of in-band interference, as it is very probable that the first step (autocorrelation) will fail, preventing the success of the second stage. The algorithm proposed in [229] uses modified cross-correlation of adjustable length to obtain coarse time synchronization. This is followed by path-timing and integer-CFO estimation according to a method from [228]. Finally, a fractional-CFO estimator uses amplitudes of Fourier-transformed cross-correlation output as in [230]. The main drawback of this algorithm is high complexity of coarse time synchronization and limited robustness to CFO. Moreover, fractional-CFO estimation utilizes the information conveyed by the strongest path component only.

3.5.2 In-Band-Interference Robust Synchronization Algorithm for an NC-OFDM System[5]

In this section, the synchronization algorithm for an NC-OFDM CRsystem robust against an LU-originating interference is presented as described in [216]. Thus, the applied preambles cannot use SC coinciding with the LU band. It is assumed that the LU transmission is protected from the CR-generated interference by appropriate spectrum sensing and managing and by the application of the guard SC or other spectrum shaping algorithm that reduces the NC-OFDM out-of-band radiation power. The algorithm is called Licensed-User Insensitive Synchronization Algorithm (LUISA). It is based on cross-correlation of the received and Reference Preamble (RP) available at the receiver. The algorithm is insensitive to false synchronization effect as the interference is uncorrelated with the NC-OFDM preamble. Additionally, the probability of synchronization error is decreased for a given SNR or Signal-to-Interference Ratio (SIR) in comparison to [148] and [229]. The algorithm works well for low SIR and SNR in the presence of LU-generated interference. It makes use of all signal path components with their phase dependencies, which allows for the decrease of the carrier-offset and timing-offset estimation MSEin comparison to [229].

Let us consider an NC-OFDM-based SU system in the presence of multipath channel distortions and interference. The transmitter uses N-point IFFT for multicarrier modulation. In each frame, P NC-OFDM symbols are transmitted, and each NC-OFDM symbol is preceded by a CP of N_{CP} samples. Let us denote a vector of complex symbols at the input of IFFT as $\mathbf{d}^{(p)} = \{d_n^{(p)}\}$, where $n = -N/2, \dots, N/2 - 1$ is the SC index, and $p = 0, \dots, P - 1$ is the NC-OFDM

---

5 © 2016 IEEE. Fragments of this subsection reprinted, with permission, from Ref. [216].

symbol index. Note that α values out of N SC are modulated by complex symbols. The indices of these SCs are elements of vector $\mathbf{I}_{DC} = \{I_{DC_c}\}$ for $c = 1, \ldots, \alpha$ and $I_{DC_c} \in \{ -N/2, \ldots, N/2 - 1\}$. The other SCs (including GS at the band edges) are modulated by zeros, that is, both real and imaginary parts of $d_n^{(p)}$ equal to zero, in order to protect the LU transmission and to allow digital–analog conversion, that is, $d_n^{(p)} = 0$ for $n \in \{ -N/2, \ldots, N/2 - 1\}\backslash\mathbf{I}_{DC}$. The noncontiguity of the considered NC-OFDM system SC results from the scenario-dependent choice of SC indices I_{DC_c}; i.e., it is not a series of consecutive integer numbers as in a typical OFDM system. The m-th sample of the p-th NC-OFDM symbol at the output of IFFT and after the CP addition is defined as

$$s_m^{(p)} = \begin{cases} \frac{1}{\sqrt{N}} \sum_{n=-N/2}^{N/2-1} d_n^{(p)} e^{j2\pi\frac{nm}{N}} & \text{for } -N_{CP} \leq m \leq N - 1 \\ 0 & \text{otherwise.} \end{cases} \tag{3.72}$$

It is assumed that $\mathbf{s}^{(0)} = \{s_m^{(0)}\}$ is a vector of preamble samples for $p = 0$ and $m = 0, \ldots, N - 1$, that is, without CP. Without the loss of generality, we narrow our considerations to one NC-OFDM frame. The transmitted signal $\tilde{s}(m)$ composed of subsequent NC-OFDM symbols is thus

$$\tilde{s}(m) = \sum_{p=0}^{P-1} s_{m-p(N+N_{CP})}^{(p)}. \tag{3.73}$$

The signal observed at the NC-OFDM receiver is distorted by the L-path fading channel with l-th path channel coefficient $h(l)$, the CFO normalized to SC spacing v, the additive interference $i(m)$, and the white noise $w(m)$. The m-th sample of the received signal $r(m)$ equals

$$r(m) = \sum_{l=0}^{L-1} \tilde{s}(m - l)h(l)e^{j2\pi\frac{mv}{N}} + i(m) + w(m). \tag{3.74}$$

Note that discrete-time representation is sufficient for our considerations. It was shown in [147] that the timing error lower than the sampling period is compensated by an NC-OFDM equalizer.

The conceptual diagram of LUISA is shown in Figure 3.28. At the output of the radio front end and the analog-to-digital converter, the baseband complex samples undergo S/P conversion. Note that in the case of no CFO, the received preamble samples are correlated with the RP $\mathbf{s}^{(0)}$. The RP is assumed to be known at the receiver as other transmission parameters, after setting the connection through a suitable *randezvous* technique or using a cognitive pilot (control) channel. The RP can be generated at the receiver from a pseudorandom generator, with the same initial state as the one at the transmitter.

Note that in the case of perfect time and frequency synchronization, the received preamble samples are correlated with RP, which results in the peak of the calculated cross-correlation. However, correlation of the received signal

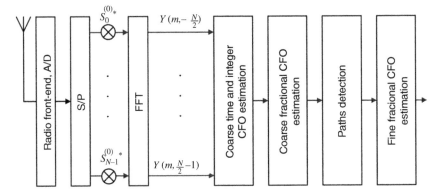

Figure 3.28 NC-OFDM receiver with the LUISA synchronization algorithm. (Based on [216].)

with the RP would not result in successful synchronization in case of relatively high CFO. Each cross-correlation component would be added with a different phase, dependent on CFO, and no peak corresponding to the optimal timing point would be observed. In order to counteract this effect, N consecutive input samples are multiplied by the complex conjugates of the RP and fed to the FFT block. Thus, at the n-th output of FFT at the m-th sampling moment we obtain

$$Y(m,n) = \sum_{k=0}^{N-1} r(m+k)s_k^{(0)*}e^{-j2\pi\frac{mk}{N}}, \tag{3.75}$$

where $(\cdot)^*$ denotes complex conjugate. $Y(m,n)$ is called the *synchronization variable* in the remainder of this section. By substituting (3.74)–(3.75), we obtain

$$Y(m,n) = \sum_{l=0}^{L-1} h(l)Y_l(m,n) + Y_{\text{in}}(m,n), \tag{3.76}$$

where

$$Y_l(m,n) = e^{j2\pi\frac{mv}{N}} \sum_{k=0}^{N-1} \tilde{s}(m+k-l)s_k^{(0)*}e^{j2\pi\frac{k(v-n)}{N}}, \tag{3.77}$$

$$Y_{\text{in}}(m,n) = \sum_{k=0}^{N-1} (i(m+k)+w(m+k))s_k^{(0)*}e^{-j2\pi\frac{nk}{N}}, \tag{3.78}$$

and the defined $Y_l(m,n)$ and $Y_{\text{in}}(m,n)$ are the l-th channel-path component and the interference-and-noise involving component of $Y(m,n)$, respectively. Before we discuss the next steps of the synchronization algorithm (as indicated in the respective blocks in Figure 3.28), let us consider some statistical properties of $Y(m,n)$.

In the case of LUISA, two preamble shapes are considered in [216]. The *S&C preamble*, as mentioned previously, is generated by the pseudorandom $d_n^{(0)}$ symbols modulating SC of even indices out of a set $\mathbf{I}_{DC}$ (SC of odd indices are modulated by zeros). In order to maintain the constant power of the transmitted signal, vector $x_m^{(0)}$ for $m = -N_{CP}, \dots, N - 1$ is multiplied by $\sqrt{2}$ for this kind of the preamble. Moreover, another preamble shape is considered in [216], herein referred to as *simple preamble*. It is generated by modulation of all α SC (of all indices in $\mathbf{I}_{DC}$) by the pseudorandom independent symbols.

According to the Central Limit Theorem (CLT), the samples $\tilde{s}(m)$ can be treated as complex random Gaussian variables with the zero mean and variance σ_s^2, approximately independent for sufficiently high α (for $1 \ll \alpha < N$), as shown in [231] (with the exception of repeatability of the CP or the S&C preamble, which will be taken into account in the latter calculations). The assumption is that if the pseudorandom preamble symbols were chosen to lower the PAPR (e.g., Zadoff–Chu sequences in LTE), the PAPR-lowering method would not change the signal probability distribution [232]. These assumptions allow for negligence of correlation between preamble samples. More accurate modeling of the preamble properties would increase the algorithm complexity, while the synchronization quality improvement would be minor.

The noise $w(m)$ is modeled as Gaussian complex random variable of the zero mean and variance σ_w^2. For the purpose of LUISA, interference observed in the NC-OFDM system is modeled in two ways: one as the random stationary complex Gaussian process of the zero mean and variance σ_i^2, here referred to as *NL interference model*, and the other as

$$i(m + k) = \frac{1}{\sqrt{N}} \sum_{n=-N/2}^{N/2-1} g_n(m) e^{j2\pi \frac{nk}{N}}, \tag{3.79}$$

where $g_n(m)$ is the frequency representation of interference at the n-th SC at the mth moment, and $k \in \{0, \dots, N-1\}$. More importantly, this representation of (3.79) holds for any interfering signal, which may or may not be orthogonal to NC-OFDM SC (in the classical sense of orthogonality over the period of N samples). As an example, the NBI of carrier frequency not aligned with any NC-OFDM FFT frequency (thus, not orthogonal to the NC-OFDM signal) has a high number of nonzero coefficients $g_n(m)$ for $n \in \{-N/2, \dots, N/2 - 1\}$. This second interference model defined by (3.79) is referred to as *GIB interference model* in the remainder of this section. Note that according to the Parseval theorem: $\sum_{n=-N/2}^{N/2-1} |g_n(m)|^2 = N\sigma_i^2$. Moreover, the LU-generated interference has usually most of the energy located in the NC-OFDM spectrum notch(es), that is, in frequency bands not utilized by NC-OFDM (in (3.79), $g_n(m) \approx 0$ for $n \in \mathbf{I}_{DC}$). As shown in [233], $g_n(m)$ can be modeled as random complex Gaussian variable with the zero mean, uncorrelated with interference at any other frequency (different from n), although interference variances for various n can be correlated.

As shown earlier, the $r(m)$ sample for each m can be treated as an approximately independent complex random variable. Based on the CLT, $Y(m, n)$ probability distribution can be modeled as complex Gaussian because it is a sum of N weighted random variables. The proposed metric adds N consecutive samples in time, which should keep the expectation in time close to the ensemble expectation. Thus, the expected value $\mathbb{E}[Y(m, n)]$ can be calculated as follows:

$$\mathbb{E}[Y(m, n)] = e^{j2\pi \frac{mv}{N}} \sum_{l=0}^{L-1} h(l) \sum_{k=0}^{N-1} \mathbb{E}[\tilde{s}(m + k - l)s_k^{(0)*}] e^{j2\pi \frac{k(v-n)}{N}}$$

$$+ \sum_{k=0}^{N-1} (\mathbb{E}[i(m + k)s_k^{(0)*}] + \mathbb{E}[w(m + k)s_k^{(0)*}]) e^{-j2\pi \frac{nk}{N}}. \tag{3.80}$$

In [216], it has been shown that with all aforementioned assumptions, for both interference models (Noise-Like (NL) and Generalized In-Band (GIB)):

$$\mathbb{E}[Y_{\text{in}}(m, n)] = 0. \tag{3.81}$$

Now, (3.80) can be rewritten as

$$\mathbb{E}[Y(m, n)] = \sum_{l=0}^{L-1} h(l)\mathbb{E}[Y_l(m, n)], \tag{3.82}$$

where

$$\mathbb{E}[Y_l(m, n)] = e^{j2\pi \frac{mv}{N}} \sum_{k=0}^{N-1} \mathbb{E}[\tilde{s}(m + k - l)s_k^{(0)*}] e^{j2\pi \frac{k(v-n)}{N}}. \tag{3.83}$$

Note that $\mathbb{E}[\tilde{s}(m + k - l)x_k^{(0)*}]$ is nonzero only when $\tilde{s}(m + k - l)$ and $s_k^{(0)*}$ are correlated. Figure 3.29 presents all cases when this is true for different values of $m - l$.

In Figure 3.29, it is assumed that preamble $\mathbf{s}^{(0)}$ is the S&C preamble. The received signal (at the top of Figure 3.29) is processed according to (3.75), where the correlation window of N samples for every time instance n is applied. As the correlation sliding window shifts in time over received signal samples (to the right in Figure 3.29), there are a few cases when $\tilde{s}(n + m - l)$ and $s_m^{(0)}$ are highly correlated. First, it happens when only the CP of the preamble is received, which is denoted as Case A. After $N/2$ sampling intervals, the second half of the RP is aligned with the first part of received preamble inside the correlation sliding window, and additionally, N_{CP} samples of received preamble CP are aligned with the end of the first part of the RP, resulting in high correlation between the received preamble and its original (reference) version. This case is denoted

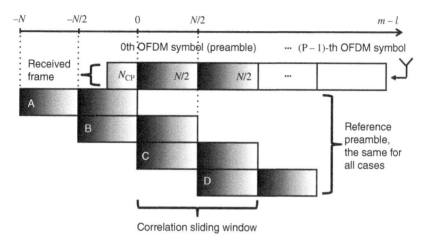

Figure 3.29 Cases of nonzero correlation when preamble consists of two repeated sequences of $N/2$ samples as in the S&C algorithm. The shade of gray represents the same preamble sample (either originally transmitted or in the RP at the receiver). (© 2016 IEEE. Reprinted, with permission, from [216].)

as B. Case C reflects the timing and the correlation window position when all N samples of the RP are aligned with the received preamble samples. The last correlation peak is obtained in Case D, over $N/2$ samples when the second part of the received preamble is aligned with the first half of the RP. In the case of the simple preamble, only cases A and C result in $\mathbb{E}[\tilde{s}(n+m-l)s_m^{(0)*}] \neq 0$. The simple preamble does not have two identical halves, so the correlation of the first half of the received preamble and the second half of the RP does have any peak values (Cases B and D).

Derivation of $\mathbb{E}[Y_l(m,n)]$ is quite complicated for both the simple preamble and the S&C preamble. These expected values $\mathbb{E}[Y_l(m,n)]$ are presented in [216] for the discussed cases.

The variance of $Y(m,n)$ is by definition

$$\mathbb{V}[Y(m,n)] = \mathbb{E}[Y(m,n)Y(m,n)^*] - \mathbb{E}[Y(m,n)]\mathbb{E}[Y(m,n)^*]. \qquad (3.84)$$

As the second term is the squared absolute value of expectation defined in (3.82), the calculation of the first term is of interest now.

The noise components are not correlated with the RP so $\mathbb{E}[w(m+k_1)w(m+k_2)^*s_{k_1}^{(0)*}s_{k_2}^{(0)}] = \mathbb{E}[w(m+k_1)w(m+k_2)^*]\mathbb{E}[s_{k_1}^{(0)*}s_{k_2}^{(0)}]$. Moreover, the noise components are correlated with each other only for $k_1 = k_2$. This is also true when NL interference is considered. Assuming NL interference model and that the noise is not correlated with the RP, and following derivations from [216], it is possible to find the formula for this component:

$$\mathbb{E}\left[Y(m,n)Y(m,n)^*\right] =$$
$$\sum_{l=0}^{L-1}|h(l)|^2 \sum_{k_1=0}^{N-1}\sum_{k_2=0}^{N-1}\mathbb{E}\left[\tilde{s}(m+k_1-l)s_{k_1}^{(0)*}\tilde{s}(m+k_2-l)^*s_{k_2}^{(0)}\right] \cdot$$
$$e^{j2\pi(k_1-k_2)\frac{v-n}{N}} + \sum_{k=0}^{N-1}\left(\mathbb{E}[|i(m+k)|^2]+\mathbb{E}[|w(m+k)|^2]\right)\mathbb{E}[|s_k^{(0)}|^2]. \qquad (3.85)$$

It should be noted that summation over expectations in the first component in (3.85) can be independently performed for each path, which allows for a simple formula for the variance of $Y(m,n)$:

$$\mathbb{V}[Y(m,n)] = \sum_{l=0}^{L-1}|h(l)|^2\mathbb{V}[Y_l(m,n)] + N(\sigma_w^2+\sigma_i^2)\sigma_s^2. \qquad (3.86)$$

Assuming the GIB interference model defined by (3.79), and using the same derivations, the variance of $Y(m,n)$ is

$$\mathbb{V}[Y(m,n)] = \sum_{l=0}^{L-1}|h(l)|^2\mathbb{V}[Y_l(m,n)] + N\sigma_w^2\sigma_s^2+$$
$$\sum_{n_1=0}^{N-1}\mathbb{E}[|g_{n_1}(m)|^2]\mathbb{E}[|d_{\overline{(n_1-n)}}^{(0)}|^2], \qquad (3.87)$$

where $x_{(\tilde{n})}$ denotes cyclic indexing of elements in $\mathbf{x}$. Let us denote the interference-related component in (3.87) as

$$V_i(m,n) = \sum_{n_1=0}^{N-1}\mathbb{E}[|g_{n_1}(m)|^2]\mathbb{E}[|d_{\overline{(n_1-n)}}^{(0)}|^2]. \qquad (3.88)$$

For interference occurring in a different band than the NC-OFDM spectrum (typical scenario), and for $n \approx 0$, $V_i(m,n)$ is small or, in the case of both signals being orthogonal, equal to zero. However, $V_i(m,n)$ can be higher if the useful signal and the interference overlap in frequency. It can reach a maximum of $N^2\sigma_i^2\sigma_s^2$ (for the theoretical case of both interference and NC-OFDM signal being complex sinusoids) [216]. Thus, (3.86) reflects the worst case (the highest value) and is used in the later LUISA stage for $\mathbb{V}[Y(m,n)]$. Expressions for $\mathbb{V}[Y_l(m,n)]$ are also quite complicated to derive, and therefore, we omit them. An interested reader can find them in [216].

Coarse Time-Offset and Integer CFO Estimation

On the basis of the defined synchronization variable $Y(m,n)$, the coarse time and frequency synchronization can be directly achieved. The search for the

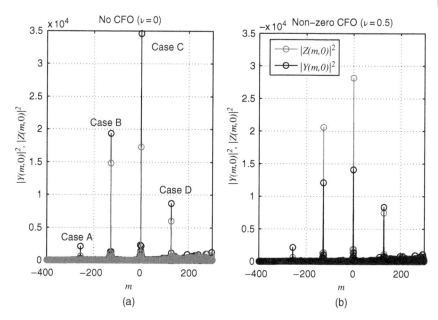

Figure 3.30 $|Y(m, n)|^2$ and $|Z(m, n)|^2$ for $n = 0$, S&C preamble, $N = 256$, $N_{CP} = N/4$, no interference, no noise, no multipath ($L = 1$), and no CFO ($v = 0$), (a) and worst case of CFO ($v = 0.5$) (b) considered. (© 2016 IEEE. Reprinted, with permission, from [216].)

peak $Y(m, n)$ power must be made in two dimensions: over frequency and time. The simplest decision method would seek $m = m_Y$ and $n = n_Y$ time- and frequency indices to maximize the squared absolute of $Y(m, n)$:[6]

$$[m_Y, n_Y] = \arg\max_{m,n} |Y(m, n)|^2. \tag{3.89}$$

An example plot of $|Y(m, n)|^2$ over m for $n = 0$ in the idealistic condition of no-noise, no-interference, and no-multipath effect can be observed in Figure 3.30a for $N = 256$, S&C preamble and $N_{CP} = N/4$. There are four peaks of $|Y(m, n)|^2$ that reflect four cases when nonzero expected value of $Y(m, n)$ is obtained, as presented in Figure 3.29.

According to the statistical properties of the random complex Gaussian variable $Y_l(m, n)$ given in [216], it is the most probable that the strongest path, that is, $m_Y = l_{max} = \arg\max_l |h_l|^2$, could be found with (3.89) because for m_Y the

---

6 Although the first stage of synchronization is achieved by maximization (as in all competitive algorithms), LUISA requires some threshold above which the maximal-value sample is searched for. This threshold eliminates situations when none of the A–D cases occurs in the observed range of time samples m. It can be obtained in a similar way as described in *Path detection* section, but the probability of false detection should be divided by the total number of possible points in 2D plane (instead of $2N_{CP}$). Moreover, at least N samples following the sample exceeding this threshold should be examined to find the highest peak (Case C).

magnitude of the expected value of $Y(m_Y, n_Y)$ should be the highest. It is also the most probable that this highest peak will be the correlation peak of Case C as it has the highest expected value among other cases. This expected value for $\sigma_w^2 = 0$, $\sigma_i^2 = 0$, and $m_Y = l_{\max}$ equals

$$\mathbb{E}[|Y(m_Y, n_Y)|^2] \approx |h_{l_{\max}}|^2 \sigma_s^4 \left(\frac{\sin(\pi(v - n_Y))}{\sin(\pi(v - n_Y)/N)} \right)^2 . \tag{3.90}$$

Note that $\mathbb{E}[|Y(m_Y, n_Y)|^2] = \mathbb{V}[Y(m_Y, n_Y)] + |\mathbb{E}[Y(m_Y, n_Y)]|^2$, but because $|\mathbb{E}[Y(m_Y, n_Y)]|^2$ is about N times higher than $\mathbb{V}[Y(m_Y, n_Y)]$, the following approximation is applied: $\mathbb{E}[|Y(m_Y, n_Y)|^2] \approx |\mathbb{E}[Y(m_Y, n_Y)]|^2$.

The search for optimum n_Y in (3.89) for $m_Y = l_{\max}$ is done over $n = -N/2, \ldots, N/2 - 1$. Let us note that there exists one FFT output for which the error of the CFO estimate is equal to or lower than half of the SC spacing, that is, $|v - n_Y| \leq 0.5$. This FFT-output index should be chosen as n_Y because it maximizes the expected value in (3.90). This is because $|v - n_Y| \ll N$, and we can approximate the denominator in (3.90) by $\sin(\pi(v - n_Y)/N) \approx \pi(v - n_Y)/N$. The function $\mathbb{E}[|Y(m_Y, n_Y)|^2]$ then reaches its maximum equal to $|h_{l_{\max}}|^2 \sigma_s^4 N^2$ for $|v - n_Y| = 0$ (example shown in Figure 3.30a). The minimum over the whole domain, that is, for $v - n_Y \in \langle -0.5; 0.5 \rangle$, is obtained for $|v - n_Y| = 0.5$, and equals $|h_{l_{\max}}|^2 \sigma_s^4 (N2/\pi)^2$ (example shown in Figure 3.30b). Note that the minimum of $\mathbb{E}[|Y(m_Y, n_Y)|^2]$ is about 40% of the maximum. This may cause a decrease in the correct time- and frequency-point detection probability for the CFO not aligned with the frequencies of the FFT used to calculate $Y(m, n)$. To increase the detection probability in this case, two adjacent FFT outputs are utilized: $Y(m, n)$ and $Y(m, n + 1)$. For $v - n \approx 0.5$, the FFT output $n + 1$ at $m = l$, as in case C, should have the absolute value close to the one for n. In order to make use of the phase dependency of the adjacent FFT outputs, we calculate $Z(m, n) = [Y(m, n) - Y(m, n + 1) \exp(-j\pi/N)]/\sqrt{2}$, and optimize the squared module of this variable:

$$[m_Z, n_Z] = \arg\max_{m,n} |Z(m, n)|^2 . \tag{3.91}$$

Note that calculation of $Z(m, n)$ is based on the $Y(m, n)$ values, resulting in a small increase in computational complexity in comparison to detection based on (3.89) only. Following the assumption that $l_{\max} = m$, and that $Y(m, n)$ and $Y(m, n + 1)$ are presumably uncorrelated [234], the mean value of $Z(m, n)$ can be obtained using the values of $\mathbb{E}[Y_l(m, n)]$ and $\mathbb{V}[Y_l(m, n)]$ (see Ref. [216]):

$$\mathbb{E}[Z(m, n)] = \frac{1}{\sqrt{2}} (\mathbb{E}[Y(m, n)]) - \mathbb{E}\lfloor Y(n, k + 1)]e^{-j\frac{\pi}{N}}) =$$

$$\frac{h_{l_{\max}} \sigma_s^2}{\sqrt{2}} e^{j2\pi \frac{lv}{N} + j\pi(v-n)(1 - \frac{1}{N})} \cdot \left(\frac{\sin(\pi(v - n))}{\sin(\pi \frac{v-n}{N})} - \frac{\sin(\pi(v - n))}{\sin(\pi \frac{v-n-1}{N})} \right) . \tag{3.92}$$

Similarly, for both simple and S&C preamble cases, the variance of $Z(m, n)$ can be calculated and is the same as for $Y(m, n)$ in the case of simple preamble. Assuming that $\mathbb{E}[|Z(m, n)|^2] \approx |\mathbb{E}[Z(m, n)]|^2$ (as in (3.90)):

$$\mathbb{E}[|Z(m, n)|^2] \approx |\mathbb{E}[Z(m, n)]|^2$$

$$= \frac{1}{2} |h_{l_{max}}|^2 \sigma_s^4 \left(\frac{\sin(\pi(v - n))}{\sin(\pi(v - n)/N)} - \frac{\sin(\pi(v - n))}{\sin(\pi(v - n - 1)/N)} \right)^2. \tag{3.93}$$

Thus, both summation components under square in (3.93) have the same phase for $v - n \approx 0.5$, which increases the detection probability. For $v - k = 0.5$, the decision variable in (3.93) can be approximated as $0.5|h_{l_{max}}|^2 \sigma_s^4 (N4/\pi)^2$, which is twice the minimum (obtained for $v - n = 0.5$) and about 80% of the maximum (obtained for $v - n = 0$) of the expected mean square of $Y(m_Y, n_Y)$. (The example of this decision metric is shown in Figure 3.30.) Therefore, in the more advanced detection case, the maximum out of both decision variables maxima is chosen:

$$[m_M, n_M] = \arg \max_{(m_Y, n_Y),(m_Z, n_Z)} (|Y(m_Y, n_Y)|^2, |Z(m_Z, n_Z)|^2). \tag{3.94}$$

Most importantly, in the aforementioned maximization problem, only two values are compared: $|Y(m_Y, n_Y)|^2$ and $|Z(m_Z, n_Z)|^2$ resulting from (3.89) and (3.91) respectively, each of them being optimization problem solutions over two arguments: (m_Y, n_Y) and (m_Z, n_Z), respectively. The variable $|Y(m_Y, n_Y)|^2$ should be greater than $|Z(m_Z, n_Z)|^2$ for CFO closer to the frequencies of the receiver FFT, otherwise $|Z(m_Z, n_Z)|^2$ should be higher. Note that if (3.91) finds the coarse time and frequency synchronization point, that is, $|Z(m_Z, n_Z)|^2 > |Y(m_Y, n_Y)|^2$, it is ambiguous whether a peak occurs for v closer to the normalized frequency m_Z or to $n_Z + 1$ ($v \in (n_Z; n_Z + 1)$). In this particular case, the following modification of n_M is performed:

$$n_M = \begin{cases} n_Z + 1 & \text{for} |Y(m_Z, n_Z + 1)|^2 > |Y(m_Z, n_Z)|^2 \\ n_Z & \text{otherwise.} \end{cases} \tag{3.95}$$

In the remainder of this description, $[m_M, n_M]$ is used as a result of the coarse time and integer CFO estimation. Note that the computational complexity of this estimation can be decreased if the receiver has the knowledge on the upper limit of the CFO v_M, where $|v| < v_M$. It allows for limiting the integer-CFO search range to $n = \{ - \lceil v_M \rceil, \dots, \lceil v_M \rceil \}$, where $\lceil \cdot \rceil$ is the *ceiling* function.

Coarse Fractional CFO Estimation

As shown earlier, $Y(m, n)$ is a Gaussian random variable with nonzero expected value when the preamble is detected. The mean value has *sinc*-like shape (over n). The output of the FFT, $Y(m, n)$, for a given m is therefore similar to the output signal after processing the complex sinusoid of an unknown frequency with AWGN [234]. The algorithm to find the frequency of this complex sinusoid presented in [234] can be used to estimate v corresponding to the peak of *sinc*-like frequency response. The difference between the FFT output defined in [234] and in the present case is the variance of this output signal. In the case of processing the complex sinusoid without noise as in [234], the variance of the variable at the FFT output is zero, which is different from our case. However, as the fine fractional CFO estimation is to be performed in LUISA afterward, this nonzero variance is not a problem at the coarse CFO estimation phase. The integer CFO estimate n_M is used to calculate the coarse CFO estimate $\widehat{v_C}$ using the following formula:

$$\widehat{v_C} = \frac{N}{\pi}\text{atan}\left(\tan\left(\frac{\pi}{N}\right)\Re\left(\frac{Q_1(m_M, n_M)}{Q_2(m_M, n_M)}\right)\right), \tag{3.96}$$

where $\Re(\ )$ is the real part of a complex-number argument,

$$Q_1(m, n) = Y(m, n-1) - Y(m, n+1), \tag{3.97}$$

$$Q_2(m, n) = 2Y(m, n) - Y(m, n-1) - Y(m, n+1), \tag{3.98}$$

and $\tan(\cdot)$ and $\text{atan}(\cdot)$ are *tangent* and *arcus tangent* functions, respectively. After this phase, the first (coarse) estimate of v is $\hat{v}_0 = \hat{v}_C + n_M$, that is, $\hat{v}_0 \approx v$. The lower index 0 of $\hat{v}_0$ denotes 0-th iteration of the CFO estimate, that is, the initial estimate.

Path Detection

Assuming that in the previous steps of LUISA, the beginning of the maximum-power channel path ($m_M = l_{\max}$) has been found and the CFO estimate is close to its actual value v, the received signal can be frequency-corrected, that is, the values of $Y(m, n)$ can be calculated for $n = \hat{v}_0$. As usual in OFDM, all channel-path components are assumed to be spread over the maximum of N_{CP} sampling intervals. Thus, we calculate

$$Y(m, \hat{v}_0) = \sum_{k=0}^{N-1} r(m+k)s_k^{(0)*}e^{-j2\pi\frac{k\hat{v}_0}{N}}, \tag{3.99}$$

where $m \in \{m_M - N_{CP}, \ldots, m_M - 1, m_M + 1, m_M + N_{CP}\}$. The CFO correction maximizes the peaks of $|Y(m, \hat{v}_0)|^2$ caused by the channel-path components that can be now more easily distinguished from noise and interference. The aim of this step is to find the first path and all other paths that can be used later for fine CFO correction. After the coarse CFO estimation, the set of detected path indices $\mathbf{D}$ consist of the strongest detected path: $\mathbf{D} = \{m_M\}$. At the LUISA path detection stage, the mth received signal sample represents a path component and is added to set $\mathbf{D}$, if

$$|Y(m, \hat{v}_0)|^2 > -\sigma_{\text{thr}}^2(m)\ln(P_{FD}/(2N_{CP})), \tag{3.100}$$

where P_{FD} is the probability of false detection over the range of $2N_{\mathrm{CP}}$ samples (note that $P_{\mathrm{FD}}/(2N_{\mathrm{CP}})$ approximates the false detection probability at a single sampling moment as in [228]). Moreover, $\sigma_{\mathrm{thr}}^2(m)$ in (3.100) is the estimated input signal variance. The false detection occurs if $|Y(m,\hat{v}_0)|^2$ exceeds the threshold despite m not representing a path-reception moment, that is, the noise or the interference is detected as a channel path. One should note that the aforementioned threshold is provided for reducing the probability that $Y(m,n)$ not aligned with any channel path ($m \neq l$) will be detected as a channel path. Assuming that the sampling time m is not aligned with any path-reception moment, $\mathbb{E}[Y(m,\hat{v}_0)] = 0$. Moreover, $Y(m,n)$ has complex normal distribution as shown previously while discussing its statistical properties. Thus, a normalized sum of its squared real and imaginary parts equal to $2|Y(m,\hat{v}_0)|^2/\sigma_{\mathrm{thr}}^2(m)$ has chi-square distribution with 2 degrees of freedom. This allows for providing the threshold, given in (3.100), using chi-square cumulative distribution function. Factor $2/\sigma_{\mathrm{thr}}^2(m)$ is needed in order to normalize the variance of both real and imaginary parts of $Y(m,\hat{v}_0)$ to 1. Furthermore, $\sigma_{\mathrm{thr}}^2(m)$ can be estimated as

$$\sigma_{\mathrm{thr}}^2(m) = \sigma_{\mathrm{s}}^2 \sum_{k=0}^{N-1} |r(m+k)|^2. \tag{3.101}$$

Decreased complexity of calculating $\sigma_{\mathrm{thr}}^2(m)$ can be obtained by subtracting $\sigma_{\mathrm{thr}}^2(m-1)$ from $\sigma_{\mathrm{thr}}^2(m)$ (both obtained using 3.101), which results in recurrence relation:

$$\sigma_{\mathrm{thr}}^2(m) = \sigma_{\mathrm{thr}}^2(m-1) + \sigma_{\mathrm{s}}^2(|r(m+N-1)|^2 - |r(m-1)|^2). \tag{3.102}$$

The mean of (3.101) can be calculated by substituting the expansion of $r(m+k)$ from (3.74) resulting in

$$\mathbb{E}[\sigma_{\mathrm{thr}}^2(m)] = \sigma_{\mathrm{s}}^2 \sum_{k=0}^{N-1} \Bigg(\mathbb{E}[i(m+k)i(m+k)^*] + \mathbb{E}[w(m+k)w(m+k)^*]+$$

$$\sum_{l_1=0}^{L-1}\sum_{l_2=0}^{L-1} h(l_1)h(l_2)^* \mathbb{E}[\tilde{s}(m+k-l_1)\tilde{s}(m+k-l_2)^*] \Bigg)$$

$$= \sigma_{\mathrm{s}}^2(\sigma_{\mathrm{i}}^2 + \sigma_{\mathrm{w}}^2)N + \sum_{l=0}^{L-1} |h(l)|^2 \sigma_{\mathrm{s}}^4 \min(\max(0, N+N_{\mathrm{CP}}+m-l), N)$$

$$\tag{3.103}$$

which is the same as $\mathbb{V}[Y(m,n)]$ when NL interference model is considered, and no channel-path is found. Interestingly, on average, only for the S&C preamble and when $m = l$ or $m - l = -N/2$ (cases C and B), the variance estimated by (3.101) is lower than the actual value calculated according to (3.86). However, according to the previous step of LUISA, m_{M} should reflect the strongest path in case C, so underestimation of $\sigma^2_{\text{thr}}(m)$ by (3.101) for case B is not possible. In case C, the actual variance higher than the one estimated increases the probability of correct path detection.

If all samples in the searched region exceed the threshold in (3.100), set **D** will consist of $2N_{\text{CP}} + 1$ elements, while the maximum delay spread can be of N_{CP} samples (CP typically covers the channel delay spread for a given OFDM system). Thus, it is proposed that the estimated channel path indices from set **D** be sorted in the descending channel-path power order, and only the strongest path components be chosen, which are detected over N_{CP} samples.

At this stage, the final estimate of the first channel-path timing $\hat{m}_{\text{M}}$ can be made. It is the earliest-arriving channel path component, that is, of the lowest index:

$$\hat{m}_{\text{M}} = \min \mathbf{D}. \tag{3.104}$$

Fine Fractional CFO Estimation

The previous stage has delivered the information on all detected signal paths components in **D**. It is therefore reasonable to use them all for fine CFO estimation. According to [216], in case C, the expected values of $Y(m, \hat{v}_0)$ for various $m \in \mathbf{D}$ differ by $h(m)e^{j2\pi\frac{v}{N}m}$. Thus, the method from [234] can be modified to use all path components coherently. The CFO estimate from the previous stages equals $\hat{v}_0 = n_{\text{M}} + \hat{v}_{\text{C}}$. In [216], it is proposed to update the CFO estimate $\hat{v}_0$ as follows:

$$\Delta\hat{v}_0 = \frac{N}{\pi}\text{atan}\left(\tan\left(\frac{\pi}{N}\right)\Re\left(\frac{\sum_{m \in \mathbf{D}}\{Q_1(m, \hat{v}_0)Y^*(m, \hat{v}_0)\}}{\sum_{m \in \mathbf{D}}\{Q_2(m, \hat{v}_0)Y^*(m, \hat{v}_0)\}}\right)\right). \tag{3.105}$$

Note that in order to add all components in the nominator and in the denominator in (3.105) coherently and according to the Maximal-Ratio Combining rule, each component is multiplied by $Y^*(m, \hat{v}_0)$. This can be done because, according to [216], all components $Y^*(m, \hat{v}_0)$ are identically affected by the CFO, while the channel impulse response $h(m)$ is different for each component under sums in (3.105). Thanks to the multiplication with their complex conjugates, all path components are added with the weighting factor proportional to their power. Let us note that for a path component of index $m \in \mathbf{D}$, three synchronization variable values have to be calculated to obtain $Q_1(m, \hat{v}_0)$ and $Q_2(m, \hat{v}_0)$. These values are $Y(m, \hat{v}_0)$, $Y(m, \hat{v}_0 + 1)$, and $Y(m, \hat{v}_0 - 1)$. The correctness of formula (3.105) can be justified by calculating the first-order approximation of the expected value of (3.105) for case C, as shown in [216]:

$$\mathbb{E}[\Delta\hat{v}_0] \approx$$

$$\frac{N}{\pi}\text{atan}\left(\tan\left(\frac{\pi}{N}\right)\Re\left(\frac{\sum_{m \in \mathbf{D}}|h(m)|^2}{\sum_{m \in \mathbf{D}}|h(m)|^2}\cot\left(\frac{\pi}{N}\right)\tan\left(\frac{\pi(v - \hat{v}_0)}{N}\right)\right)\right) =$$

$$v - \hat{v}_0, \tag{3.106}$$

where cot($\cdot$) is the *cotangent* function. The updated frequency estimate is thus

$$\hat{v}_1 = \hat{v}_0 + \Delta\hat{v}_0. \tag{3.107}$$

This estimate can be further improved by iterative calculation of (3.105) and (3.107) in the γ-th iteration for $\hat{v}_\gamma$ based on $\hat{v}_{\gamma-1}$ and $\Delta\hat{v}_{\gamma-1}$. The rationale behind this approach is given by the MSE of the basic estimator shown in [234]. The higher the difference between $\hat{v}_\gamma$ and v, the higher the MSE.

The MSE of frequency estimation is limited in our case by the nonzero variance of the useful NC-OFDM signal. However, in the case of S&C preamble, the proposed method of calculating the CFO update according to (3.105) utilizes the fact that the variance of $Y_l(m, n)$ at the time sample aligned with the channel-path component l and with the frequency distanced by one SC spacing from CFO equals zero ($\mathbb{V}[Y_l(l, v + 1)] = \mathbb{V}[Y_l(l, v - 1)] = 0$). As γ increases, $\hat{v}_\gamma$ approaches v, the variance of $Q_1(m, \hat{v}_{\gamma-1})$ decreases, increasing the CFO estimation quality, specially at high SNR and SIR. In this case, the variance of this estimator in the AWGN channel is

$$\mathbb{V}[v - \hat{v}_\gamma] = \frac{1}{4N \cdot \text{SNR}}, \tag{3.108}$$

the same as in [234]. It is also 1.6 times higher than the Cramér –Rao lower bound of $\frac{6}{4\pi^2 N \cdot \text{SNR}}$ [234].

3.5.3 Performance Evaluation

Let us evaluate the results of LUISA against the commonly used S&C algorithm [148] and the algorithm proposed by Abdzadeh-Ziabari and Shayesteh [229], which does not use autocorrelation for fractional CFO estimation but the method from [230]. In the considered NC-OFDM system, $N = 256$-order IFFT/FFT and the CP of $N_{\text{CP}} = N/16$ samples are applied. Only SC of indices in vector $\mathbf{I}_{\text{DC}}$ are modulated by data symbols. Other SCs are not used because the LU signal spectrum is observed in this range. Let us consider three system scenarios:

- Static Spectrum Allocation (SSA) scenario, when vector $\mathbf{I}_{\text{DC}}$ is fixed for each frame (both the CR's and LU's PSD remain constant), and GSs are not applied (because either there is no interference or static LU's PSD allows for appropriate NC-OFDM spectrum shaping techniques to be practically applied),
- SSA scenario with the application of GS, and
- Dynamic Spectrum Access (DSA) scenario, when the gap in the NC-OFDM spectrum required for the LU protection has a random, dynamically changing position in frequency for each transmitted frame and the GS are additionally applied for this protection.

The LU-generated interference is simulated in two ways: (i) as the complex sinusoid (as in Ref. [219]) observed in the NC-OFDM spectrum notch or (ii) as an OFDM-like interfering signal modulated by random QPSK symbols

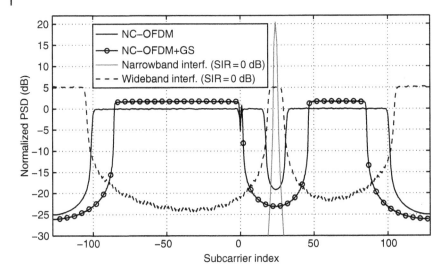

Figure 3.31 Power spectral density of the NC-OFDM signal (with and without GS), the in-band NBI at center frequency 24, and the in-band OFDM-like WBI occupying bands unused by the secondary NC-OFDM signal; SIR = 0 dB, no CFO. (© 2016 IEEE. Reprinted, with permission, from [216].)

occupying the band not used by our NC-OFDM signal, and using SC not orthogonal to our NC-OFDM SC. Note that both simulated interfering signals can be modeled with the GIB interference model. Moreover, the first one (i) can be considered as strictly NBI, while the latter (ii) is closer to WBI. The example PSD plots of the NC-OFDM signal and the interference having equal power (SIR = 0 dB) are depicted in Figure 3.31. The SNR or SIR is defined as the ratio of the mean NC-OFDM signal power to the mean noise or interference power, respectively, over the whole NC-OFDM receiver band (including the frequencies inside the spectrum gap required for the LU protection).

Both Ziabari and the S&C methods use the S&C preamble. LUISA is tested both with the S&C- and with the simple preamble. The preamble and the data symbols use QPSK mapping with random symbols for each simulation run. The S&C preamble has the same energy as the simple preamble as explained previously. For each case, 10^5 independent NC-OFDM frames have been tested, each consisting of 11 NC-OFDM symbols, preceded by two empty OFDM symbols (interframe period). For LUISA and Ziabari algorithm, only the first (preamble) symbol is used for synchronization purposes, while the S&C method also uses the second symbol for integer CFO estimation. For each NC-OFDM frame, a random instance of nine-path Rayleigh fading channel characteristic is generated according to Extended Vehicular A model [198], as well as random CFO uniformly distributed over $(-3; 3)$ normalized to SC spacing of 15 kHz (as in LTE).

For the considered S&C algorithm, the beginning of a preamble is found as the point in the middle between two points achieving 90% of the maximum of timing metric, as in [148], while integer CFO is searched over the range of $\{-20, \ldots, 20\}$. The Ziabari method uses its own type of correlation consisting of $4N$ values. It scans the set of possible integer CFO values (N) while its threshold is calculated for the probability of false detection equal to 10^{-5}. The duration of the window used in the last stage of this method for the first channel path detection is five samples. In LUISA, the probability of false detection P_{FD} in (3.100) is also set to 10^{-5}, and the number of iterations of fine CFO estimation using (3.105) and (3.107) equals 2.

SSA Scenario, No GS – example Results

In this scenario, $I_{DC} = \{-100, \ldots, -1, 1, \ldots, 16, 32, \ldots, 100\}$. In the first round of simulations, the presence of AWGN, CFO, and channel fading is considered and no interference. Because the LU transmission is not present, no GS are necessary. The estimated probability of the frame synchronization error is depicted in Figure 3.32. The frame is assumed to be roughly synchronized if the error in time is lower than N_{CP} samples ($\hat{m}_M \in \{-N_{CP} + 1, \ldots, N_{CP} - 1\}$), and the error in frequency is smaller than half of the SC spacing ($|\hat{\nu}_2 - \nu| < 0.5$). Note

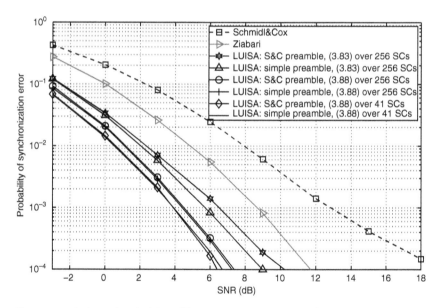

Figure 3.32 Estimated probability of frame synchronization error for LUISA configurations: preamble type, coarse time–frequency point detection method ((3.89) or (3.94)), the range of integer CFO search (in the number of SC); SSA scenario, no GSs, no interference. (Based on [216].)

that LUISA in all configurations outperforms two other reference algorithms. Performance deterioration of LUISA using the S&C preamble with respect to the simple preamble is caused by the increased variance $\mathbb{V}[Y(m,n)]$. Moreover, an improvement of about 1.5 dB SNR gain can be achieved by LUISA when detecting the frame-beginning time with formula (3.94) instead of (3.89), which is more robust against noninteger CFO.

In [216] the time and frequency MSEs are presented for all simulated frames. Interestingly, Ziabari method achieves both time and frequency MSEs higher than that from the S&C method although it has lower probability of synchronization error. This is because typically, an error for erroneously synchronized frame in the S&C algorithm is smaller in time and frequency than in the Ziabari method due to much wider range of integer CFO estimation in the latter. Therefore, it is reasonable to analyze MSE only for the synchronized frames. In Figure 3.33, it can be observed that LUISA using the simple preamble has a frequency MSE lower than that in the Ziabari method, which reaches the MSE floor of above 10^{-4}. As explained previously, it is caused by nonzero variance of $Y_l(m,n)$. All LUISA configurations utilizing the S&C preamble obtain similar frequency MSE, which requires about 1 dB higher SNR than the S&C algorithm. This is because LUISA, during the path detection stage, may not find low-power paths that could be used for the CFO estimation. In the S&C algorithm, all paths are inherently included in the autocorrelation result. The S&C algorithm has theoretical MSE lower than LUISA (3.108).

The plot in Figure 3.33b shows time MSE. For all LUISA configurations, the performance is similar, and typically, this MSE is lower than one squared sample. A bit higher timing MSE is obtained for the Ziabari method, while the S&C algorithm obtains timing MSE a bit lower than 10^2. However, such an MSE of the S&C algorithm might not mean synchronization degradation as this algorithm can choose the beginning of the frame in the middle of CP, and in case of relatively short path delay spread, its performance may be acceptable.

For the next runs of simulations, a LUISA configuration showing the most of its advantages is used. The S&C preamble has been applied, and the detection method using formula (3.94) operates over 41 FFT outputs. Now, the NBI and WBI interference is simulated, as described earlier. The NBI normalized center frequency is 24 in the NC-OFDM spectrum notch (see PSD in Figure 3.31). Despite the presence of the LU transmission, GSs are not applied. It is assumed that in the static scenario, NC-OFDM spectrum shaping methods can be practically incorporated.

The probability of frame detection error has been presented in Figure 3.34. It can be observed that for the case of strong NBI (low SIR), the S&C algorithm fails, losing nearly 100% of the transmitted preambles. Interestingly, it is visible for NBI that when SNR increases, the ratio of interference to noise power increases and the false synchronization occurs. This means that the S&C synchronization is not suitable for CR NC-OFDM receivers in the presence of

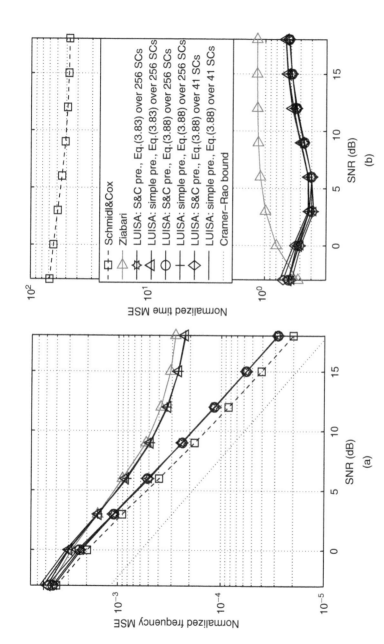

Figure 3.33 Frequency (a) and time (b) normalized MSE for successfully synchronized frames; SSA scenario, no GSs, no interference. (Based on [216].)

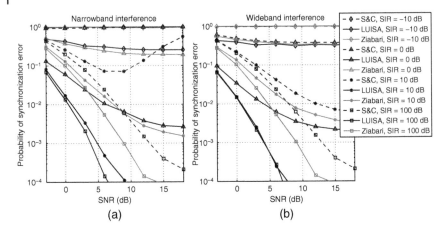

Figure 3.34 Probability of synchronization error for an NC-OFDM system in the presence of NBI (a) and WBI (b); SSA scenario, no GSs. (Based on [216].)

high-power LU-generated narrowband interference. The Ziabari method is not based on the autocorrelation, and as such, it is not prone to this false synchronization. More importantly, in case of WBI, Ziabari method is only a bit better than S&C algorithm. However, LUISA can work in both environments more efficiently. For example, for SNR = 15 dB and NBI of SIR = 0 dB, the Ziabari method loses about 80% of NC-OFDM frames, while LUISA has the probability of frame synchronization error of only 0.3%, which results in about 70 times less erroneously synchronized frames. In the considered scenario, LUISA has even better performance under WBI; that is, it provides about 200 times more correctly synchronized frames in comparison to Ziabari method. It is mostly the effect of the first detection stage, which significantly rejects the interference generated in frequencies not used by the NC-OFDM system, and takes the amplitude and the phase relations derived in [216] into account, which is robust against high CFO values.

SSA Scenario with GSs

It has been proved previously that the variance of $Y(m, n)$ for relatively low value of n is the lower, the higher the distance in frequency between the interference and the NC-OFDM spectra. Therefore, another test case is defined assuming a guard band of 15 GS on each side of occupied frequency band, that is, $I_{DC} = \{ -85, \dots, -1, 1, 47, \dots, 85 \}$. The GSs are a simple solution to decrease the effective interference power in both the LU and the secondary NC-OFDM receiver. The resultant probabilities of synchronization error are shown in Figure 3.35. Interestingly, in comparison to the previous scenario, this probability is slightly increased for both the S&C and Ziabari methods.

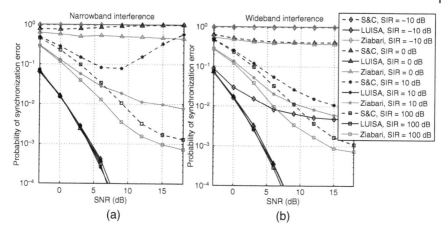

Figure 3.35 Probability of synchronization error for an NC-OFDM system in the presence of NBI (a) and WBI (b); SSA scenario, GSs applied. (Based on [216].)

Although the number of complex symbols modulating the preamble decreases, the performance of LUISA improves, especially for low SIR values. It is visible that only for wideband interference of SIR = −10 dB, the probability of synchronization error floor occurs. However, more than 99% of NC-OFDM frames are synchronized correctly for usable SNR values. In other cases of LUISA, interference does not increase the probability of synchronization error in comparison to the system without interference. The perfect operation of LUISA under NBI is caused by integer CFO search range limited to $n \in \{-20, \ldots, 20\}$ in $Y(m, n)$ and minimum distance in frequency between NBI (at normalized frequency 24) and occupied NC-OFDM. According to (3.87), it results in zero variance of interference-based component in $Y(m, n)$ over the observed output range. In case of the wideband interference, this component is nonzero and adds to $Y(m, n)$. Thus, LUISA is highly suitable for NC-OFDM with the LU signal occupying a different band.

DSA Scenario with GSs

In this scenario, it is assumed that the frequency allocation of the secondary NC-OFDM and LU systems changes dynamically and randomly and remains constant only within a single frame. Still, it requires the knowledge of the used set of subcarriers at the NC-OFDM receiver. The potentially available initial SC set is $\mathbf{I}_{DC} = \{-100, \ldots, -1, 1, \ldots, 100\}$. The NBI of normalized frequency (to the SC spacing) drawn from the uniform distribution over $(-128; 127)$ is generated for each frame. This requires dynamic reconfiguration of NC-OFDM-used SC, turning off SC around the NBI carrier frequency, and application of GS. Similarly, as earlier, the spectral notch of about 45 subcarriers (including 15

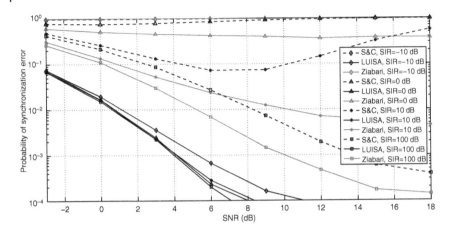

Figure 3.36 Probability of synchronization error for an NC-OFDM system in the presence of NBI changing its center frequency for each frame; DSA scenario, GAs applied. (Based on [216].)

GS on each NC-OFDM spectrum fragment edge) is created to protect the LU transmission.

The resultant probability of synchronization error is shown in Figure 3.36. It is visible that the performance of the S&C and Ziabari methods for high SIR is slightly increased in comparison to the results presented in Figure 3.35. However, stronger interference causes nearly the same probability of synchronization error as in the SSA scenario. Similarly, the randomly changing NBI center frequency still causes LUISA to outperform all other schemes. A slight increase of LUISA synchronization error probability for lower SIR is observed as a result of the noninteger NBI normalized frequency, that is, some interference-related components of $\mathbb{V}[Y(m, n)]$ are observed in the considered range of n according to (3.87).

3.5.4 Computational Complexity

In the considered synchronization algorithms, the operation dominating the complexity is the computation of the squared module of the synchronization variable ($|Y(m, n)|^2$ in the case of LUISA). For each NC-OFDM frame of K samples ($K \geq P(N + N_{CP})$), this value is calculated at least K times until a peak is found. The complexity evaluation of the considered S&C, Ziabari algorithm, and LUISA is presented in Table 3.2. For LUISA, FFT using Radix-2 method is assumed, preceded by N complex multiplications. The results presented for LUISA concern the method with increased robustness against CFO using (3.94) over a limited range of integer CFO, where $\lceil v_M \rceil = 20$. Note that LUISA has the complexity about three orders of magnitude higher than that in the S&C algorithm for $N = 256$ (as in the simulations). However, assuming that

Table 3.2 Number of operations per single input sample n for $N = 256$.

Algorithm	Real additions/subtractions	Real multiplications
S&C	7	6
Ziabari	$16N + 11 = 4107$	$16N + 22 = 4118$
LUISA	$2N + 6(2\lceil v_M \rceil + 1) +$ $3N\log_2 N = 6902$	$4N + 8(2\lceil v_M \rceil + 1) +$ $2N\log_2 N = 5448$
LUISA (optimized, approx.)	$3N + 3N\log_2(2\lceil v_M \rceil + 2)$ $+4(2\lceil v_M \rceil + 1) \approx 5073$	$2N + 2N\log_2(2\lceil v_M \rceil + 2)$ $+8(2\lceil v_M \rceil + 1) \approx 3601$

the processors' speed doubles every 18 months, the increase in the processing power since 1997, when the S&C algorithm was proposed, should cover the computational complexity increase of LUISA over the S&C algorithm complexity. Moreover, when comparing LUISA against Ziabari algorithm, the increase in complexity is acceptable.

LUISA's complexity can be further decreased by utilization of FFT pruning (because only $2\lceil v_M \rceil + 2\ Y(m, n)$ samples for a given m are needed as a result of FFT calculation). The Skinner approach (Decimation in Frequency [235]) can be applied that allows for embedding half of the correlation multiplications into the FFT block for the application of the S&C preamble. The approximate saving in real additions and multiplications is about 26% and 34%, respectively, in comparison to basic LUISA.

To summarize, the synchronization algorithm for NC-OFDM, named LUISA, works efficiently in a secondary CR system affected by the strong LU-generated in-band interference. This effectiveness is obtained at the cost of increased computational complexity when compared with the S&C algorithm.

3.6 Summary: Potentials and Challenges of NC-OFDM

The NC-OFDM is considered to be a modulation/multiplexing technique for future applications, allowing for high transmission flexibility and spectral efficiency while preserving relatively low computational complexity. Additionally, it can adapt most of the algorithms designed for standard OFDM, and utilize these well-established algorithms. Therefore, NC-OFDM has a great potential to serve as the radio interface for future communications with opportunistic spectrum sharing [36, 38]. Let us summarize the merits and challenges of NC-OFDM for this type of future communication.

Environmental Awareness

Recognition of the available spectrum with high reliability is one of the issues of making an opportunistic use of the frequency resources. In the NC-OFDM system, the symbol detection is performed in the frequency domain, at the output of the FFT block. Thus, the same reception setup can be used for the PU signal detection with the appropriately high frequency resolution. For sensing purposes, the radio front end should cover a wide bandwidth and have a high dynamic range (limited by the resolution of an A/D converter) in order to be able to detect PUs transmitting with variable powers. Some modifications to reduce the required dynamic range have been proposed in [212]. The first concept is to use a notch filter before the A/D converter to reduce strong signal power. However, this requires tunable filters that can be too complex to implement. To overcome this problem, spatial filtering is proposed. It is based on multiantenna receiver that forms its beam to have its antenna array characteristics minima in the direction from which a strong signal is received. Importantly, Multiple Input, Multiple Output (MIMO) techniques can be easily utilized in NC-OFDM, similarly as in standard OFDM. The drawback is that in the beginning, a few sweeps in many directions are needed to estimate the direction of the strong signal.

Location awareness is also crucial as the number of location-based services increases [38]. It can be used to optimize the network traffic, especially when channel information is sent to a base station marked with a measurement location. An example of an OFDM system that locates itself based on pilot subcarriers, without any need to use a Global Positioning System (GPS) receiver, is shown in [38]. The result can be obtained by calculations either in the time domain or in the frequency domain.

Flexibility

One of the most advantageous features of employing NC-OFDM in a CR system is its adaptability [38]. The NC-OFDM system can be highly flexible and can adapt the set of utilized subcarriers and allocated power in order to limit interference observed at the PU receiver (details will be discussed in Chapter 6). Parameters such as modulation constellation order or coding gain can flexibly be assigned to each subcarrier separately or to a block of subcarriers can be flexibly adapted to maximize the throughput and performance quality, to increase of the energy efficiency depending on the system goals and constrains. Moreover, the NC-OFDM system can be configured to use variable number of subcarriers (including DC, GS, CC, EC) to control the OOB emission power. Importantly, in the case of relatively low number of utilized subcarriers, it is possible to optimize FFT and IFFT operation by *pruning* [236].

A big advantage of NC-OFDM over OFDM is its ability to aggregate available spectrum over the whole considered transmitter bandwidth. Spectrum aggregation to increase the system throughput has already been applied in the Long Term Evolution Advanced (LTE-A) system where CA technique has been introduced. However, its flexibility in aggregating any kind of spectrum fragments (available frequency bands) is limited, because only a limited and integer number of resource blocks can be used for the data communication. Moreover, the proposed protocols do not allow for highly dynamic spectrum access and aggregation. As opposed to this, the small-scale and more flexible spectrum aggregation is envisioned in the NC-OFDM-based CR, where the aggregated fragments of spectrum are interleaved with other systems' spectrum.

Spectrum Shaping

As emphasized earlier in this chapter, in the future system coexistence scenarios, the legacy licensed PU systems have to be protected from the SU system-originating interference. In NC-OFDM, subcarriers coinciding with a PU band are modulated by zeros. The other subcarriers can be used to carry data symbols. However, simple turning off the subcarriers within a given band results in the interference generated in this band at the PSD level of -20 to -30 dB (relative to in-band PSD), which may not be sufficient for the protection of coexisting systems operating in this band. There is a number of techniques to fight this phenomenon. Some utilize correlation between different subcarriers sidelobes in order to null the spectrum components in the PU band [43, 200]. Other methods focus on time-domain processing, preserving smooth transition between consecutive symbols [161, 182]. A detailed presentation of this problem and related techniques has been provided in this chapter. One very promising technique to reduce the OOB power is OCCS as well as EC method that additionally accounts for the HPA nonlinear distortions.

Multiple Accessing

Another merit of NC-OFDM employed in the future radio communication systems is its ability to be combined with medium access methods, the same as considered for OFDM [38]. The simplest served protocols are Time-Division Multiple Access (TDMA) and Frequency-Division Multiple Access (FDMA). In wireless local area networks (IEEE 802.11 a/g), a protocol called Carrier-Sense Multiple Access (CSMA) is used while Physical Layer (PHY) layer is based on OFDM. There is also some research on combining CDMA with OFDM. Moreover, the Orthogonal Frequency-Division Multiple Access (OFDMA) where inherent features of OFDM are used, can also be considered for this purpose. The main principle of this method is that distinct users have distinct sets of subcarriers assigned to their transmission. Additionally, by maintaining time and frequency synchronization, orthogonality between different users signals can be ensured.

Robustness to the PU-Originating Interference

The nature of NC-OFDM allows for separation of the PU and the SU signal frequency components. The protection of PU is achieved by means of some spectrum shaping algorithms and power control mechanisms. However, for the NC-OFDM secondary system to be efficient in terms of the Quality of Service (QoS) performance, the NC-OFDM SU receiver has to be able to reject the interference generated by the PU transmission. It is achieved by the nature of FFT-based NC-OFDM receiver. However, as the interference signal may not be orthogonal to the complex sinusoids used in the receiver FFT, some interference leakage occurs in the SU band.

Moreover, before data symbol detection in the frequency domain is carried out, time and frequency synchronization has to be achieved. Although PU and SU signals are separated in frequency, typically the time and frequency synchronization is achieved by the time-domain processing, where the whole PU signal at the SU receiver constitutes interference, that is, it overlaps the SU useful signal in the time domain. It is an important issue, as NC-OFDM transmission is very sensitive to CFO and relatively sensitive to timing offset. If the synchronization is not achieved, ICI and ISI occur. The main problem in the NC-OFDM-based CR is high-power LU-originating interference in-band of the SU receiver. Here, a promising synchronization algorithm for this type of NC-OFDM system has been presented, that is, LUISA, which is robust to the LU-originating narrowband and wideband interference. Other well-known synchronization algorithms applied typically in OFDM systems can also be applied in other less-demanding cases of interference occurring in the NC-OFDM receiver. Finally, most of the other standard channel estimation, equalization, and detection methods developed for OFDM can be efficiently adopted to future NC-OFDM systems.

4

Generalized Multicarrier Techniques for 5G Radio

As discussed in the previous chapter, enhanced Orthogonal Frequency-Division Multiplexing (OFDM) and Non-contiguous Orthogonal Frequency-Division Multiplexing (NC-OFDM) techniques have evident advantages that make them suitable for future radio communication systems. However, some major drawbacks also exist, for example, a high value of the Peak-to-Average Power Ratio (PAPR) or a high impact of the frequency offset on the Bit Error Rate (BER) observed at the receiver. Moreover, considering the efficiency of spectrum utilization, the Cyclic Prefix (CP)-based multicarrier techniques have their limitations as well. The necessity of the usage of CP in order to avoid Inter-Symbol Interference (ISI), rectangular pulse shape of the frequency-domain symbols modulating the subcarriers, and strictly defined subcarrier spacing (equal to the inverse of the orthogonality time of a single OFDM symbol) results in the fact that the spectral efficiency can be improved only by increasing the number of bits transmitted in a subchannel. This in turn requires application of high-order constellations of digital data modulation and has negative impact on the BER performance. It is worth mentioning that such limitations are not set in general in Multi-carrier (MC) systems using nonorthogonal subcarriers. In another subclass of multicarrier systems, for example, in the Filter-Bank Multi-Carrier (FBMC), CP is no longer required. Contrary to the OFDM, NC-OFDM and Filtered Multi-Tone (FMT) techniques, nonorthogonal multicarrier systems [237, 238] have higher degree of freedom in the radio air-interface definition. In these systems, the pulse shape can be flexibly defined, and its duration is not related to the frequency spacing. Furthermore, it is allowed that the neighboring pulses overlap both in the time and in the frequency domain. Owing to this, the spectral efficiency of MC systems with nonorthogonal subcarriers can be improved not only by means of increasing the number of bits assigned to particular subchannels but also by discarding the cyclic prefix and reducing the distances between adjacent waveforms on the Time-Frequency (TF) plane. The denser the allocation of the pulses on the TF plane, the higher the theoretical spectral efficiency and possible channel capacity.

Advanced Multicarrier Technologies for Future Radio Communication: 5G and Beyond, First Edition.
Hanna Bogucka, Adrian Kliks, and Paweł Kryszkiewicz.
© 2017 John Wiley & Sons, Inc. Published 2017 by John Wiley & Sons, Inc.

One may categorize two classes of multicarrier signals: the ones with orthogonal and nonorthogonal subcarriers. However, any kind of multicarrier signals can be classified into a broad class of Generalized Multicarrier (GMC) signals, as suggested in [51, 84]. The sets of orthogonal and nonorthogonal signals are disjoint and mutually complement, creating the aforementioned set of generalized multicarrier signals. (However, as it is described later in this chapter, also the single carrier signals can be interpreted as a special case of a GMC signal and thus can be counted into this set.) The term *generalized* indicates that any kind of signal can be defined by a set of parameters (such as the pulse shape, the density of the TF grid). In other words, there are no restrictions in designing a GMC system but one – the GMC receiver has to be able to recover the transmitted data at any given BER level, ensuring the fulfillment of the Quality of Service (QoS) requirements. Within this chapter, theoretical background of GMC systems is provided, with a particular focus on the subcarriers orthogonality and the lack of this orthogonality in multicarrier signals.

4.1 The Principles of GMC

In the following discussion, typical symbols for denoting the set of natural, integer, real, and complex numbers are used, that is, the set of natural numbers is denoted as $\mathbb{N}$, the set of integers as $\mathbb{Z}$, the set real and complex numbers as $\mathbb{R}$ and $\mathbb{C}$. The Lebesgue space $\mathbf{L}^p(\mathbb{R})$ [239] is a space that contains all measurable functions f for which the p-th norm $\|f\|_p$ for $p \in \langle 1, \infty \rangle$ is finite, that is, $\|f\|_p = \left(\int_{-\infty}^{\infty} |f(x)|^p \, dx \right)^{\frac{1}{p}} < \infty$. We also assume that the continuous signal $s(t)$ belongs to $\mathbf{L}^2$, that is, $s(t) \in \mathbf{L}^2(\mathbb{R})$. At the same time, it means that the signal belongs to the Hilbert space S (a space with strictly defined inner product [239]) and the integral over the considered space S of the second power of the absolute value of $s(t)$ is finite, that is, $\int_S |s(t)|^2 \, dt < \infty$. In other words, it can be stated that signal $s(t)$ is *square integrable*. In practice, it usually refers to the finite-energy signals. In the case of discrete signals $s[k]$, it is assumed that it belongs to the sequence space, that is, $s[k] \in l^2(\mathbb{Z})$ and the square sum of the absolute values of the sequence is finite, that is, $\sum_S |s[k]|^2 < \infty$. In other words, it can be stated that signal $s[k]$ is *square-summable signal*.

One can represent any signal $s(t)$ in the form of a series [51, 81, 82, 239, 240]:

$$s(t) = \sum_{n,m \in \mathbb{Z}} d_{n,m} g_{n,m}(t) = \sum_{n,m \in \mathbb{Z}} d_{n,m} g(t - mT) \exp(j2\pi nFt), \qquad (4.1)$$

where j is an imaginary unit ($j^2 = -1$), n, m are integers ($n, m \in \mathbb{Z}$), $\{d_{n,m}\}$ are the series coefficients, and $\{g_{n,m}(t)\}$ creates a set of elementary functions (also known as *atoms* [84, 241]) used for signal expansion. Note that it is customary to neglect the phase value $\exp(-j2\pi nFmT)$ in $s(t)$, that is, $g_{n,m}(t) = g(t - mT) \exp(j2\pi nFt)$ rather than $g_{n,m}(t) =$

$g(t - mT) \exp(j2\pi nF(t - mT))$ [242]. Since the TF representation of the signal is considered, the set of functions is often called a Gabor set [239]. This set is created by means of translation (which shifts pulse $g(t)$ in the time domain by the time period T) and modulation (which shifts signal $g(t)$ in the frequency domain by the frequency distance F) of the elementary Gabor function $g(t)$ of duration L_g expressed in the assumed time units. Using the operator notation, one can formulate the aforementioned statement as

$$g_{n,m}(t) = \dot{T}_{mT}\dot{F}_{nF}g(t) = \dot{T}_{ma}\dot{F}_{nb}g(t), \tag{4.2}$$

where $\dot{T}_{ma}$ and $\dot{F}_{nb}$ denote the translation by a period ma and modulation by the frequency distance nb, respectively. In other words, elementary function $g(t)$ is shifted along the specific *lattice* $\Lambda = a\mathbb{Z} \times b\mathbb{Z} = T\mathbb{Z} \times F\mathbb{Z}$ in the time–frequency plane. Parameters $(a, b) = (T, F)$ are called *time–frequency shift parameters* and define the distance between consecutive waveforms in the time and in the frequency domain, accordingly. Thus, one can say that waveform $g_{n,m}(t)$ is localized at a specific point (mT, nF) in the TF plane. The exemplary set of functions and the graphical representations of the set of elementary functions $\{g_{n,m}(t)\}$ are presented in Figure 4.1. In the discrete-time form, formula (4.1) can be written as follows:

$$s[k] = \sum_{n,m\in\mathbb{Z}} d_{n,m}g_{n,m}[k] = \sum_{n,m\in\mathbb{Z}} d_{n,m}g[k - mM_\Delta]\exp[j2\pi nN_\Delta k], \tag{4.3}$$

where k denotes the sample's index in the time domain. In such a case, the distance between the adjacent pulses is expressed in terms of *samples*, that is, two consecutive pulses in time domain are spaced by $M_\Delta [samples]$ and two neighboring pulses in frequency domain are spaced by $N_\Delta [samples^{-1}]$ in frequency domain, respectively. Finally, the duration of the pulse $g[k]$ is equal to L_g samples.

Since in practical realizations of multicarrier signals, the transmit signals are organized in the form of transmission frames (or bursts) of the size $N \times M$ (i.e., one burst consists of M blocks of N subcarriers giving in total $N \cdot M$ pulses in one frame), the formulas (4.1) and (4.3) can be rewritten in the following way [237, 239]:

$$s(t) = \sum_{m=0}^{M-1}\sum_{n=0}^{N-1} d_{n,m}g_{n,m}(t) = \sum_{m=0}^{M-1}\sum_{n=0}^{N-1} d_{n,m}g(t - mT)\exp(j2\pi nFt), \tag{4.4}$$

$$s[k] = \sum_{n=0}^{N-1}\sum_{m=0}^{M-1} d_{n,m}g_{n,m}[k] = \sum_{n=0}^{N-1}\sum_{m=0}^{M-1} d_{n,m}g[k - mM_\Delta]\exp[j2\pi nN_\Delta k]. \tag{4.5}$$

The bandwidth of the frequency band occupied by one GMC-frame of duration T_{GMC} is equal to F_{GMC}.

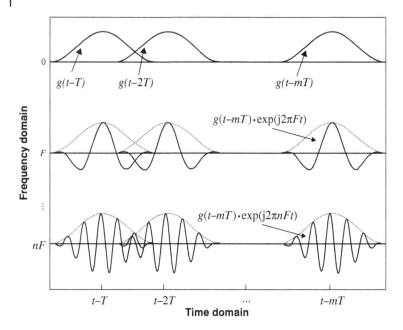

Figure 4.1 Gabor's elementary functions localized in the time–frequency plane.

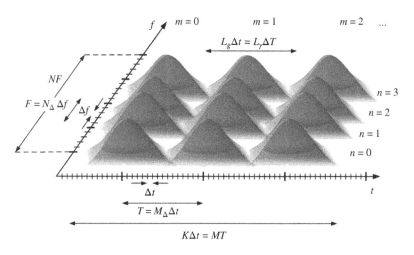

Figure 4.2 Exemplary time–frequency representation of one GMC-frame.

The exemplary representation of one GMC-frame (or GMC-burst) on the time–frequency plane is shown in Figure 4.2, where the meaning of all crucial variables used in this chapter is illustrated. Note that we slightly change the meaning of some variables used in Chapter 3 in order to better reflect the time and frequency relations of the transmitted pulses on the TF plane. Thus,

- M defines the number of blocks of N parallel pulses in one GMC-frame in the time domain;
- N defines the number of parallel pulses in the frequency domain transmitted in one block of the GMC-frame, which is equivalent to the number of subcarriers and usually the (I)FFT size;
- $N \cdot M$ is the total number of pulses in one GMC-frame;
- T is the time distance between the consecutive pulses;
- F is the frequency distance between the adjacent pulses in the frequency domain (subcarrier frequency distance, formerly denoted as Δf in Chapter 3);
- M_Δ is the time distance between the consecutive pulses (in samples, i.e., in the case of discrete representation of the signals);
- N_Δ is the frequency distance between the adjacent pulses (in discrete representation of the signal spectra);
- Δt is the time distance between the consecutive samples in the time domain;
- Δf is now the frequency distance between the consecutive spectrum samples in the frequency domain;
- L_g, L_γ denote the duration of pulse g or γ (introduced later in this chapter) and expressed in samples;
- T_{GMC} denotes the duration of one GMC-frame;
- F_{GMC} denotes the bandwidth of the occupied frequency band of one GMC-frame.

The series coefficients $d_{n,m} \in l^2(\mathbb{Z} \times \mathbb{Z})$ are obtained by means of projection of the time-domain signal on the two-dimensional (time–frequency) plane, or more precisely on the set of the elementary functions $\{g_{n,m}\}$. Such an operation can be realized by the Gabor transform or Short-Time Fourier Transform (STFT) [238, 239, 241]. In digital communications, these coefficients can be interpreted as data symbols (e.g., QAM, PSK symbols). In such a case, the set of elementary functions is the set of $N \cdot M$ information-bearing pulses *spread over* the time–frequency plane. Hence, $s(t)$ (or its sampled version $s[k]$) is the transmitted signal representing a set of data symbols. The received signal, $r(t)$ (or $r[k]$), has to be filtered appropriately in the bank of analysis filters $\{\gamma_{n,m}\}$ in order to recover the transmitted data $d_{n,m}$. The duration of pulse $\gamma(t)$ equals L_γ of the assumed time units (or in a case of discrete representation of this pulse $\gamma[k]$ – L_γ samples). Further information regarding the issues related to the design of the pulse pair $(g(t), \gamma(t))$ is provided in the next sections.

4.1.1 Frame Theory and Gabor Transform

In this section, some theoretical background of the frame theory and Gabor transform is provided [80–82, 84, 93, 238–240, 243, 244]. In the most popular approach (i.e., in the engineering applications), the orthogonal (or even orthonormal) bases of functions are used for signal expansions. The basis is defined as a set of functions $\{g_i(t)\}$, where $i \in \mathbb{I}$ and $\mathbb{I}$ is the countable set of indices $\mathbb{I} = \mathbb{Z}$ (or $\mathbb{I} = \mathbb{Z}^2$ in a sense of two-dimensional, *Weyl–Heisenberg system*) for which two main properties of the signal expansion can be defined [82]:

- the possibility of signal expansion (it is also the sufficient condition of the *completeness* of the considered basis in the Hilbert space)

$$s(t) = \sum_{i \in \mathbb{I}} \langle s(t), g_i(t) \rangle \cdot g_i(t), \tag{4.6}$$

- the fulfillment of the generalized Parseval formula expressed as

$$\|s(t)\|^2 = \sum_{i \in \mathbb{I}} |\langle s(t), g_i(t) \rangle|^2, \tag{4.7}$$

where $\langle s(t), g(t) \rangle = \langle s, g \rangle = \int s(t)\overline{g(t)} \, dt$, and $\|s\| = \sqrt{\langle s, s \rangle}$ define the inner product and the norm, respectively.

Usually, in most practical realizations, the *completeness* of the basis is required, that is, the set of basis functions generates the linear (sub)space, and in other words, it is possible to recover the expanded signal without any information loss. Although the orthonormal bases are very useful from a practical point of view, the *nonorthogonal basis, spans*, and *frames* are of high importance in the context of GMC signaling, which are explained as follows:

- in general, the *nonorthogonal* basis is a set of functions that are linearly dependent (thus nonorthogonal) but still span the considered (Hilbert) space;
- assuming that $\mathcal{V}^D$ is a D-dimensional vector space and $[\mathbf{v}_1, \mathbf{v}_2, \dots, \mathbf{v}_D]$ is a set of elements in $\mathcal{V}$, and if every vector in $\mathcal{V}$ can be expressed as a linear combination of all the vectors $[\mathbf{v}_1, \mathbf{v}_2, \dots, \mathbf{v}_D]$, then $[\mathbf{v}_1, \mathbf{v}_2, \dots, \mathbf{v}_D]$ *spans* $\mathcal{V}$. In other words, span$(\mathcal{V})$ = span$(\mathbf{v}_1, \dots, \mathbf{v}_D)$ = $\{\lambda_1 \mathbf{v}_1 + \cdots + \lambda_D \mathbf{v}_D \mid \lambda_1, \dots, \lambda_D \in \mathbf{K}\}$, and λ_i are the scalars in the considered field $\mathbf{K}$; one can state that the span of $\mathcal{V}$ can be defined as the set of such elements of $\mathcal{V}$, which are combinations of vectors $[\mathbf{v}_1, \mathbf{v}_2, \dots, \mathbf{v}_D]$;
- if the set of basis functions is not linear, it is *overcomplete*; the overcomplete set of functions $\{g_i\}_{i \in \mathbb{I}} \subseteq \mathcal{L}^2$ is called a *frame* if and only if there exist two constants (known as the *lower* and the *upper frame bound*) A and B such that $0 < A \le B < \infty$, and

$$A\|s\|^2_{\mathcal{L}^2} \le \sum_{i \in \mathbb{I}} |\langle s, g_i \rangle_{\mathcal{L}^2}|^2 \le B\|s\|^2_{\mathcal{L}^2}. \tag{4.8}$$

When analyzing the aforementioned definition of a frame, it can be concluded that a frame can be treated as a generalization of the basis. The mid-term in the formula (4.8) represents the power of the signal represented by the frame elements (in our case, functions g_i). Thus, the energy of a signal represented by means of the frame elements is greater than $A > 0$ and lower than $B < \infty$. This formula can also be treated as a special form of the generalized Parseval formula [241]. The frame is called *tight* if the lower and upper bounds of the frame are equal [239, 241, 245]. Additionally, if $A = B = 1$, the frame is called normalized frame and constitutes an orthonormal system. Detailed information about the frames in the context of telecommunication can be found in [239, 241, 245, 246]. A tutorial on the frame theory has been presented in [247].

Typically, the term *basis* refers to the *linearly independent* and *complete* set of functions. If one removes some elements from such a set, the completeness is lost. On the other hand, if one adds some elements to such a defined set, the linear independence is lost.

In a case of the nonorthogonal signals, a so-called *dual* frame is required to recover the original signal from the expanded one by means of the original frame functions. The set of functions $\{\gamma_i\}_{i \in \mathbb{I}}$ constitutes the dual frame if the following relation is fulfilled:

$$s(t) = \sum_{i \in \mathbb{I}} \langle s(t), g_i(t) \rangle \gamma_i(t). \tag{4.9}$$

One can state that formula (4.9) generalizes the relation (4.6).

Very often, mainly in the case of time-discrete signals, one has to consider the square-summable sequences of data symbols of the form $\{d_i\}_{i \in \mathbb{I}}$. As a consequence, the Riesz basis has to be introduced. Formal definition of the Riesz basis is as follows: a sequence of vectors $\{g_i\}$ constitute the Riesz basis if for any sequence d_i two positive constants A and B exist such that $A \sum_{i \in \mathbb{I}} |d_i|^2 \leq \|\sum_{i \in \mathbb{I}} d_i \cdot g_i(t)\|^2 \leq B \sum_{i \in \mathbb{I}} |d_i|^2$. It means that all possible vectors obtained by means of the basis vectors are of finite energy if and only if coefficients d_i are of finite energy. Thus, Riesz basis is sometimes called a *stable* basis. More information about the Riesz basis can be found in [248].

In the context of the signal representation utilizing the nonorthogonal basis functions, two additional terms have to be introduced. The first one is the definition of the *Weyl–Heisenberg* frames. Based on (4.2), one can explicitly define the translation $\dot{T}_m a$ and modulation $\dot{F}_n b$ operators over the function $g(t)$ as follows:

$$\dot{T}_{ma} g(t) = g(t - ma), \quad \text{for } a \in \mathbb{R}, \tag{4.10}$$

$$\dot{F}_{nb} g(t) = \exp(j2\pi t n b) g(t), \quad \text{for } b \in \mathbb{R}. \tag{4.11}$$

In the case of GMC systems introduced in the previous section, $a = T$ and $b = F$. Thus, the set of functions $\{\dot{T}_{mT} \dot{F}_{nF} g(t)\} = \{g_{n,m}(t)\}$ (and accordingly in the discrete form) generates the *Weyl–Heisenberg* frame (also known as

Gabor frame) if two constants A and B exist such that $0 < A \le B < \infty$, and

$$A\|s\|^2 \le \left\|\sum_{(m,n)\in\mathbb{Z}^2} \langle s, g_{n,m}\rangle\right\|^2 \le B\|s\|^2. \tag{4.12}$$

It can be observed that formula (4.12) is a special case of (4.8). Comprehensive tutorials about the Weyl–Heisenberg frames can be found in [249, 250].

The term that must also be defined is the *biorthogonality* relation of two bases. Let $\{g_i\}_{i\in\mathbb{I}}$ and $\{\gamma_i\}_{i\in\mathbb{I}}$ be the basis in a Hilbert space. Two sets of sequences (vectors) are in the biorthogonal relation if and only if $\forall_{i,j\in\mathbb{Z}}\langle g_i, \gamma_j\rangle = \delta[i-j]$, where $\delta(\cdot)$ is the Kronecker delta function. It can be proved that the basis dual to the Riesz basis is also a Riesz basis with the inverse frame bounds. The necessary condition for the existence of (bi)orthogonality between two function sets is that the sampling grid cannot be "too dense," that is, $TF \ge 1$ [82, 239].

In most of the practical realizations of the multicarrier systems (such as OFDM), the so-called rectangular lattice is applied, which means that the centers of the transmit pulses (nF, mT) (or (nN_Λ, mM_Λ) in case of discrete representations of signals) create a rectangular grid as shown in Figure 4.3.a. In other words, it means that the localization of pulses on the time–frequency plane is identical in every time slot. However, as it will be shown in the following sections, the rectangular lattice creates some limits with regard of the overall system capacity.

Every lattice, denoted as Λ, is defined by its generator matrix $\mathbf{L_R}$ as follows [244, 251]:

$$\mathbf{L_R} = \begin{pmatrix} a_{11} & a_{12} \\ a_{21} & a_{22} \end{pmatrix}, \tag{4.13}$$

where $a_{11}, a_{12}, a_{21}, a_{22}$ denote the distances between the neighboring points (waveform centers) of the lattice. For example, the generator matrix for the

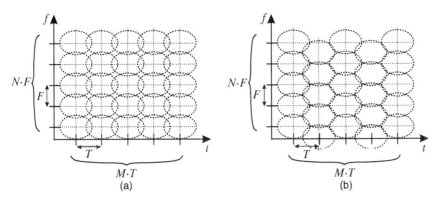

Figure 4.3 (a) Rectangular lattice, (b) Hexagonal lattice.

rectangular lattice with time distance T and frequency spacing F is defined as

$$\mathbf{L}_R = \begin{pmatrix} F & 0 \\ 0 & T \end{pmatrix}, \tag{4.14}$$

whereas the hexagonal lattice (see Figure 4.3b) is defined as

$$\mathbf{L}_R = \begin{pmatrix} \rho F & 0.5\rho F \\ 0 & \rho^{-1}T \end{pmatrix} \tag{4.15}$$

and $\rho = \sqrt{2}/\sqrt[4]{3}$. Lattices are also often defined by its density, $\delta(\Lambda)$:

$$\delta(\Lambda) = \frac{1}{\det(\mathbf{L}_R)}. \tag{4.16}$$

Thus, the density of the rectangular lattice (used for OFDM) is equal to

$$\delta(\Lambda) = \frac{1}{TF}. \tag{4.17}$$

As it has been stated, the system defined by the set of functions: $\{\dot{T}_{mT}\dot{F}_{nF}g(t)\} = g_{n,m}(t)$ (for $T = M_\Delta \Delta t$ and $F = N_\Delta \Delta f$) constitutes the Weyl–Heisenberg frame (or Gabor frame). One can observe that the relation between distances on time–frequency grid (i.e., M_Δ and N_Δ) defines the lattice size. In other words, the lattice size can be characterized by the product of these numbers, that is, $N_\Delta \cdot M_\Delta$. This product value is one of fundamental properties of the considered set of functions $g_{n,m}(t)$ since it defines the completeness and linear dependence of its elements. One can distinguish three separate classes of these functions [80–82, 239]:

- when the product of T and F is higher than 1 ($TF = N_\Delta \Delta t \cdot M_\Delta \Delta f > 1$), the grid is called *undercritical* grid, which means that there cannot exist complete Riesz basis for the square-integrable functions; in such a case, the linear independence of the basis functions is guaranteed;
- when the product of T and F is lower than 1 ($TF = N_\Delta \Delta t \cdot M_\Delta \Delta f < 1$), the grid is called *overcritical* grid, which means that there cannot exist Riesz basis for the square-integrable functions; however, the existence of frames (nonorthogonal overcomplete set of functions) is possible. Moreover, frames with very good time–frequency localizations exist, which means that $g_{n,m}(t)$ have concentrated energy over the time–frequency plane;
- when the product of T and F is equal to 1 ($TF = N_\Delta \Delta t \cdot M_\Delta \Delta f = 1$), the frame is called *critical* grid, which means that the complete Weyl–Heisenberg frame exists for the square-integrable functions; however, the localization of the pulses is very bad (see the Balian–Low theorem [82, 238, 239, 241]). The well-known examples of the orthonormal basis $\{g_{n,m}(t)\}$ are the set of rectangular functions and the set of *Sinc* functions used in OFDM. The frequency localization (expressed by the means of the second moment of the frequency-domain representation $G_{n,m}(f)$ of a signal, that

is, $\int_{-\infty}^{\infty} f^2 |G_{n,m}(f)|^2 \, df = \infty$) in the case of the rectangular function and the time localization (expressed by means of the second moment of the time-domain representation $g_{n,m}(t)$ of a signal, i.e., $\int_{-\infty}^{\infty} t^2 |g_{n,m}(t)|^2 \, dt = \infty$) in the case of the *Sinc* function are extremely bad. The bad time–frequency localizations means, as a consequence, poor energy concentration on the time–frequency plane.

It is known that for a large number of subcarriers, the spectral efficiency of multicarrier systems can be approximated as $\eta = \frac{\beta}{TF}$ [252], where β denotes the number of bits per one data symbol. Assuming that β is constant, one can state that the spectral efficiency depends on the parameters of the time–frequency grid, that is, distance between pulses in the time domain T and in the frequency domain F. In other words, spectral efficiency for the rectangular grids can be defined as the density of the lattice, that is, $\eta = \beta \delta(\Lambda)$. In addition, the classification of the Gabor systems can be done with regard to the lattice type.

In all existing multicarrier communication systems, it is decided to ensure the linear independence (regarding of the orthogonality property) of the pulses on the time–frequency plane. In other words, the value of TF cannot be lower than 1. It means that the spectral efficiency of these multicarrier systems cannot be higher than β [bits/Hertz/second] and is reached only in the case of critical sampling. In this context, the maximum spectral efficiency is achieved in the standard OFDM transmission, where $TF = 1$. However, in order to limit the influence of Intersymbol Interference (ISI), the cyclic prefix is added to every transmission frame, thus increasing the TF product and decreasing the efficiency.

Another approach is to increase the distance between the pulses and "orthogonalize" them in order to ensure perfect signal reconstruction at the receiver. In such a case, there is no need for the transmission of the cyclic prefix. In addition, the idea of transmission of nonorthogonal and biorthogonal pulses (if only the distance between the neighboring pulses on the time–frequency plane is large enough to limit the self-interference) has been considered [82].

In all the aforementioned examples, the main assumption was to use the linearly independent set of basis functions, which translates to the relation $TF \geq 1$. However, as it has been shown in some recently published works, that this constraint can be relaxed [81, 84, 85, 240]. As a consequence, the spectral efficiency can be increased by proper pulse shaping (and lattice shaping), since TF can be lower than 1. Accordingly, the transmission is organized by utilization of the overcomplete, linearly dependent set of functions that constitute the Weyl–Heisenberg frame.

4.1.2 Short-Time Fourier Transform and Gabor Transform

From (4.3) and (4.6) one can conclude that the expansion coefficients $\{d_{n,m}\}$ (or in the sense of the wireless communications – user data symbols) can be calculated by means of the STFT of function (signal) $s(t)$. In case of continuous signals, STFT of $s(t)$ equals:

$$d(f, t) = \int_{-\infty}^{\infty} s(\tau)\gamma^*(\tau - t) \exp(-j2\pi f\tau) \, d\tau, \tag{4.18}$$

and accordingly in the case of discrete signals:

$$d[f_n, m] = \sum_k s[k]\gamma^*[k - mM_\Delta] \exp[-j2\pi f_n k]. \tag{4.19}$$

Here, $(\cdot)^*$ denotes complex conjugate. The sampled version of the STFT (i.e., decimated in time and in frequency) is the so-called Gabor transform defined as

$$d_{n,m} = \int_{-\infty}^{\infty} s(\tau)\gamma^*(\tau - mT) \exp(-j2\pi nF\tau) \, d\tau, \tag{4.20}$$

and accordingly in the discrete representation of signals:

$$d_{n,m} = \sum_k s[k]\gamma^*[k - mM_\Delta] \exp[-j2\pi nN_\Delta k]. \tag{4.21}$$

4.1.3 Calculation of the Dual Pulse

As it was mentioned in the previous section, in order to recover the user data from the transmit signal (represented by means of frame $\{g_{n,m}(t)\}$), one has to use the dual frame $\{\gamma_{n,m}(t)\}$. The assumption of the perfect recovery of coefficients $d_{n,m}$ in the case of nondispersive and noise-free channel [64, 82] is fulfilled when the dual-pulse prototype satisfies the biorthogonality condition. In such a case, a problem of efficient calculation of the dual pulse arises. Obviously, the Gabor expansion exists if and only if the following relation is fulfilled:

$$\sum_n \sum_m \gamma_{n,m}^*(t')g_{n,m}(t) = \delta(t - t'), \tag{4.22}$$

or equivalently – after applying the Poisson-sum formula [253]:

$$\frac{2\pi}{TF} \int_{\mathbb{R}} g(t)\gamma^* \left(t - m\frac{2\pi}{T}\right) \exp\left(\frac{-j2\pi n}{F}\right) \, dt = \delta(n)\delta(m). \tag{4.23}$$

When considering the discrete signals, the aforementioned relations shall be rewritten as follows:

$$\sum_n \sum_m \gamma^*[k' - mM_\Delta]g[k - mM_\Delta] \exp\left\{\frac{j2\pi(k - k')}{N}\right\} = \delta[k - k'], \tag{4.24}$$

and equivalently – after applying the Poisson-sum formula:

$$\sum_{k=0}^{K-1} g[k + qN]\gamma^*[k] \exp\left[\frac{-j2\pi pk}{M_\Delta}\right] = \frac{M_\Delta}{N}\delta[p]\delta[q] \tag{4.25}$$

for $0 \leq p \leq M_\Delta - 1$ and $0 \leq q \leq N_\Delta - 1$ and where K denotes the signal period (in samples). The relations (4.23) and (4.25) are known as the *Wexler–Raz identity* [82, 238, 239]. It is assumed that the so-called *oversampling ratio* $\tilde{\alpha} = \frac{K}{N_\Delta M_\Delta}$ is chosen in such a way that perfect reconstruction is guaranteed, that is, $\tilde{\alpha} \geq 1$. It is also required that $N_\Delta \cdot N = M_\Delta \cdot M = K$. The formula (4.24) can be rewritten in a matrix form, thus allowing for efficient calculation of the dual pulse:

$$\mathbf{U}\overline{\gamma}^* = \overline{\mu}, \tag{4.26}$$

where $\overline{\gamma}^*$ is the vector whose elements are the conjugate samples of dual pulse $\gamma[k]$, $\mathbf{U}$ is the $N_\Delta M_\Delta$-by-K matrix, whose elements are defined as $u_{pM_\Delta+q,k} \equiv g[k + qN] \exp[\frac{-j2\pi pk}{M_\Delta}]$ and μ is the $N_\Delta M_\Delta$ dimensional vector given by $\overline{\mu} = [\frac{M_\Delta}{N}, 0, 0, \dots, 0]^{\mathrm{T}}$.

However, it must be stressed that if a pair of functions $g[k]$ and $\gamma[k]$ are not well-localized (have low value of the second moment) either in the time or in the frequency domain, the Gabor coefficients (i.e., the inner product of $s[k]$ and $\gamma[k]$) do not characterize the signal's local variation. Moreover, for a selected pulse shape and the certain time–frequency sampling grid, the dual pulse would not be unique. Thus, the problem arises on how to choose the best shape of the dual pulse. The popular (in the literature) approach is to select a dual pulse that is closest to $g[k]$ in the sense of the least squared error $\epsilon_{\min}$, that is:

$$\epsilon_{\min} = \min_{\mathbf{U}\overline{\gamma}^*=\overline{\mu}} \sum_{k=0}^{K-1} \left| \frac{\gamma[k]}{\|\gamma\|} - g[k] \right|^2 = \min_{\mathbf{U}\overline{\gamma}^*=\overline{\mu}} \left(1 + \frac{M_\Delta}{\|\gamma\|N} \right), \tag{4.27}$$

where

$$\|\gamma\| = \sqrt{\sum_{k=0}^{K-1} |\gamma[k]|^2}. \tag{4.28}$$

If matrix $\mathbf{U}$ is of full rank, the dual pulse of the minimum energy is defined as

$$\overline{\gamma} = \mathbf{U}^{\mathrm{T}}(\mathbf{U}\mathbf{U}^{\mathrm{T}})^{-1}\overline{\mu}. \tag{4.29}$$

The straightforward application of the pseudoinverse operation seems to be impractical. Thus, efficient algorithms (i.e., such that could be applied in real time) should be used instead of the direct application of (4.29). One can find some computationally efficient algorithms for calculating the dual pulse in the literature. Four algorithms for calculation of the dual pulses should be mentioned, mainly the one using the straightforward formula described earlier, the fast algorithm described in [254] (where the QR decomposition has been applied), the one described in [255] (where some specific properties of matrix $\mathbf{U}$ are considered), and the one proposed in [256].

It has to be mentioned that several procedures of calculation of the dual pulse have been proposed in [81, 85] that are dedicated to the case when $N_\Delta M_\Delta < 1$, and some of them take the channel's characteristics into account.

4.1.4 GMC Transceiver Design Using Polyphase Filters

It is well-known that the general transceiver of the multicarrier system can be realized by means of the transmultiplexer, presented in Figure 4.4 [82, 239]. According to [49, 64, 84, 257, 258], one can think of the efficient realization of the Gabor transform used for the GMC signal generation by means of the DFT perfect-reconstruction filter bank (see Figure 4.5) [64]. In such a case, a set of user data (i.e., constituting the Gabor coefficients) $\{d_{n,m}\}$ is the input to the synthesis filter bank (N filters defined by $G_n[z]$, that is, the Z transform of function $g_n[k]$). Then, it passes through the transmission channel, and the received signal $r[k]$ is analyzed in a so-called analysis filter bank ($\Gamma_n[z]$ denotes the Z transform of the dual pulse $\gamma_n[k]$). By changing the oversampling factor M_Δ, one defines the time distance between corresponding samples of the consecutive waveforms in the time domain.

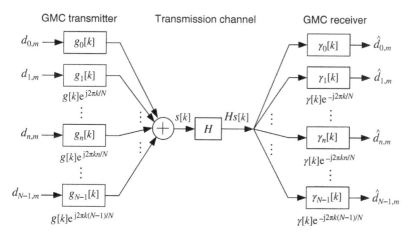

Figure 4.4 The structure of the GMC transmultiplexer.

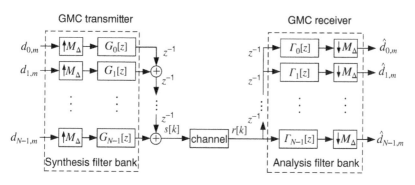

Figure 4.5 The structure of the GMC transceiver realized by means of the DFT perfect-reconstruction filter bank.

Furthermore, the idea of a so-called polyphase decomposition [64] can be applied in order to make the realization of the DFT filter bank even more effective (in terms of the structure complexity and the number of required operations per one sampling period). In such a case, the analysis and the synthesis prototype filters (or simply prototypes) are represented by means of polyphase filters:

$$G[z] = \sum_{n=0}^{N-1} z^{-(N-1-n)} \tilde{G}_n[z^N], \qquad (4.30)$$

$$\Gamma[z] = \sum_{n=0}^{N-1} z^{-n} \tilde{\Gamma}_n[z^N], \qquad (4.31)$$

where $\tilde{G}_n[z]$ and $\tilde{\Gamma}_n[z]$ are the Z transforms of the polyphase components of the synthesis and the analysis pulses. The general structure of the transmultiplexer realized through the IFFT and FFT blocks and the polyphase filter banks is illustrated in Figure 4.6. There, $\tilde{g}_n[k]$ and $\tilde{\gamma}_n[k]$ denote the impulse responses of the synthesis and analysis polyphase filters, respectively.

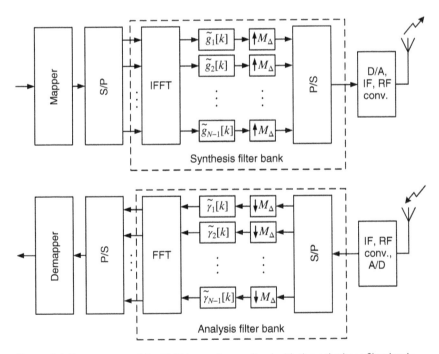

Figure 4.6 The structure of the GMC transceiver realized with the polyphase filter bank; applied standard blocks: Serial-to-Parallel (S/P), Parallel-to-Serial (P/S), Digital-to-Analog (D/A), Analog-to-Digital (A/D), Intermediate Frequency (IF), and Radio Frequency (RF) conversion.

4.2 Peak-to-Average Power Ratio Reduction in GMC Transmitters

In this section, we analyze the problem of high PAPR, in GMC systems [130]. As discussed in Chapters 2 and 3, this is the typical problem of all MC systems including OFDM and NC-OFDM transmissions. Similar issues occur in other MC techniques, for example, in the ones using nonorthogonal subcarriers or filtered subcarrier signals. The PAPR metric, or its square root known as CF, and the CM are usually used to reflect the influence of a nonlinear element on the transmitted signal. High values of these metrics indicate that the probability of nonlinear distortions in the transmit signal is also high when a nonlinear element, such as Power Amplifier (PA) (or High Power Amplifier (HPA)), is placed in the transmission chain. These nonlinear distortions usually result from clipping of high-amplitude samples in HPA and have repercussions in the transmit (amplified) signal, such as the in-band (self) interference and out-of-band radiation. They lead to the BER degradation at the receiver and to the fact that the spectral mask (especially the dynamic Spectrum Emission Mask (SEM)) is not ensured as intended.

4.2.1 Optimization of the Synthesis Pulse Shape for Minimization of Nonlinear Distortions

Let us consider whether PAPR could be reduced in GMC systems by finding the advantageous (or optimal) shape of the pulse. The aim of the optimization is to find the pulse shape $g(t)$ that minimizes the maximum PAPR value denoted as PAPR_{max}. (Note that the actual PAPR may vary for each OFDM symbol and depends on the data symbols modulating subcarriers and the pulse shape, while PAPR_{max} is the maximum value that can never be exceeded.) Following [259], one can define the PAPR_{max} for the frame-based transmission as

$$\text{PAPR}_{\text{max}} = \frac{1}{N} \max_{0 \leq k \leq L_g - 1} \left(\sum_{n=0}^{N-1} |g_n[k]| \right)^2. \tag{4.32}$$

The optimization considerations should be carried out under the energy constraint – the amount of the energy carried by the pulse $g[k]$ should be normalized. Thus, let us consider two theoretical, radically different pulse shapes – the rectangular pulse (defined as in (4.33)) and the delta pulse (or unit impulse defined for discrete signals as in (4.34), analogously as the Dirac delta for continuous signals) that carry the same amount of energy:

$$g_a[k] = A\Pi \left(\frac{k}{N} \right), \tag{4.33}$$

$$g_b[k] = \sqrt{N} \cdot A\delta[k]. \tag{4.34}$$

In the aforementioned equations, A is the amplitude of the rectangular pulse, N denotes the IFFT size, $\Pi\left(\frac{k}{N}\right)$ is the sampled rectangular function of duration of N samples, and $\delta[k]$ is the delta function (unit impulse). In fact, the PAPR value does not depend on the amplitude of the pulse, so A can be omitted. In the case of a multicarrier signal without overcritical sampling and with the pulse duration L_g (in samples) equal to the IFFT size, the maximal PAPR for rectangular pulse is equal to $\text{PAPR}_{\max} = N$, whereas for the delta pulse, $\text{PAPR}_{\max}$ is equal to $\text{PAPR}_{\max} = \sum_{k=0}^{N-1} N = N^2$. It seems that any pulse shape between the delta pulse and the rectangular pulse results in $N \leq \text{PAPR}_{\max} \leq N^2$ and would deteriorate the PAPR characteristic.

Now, let us consider the overcritical sampling case. Contrary to the OFDM case, where there is no overlapping of subsequent data blocks ($N = M_\Delta$, where M_Δ is the distance between consecutive pulses in samples), any two adjacent atoms overlap in the time domain. When $M_\Delta = 1$, there are N overlapping data blocks at the output of the modulator. The maximal PAPR for rectangular pulse is given by

$$\text{PAPR}_{\max} = \left\lceil \frac{N}{M_\Delta} \right\rceil \cdot N, \tag{4.35}$$

where $\lceil \cdot \rceil$ is the *ceil* function, that is, the nearest integer not lower than the argument. For the delta pulse, the maximum value of the PAPR metric is the same as for the critical sampling case. One can observe that only for small distances M_Δ (expressed in the number of samples) between the adjacent atoms in the time domain, the maximal PAPR values for the two considered pulse shapes are comparable (see Figure 4.7). The higher M_Δ, the lower $\text{PAPR}_{\max}$ for the rectangular pulse.

Let us now consider the $g[k]$ pulse function of longer duration, that is, $L_g > N$. For this situation, the theoretical maximum value of PAPR for rectangular pulse is defined as

$$\text{PAPR}_{\max} = \left\lceil \frac{N}{M_\Delta} \right\rceil \cdot N \cdot \left\lceil \frac{L_g}{N} \right\rceil. \tag{4.36}$$

For fair comparison, the pulse amplitude of the delta pulse must be increased to preserve the energy of the pulse function:

$$g_b[k] = \sqrt{L_g} \cdot g_b[k] = \sqrt{L_g N} \cdot A\delta[k], \tag{4.37}$$

and thus,

$$\text{PAPR}_{\max} = L_g \cdot N. \tag{4.38}$$

As discussed earlier, the theoretical maximal PAPR value for rectangular pulse is in most cases (i.e., except for small M_Δ) lower than that for the delta pulse. This effect can be observed in Figure 4.8. Only in the case where the

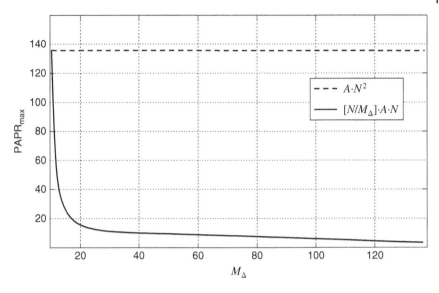

Figure 4.7 The comparison between the maximum PAPR values for two different pulse shapes: rectangular (solid line) and delta pulse (dashed line) versus the distance between atoms M_Δ (in samples) in the time domain.

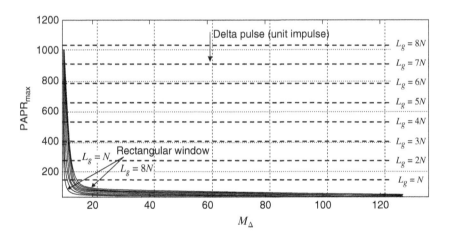

Figure 4.8 The comparison between the maximum PAPR values for two pulse shapes of long duration: rectangular pulse (solid lines) and delta pulse (dashed lines) versus the distance between atoms M_Δ in the time domain (in samples).

number of samples of the pulse function is slightly higher than the IFFT size, the theoretical maximal PAPR value for the rectangular pulse for low N is equal to that value for the delta pulse.

Example Results

Let us verify the analysis presented earlier by computer simulations. In order to find the optimum pulse shape that minimizes the PAPR and the CM metric, the genetic algorithms have been applied. Every pulse sample was represented by an 8-bit gene. For 200 different random chromosomes, at least 10 000 generations have been produced using the roulette wheel selection method. The obtained results show that regardless of the relations between N (the IFFT size), M_Δ (the distance between two adjacent pulses in the time domain), and L_g (the number of samples of the pulse-shaping filter impulse response), the best pulse shape to minimize the PAPR value is always the rectangular one. The same results have been obtained with reference to CM.

Another set of simulation experiments aimed at finding the optimal N and M_Δ parameters of the Gaussian pulse shape:

$$g[k] = \exp\left(-\frac{\pi k^2}{N M_\Delta}\right). \tag{4.39}$$

In the simulations, the same genetic algorithm has been used as the one described earlier. The parameters N and M_Δ have been optimized in the range from 0 to 1023. The obtained results show that the optimal pulse shape is for $N = M_\Delta = 1023$, that is, the highest values in the considered range. It can be anticipated that the optimal pulse shape should be described by the relation $g[k] \to \exp(0) = \Pi\left\lceil\frac{k}{M}\right\rceil$, that is, by the rectangular pulse.

Finally, CCDF of PAPR has been obtained for some exemplary selected pulse shapes. They can be observed in Figure 4.9. When looking at the graphs presented for the GMC signal with various pulse shapes applied for synthesis, one can observe that the lowest values of CCDF(PAPR_0) are obtained for the rectangular pulse, that is, for the conventional OFDM transmission without subband filtering. Similar results have been obtained for the Cubic Metric.

Let us illustrate the impact of the pulse shape and its duration on the PAPR value (see Figure 4.10, where the CCDF of PAPR is presented for pulses of long duration and for $N = 256$). It is easy to observe that when the pulse duration in samples L_g is higher than the IFFT order N, the CCDF of PAPR is higher than that in the case where L_g is equal to N. The line with triangle markers represents the PAPR characteristic for rectangular pulse and $L_g > N$ (note that this is not the case of OFDM). Note that this curve is shifted right in comparison with the reference OFDM curve, where $L_g = N$. The CCDF of PAPR for Gaussian pulse (solid line) and that for the dual Gaussian pulse (line with circle markers) show that changing the pulse shape from rectangular to the Gaussian leads to PAPR

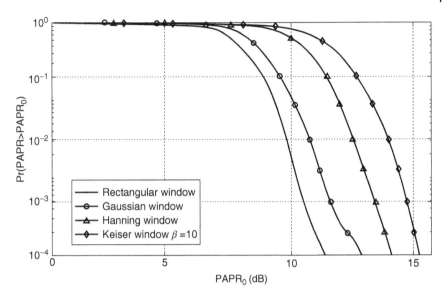

Figure 4.9 The CCDF($PAPR_0$) for various pulse shapes (rectangular, Gaussian, Hanning, and Keiser with $\beta = 10$).

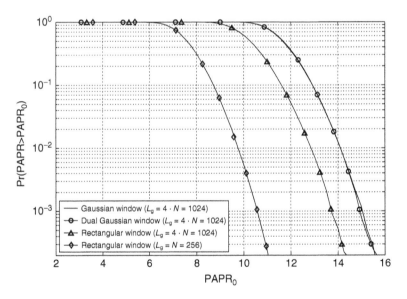

Figure 4.10 The CCDF of PAPR in case of pulses of long duration for various pulse shapes and $L_g = 4 \cdot N = M_\Delta = 1024$; no PAPR reduction method applied.

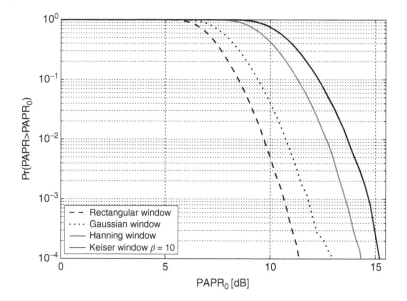

Figure 4.11 The CCDF of PAPR for various pulse shapes and $L_g = N = M_\Delta = 128$; no PAPR reduction method applied.

performance degradation. Thus, regardless of the relation between L_g, N, and M_Δ, the best pulse shape in terms of resulting PAPR is rectangular.

Similar conclusions can be drawn when analyzing the results presented in Figure 4.11, where the CCDF of PAPR has been presented for various pulse shapes of the same duration (in samples), that is, $L_g = N = M_\Delta = 128$. One can observe that changing the pulse shape from the rectangular one to any other pulse (that carries the same amount of power) leads to PAPR increase.

4.2.2 Active Constellation Extension for GMC Signals

Let us now focus on the Active Constellation Extension (ACE) method, as described in Chapter 2. This method seems to be particularly suitable for application in the future GMC-based multistandard terminals due to its advantages over other methods, that is, BER improvement or at least no degradation of the BER performance due to symbol predistortion and no transmission overhead (side information) nor additional operations at the receiver. Some interesting improvements can be proposed for this method, taking the specific properties of the generic GMC-based multistandard terminal into account.

Application of the ACE method to GMC signals requires incorporation of the pulse shape, its duration L_g (in samples), and the TF distance between atoms in the metric (PAPR or other) definition. The reason is that the GMC-modulated pulses interfere with the neighboring pulses (on the TF plane). The degree of this overlapping depends on the pulse shape, and in the time domain, it is often called the *overlapping factor*. The decision concerning predistortion of data symbols should be made for the whole block of symbols consecutively arriving in the considered GMC symbol and modulating the subcarriers (unlike in the case of OFDM and conventional ACE, where this decision can be made for each OFDM symbol separately). Thus, in the considered case, the ACE metric is defined as [79]

$$\mu_{n,m} = -\sum_{k \in \mathbf{I}} \cos(\phi_{n,m,k}) \cdot |s_k|^{p_{\text{ACE}}}, \tag{4.40}$$

where $\mathbf{I}$ is the set of indexes of peaks (samples s_k that exceed a predefined threshold), p_{ACE} is the parameter of the method, and ϕ_{mnk} is the difference between the phase of the kth output peak sample $s[k]$ and the phase of $d_{n,m}g_{n,m}[k]$, which is the contribution of the input data symbol $d_{n,m}$ to this peak. The value of $\mu_{n,m}$ indicates to what degree modification of symbol $d_{n,m}$ can positively affect the decrease of peaks. Note that theoretically there may be infinite number of elements in the aforementioned summation (in general, $k \in \mathbb{Z}$). However, in the frame-based transmission, the finite range of the time samples ($k \in \mathbf{I} \subset 0, \dots, K - 1$) is considered. However, unlike for the standard OFDM, when the ACE method can be implemented for distinct OFDM symbols, in case of GMC and overlapping atoms, the $\mu_{n,m,k}$ metric is calculated for all symbols transmitted in the GMC time–frequency block (or frame), that is, for all pairs of (n, m) [79]. In practice, the duration of the pulse shape $g[k]$ and atoms $g_{n,m}[k]$ may be finite and (4.40) can be implemented for one GMC time-domain symbol in a symbol-based transmission [51]. Moreover, overlapping of GMC time-domain symbols can be neglected if an appropriate gap between consecutive GMC frames can be assumed, analogously to the guard period between symbols in OFDM scheme. By inserting a gap between consecutive GMC symbols, we do not destroy the main advantageous feature of GMC transmission, that is, overlapping of the neighboring atoms in time–frequency domain. Indeed, the neighboring pulses on TF plane overlap each other, whereas there is no overlapping between the border pulses of two subsequent GMC frames [79].

Some further modifications to the ACE method have been proposed in [79], which improve its performance in terms of PAPR and CM reduction for the GMC signals. One of them is based on the application of a flexible parameter α, which scales the constellation outer symbols depending on the value of metric $\mu_{n,m}$ defined by (4.40). Another one incorporates the so-called *near-threshold samples* in the $\mu_{n,m}$ metric definition to avoid the peak regrowth as a consequence of the modification of the constellation shape, which may be the case particularly in a GMC transmitter with overlapping symbols. Moreover, means to reduce computational complexity of the ACE method applied in these transmitters have been proposed and evaluated in [79]. One of them uses the Inverse

Discrete Gabor Transform (DGT) for the computation of $\mu_{n,m}$, which is relatively low complex because it may use FFT preceded by the bank of polyphase filters.

Let us also consider application of other PAPR reduction methods to the GMC transmitted signals, for example, the ones summarized in Table 2.1. In order to assess the effectiveness of any PAPR reduction method, let us also introduce a new *PAPR reduction gain* parameter. The proposed PAPR reduction gain is defined as follows:

$$G_{PAPR}^{prob\_level} = PAPR_{unreduced}^{prob\_level} - PAPR_{reduced}^{prob\_level}, \tag{4.41}$$

where $G_{PAPR}^{prob\_level}$ is the *PAPR reduction gain* at the *defined probability level* and expressed in [dB], $PAPR_{unreduced}^{prob\_level}$ is the PAPR value at the given *probability level* of the system without PAPR reduction method, and $PAPR_{reduced}^{prob\_level}$ is the PAPR value at the given *probability level* of the system after applying the selected PAPR reduction method.

Example Results

When the modified ACE method using the formula (4.40) is applied for the cases of GMC signals described above, the resulting PAPR reduction gain is similar to the effectiveness of the conventional ACE method applied for OFDM. Figure 4.12 shows that the highest PAPR reduction gain is reached for the pulse

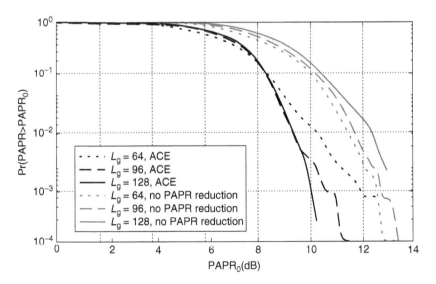

Figure 4.12 CCDF(PAPR$_0$) for rectangular pulse of long duration ($L_g > N$) – modified ACE method applied; $N = M_\Delta = 32$.

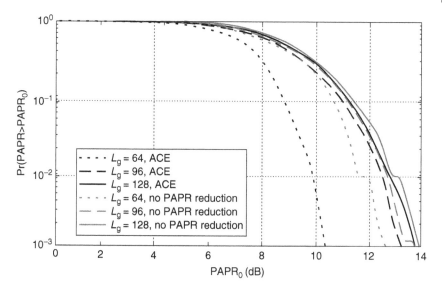

Figure 4.13 CCDF($PAPR_0$) for Gaussian pulse of duration $L_g > N$ – modified ACE method applied; $N = M_\Delta = 32$.

of longest duration ($L_g = 4 \cdot N = 128$). At the same time, the higher the number of samples of $g[k]$, the higher the PAPR value.

Different results are obtained for Gaussian pulses of long duration (see Figure 4.13). Here, the most significant improvement is achieved for the pulses with the shortest impulse response (small L_g). Moreover, the PAPR reduction gain is always lower than the one obtained for rectangular pulse (see Figure 4.12). In both cases of the rectangular pulse and the Gaussian pulse, 20% of input symbols have been predistorted.

Analogous results to these presented earlier have been presented in Figures 4.14 and 4.15 for overcritical sampling case with the number of samples of pulse $g[k]$ equal to the IFFT size ($L_g = N = 32$). The PAPR reduction gain is dependent on the overcritical sampling ratio $\frac{N}{M_\Delta}$. For fixed N, the higher the M_Δ value, the higher the PAPR reduction gain.

These results are in compliance with the derivations concerning the pulse-shape optimization presented earlier. When analyzing the obtained plots, it can be stated again that higher PAPR gain is reached for rectangular pulse rather than for the Gaussian pulse. Finally, simulation results presented in Figures 4.12–4.15 show that the effectiveness of the modified ACE method applied to the GMC signals is similar to the effectiveness of unmodified ACE method for OFDM, that is, the PAPR reduction gains are comparable.

Further simulations have been carried out to compare the effectiveness of other PAPR reduction methods that have been enumerated in Table 2.1.

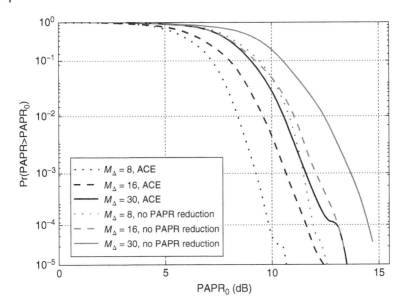

Figure 4.14 CCDF(PAPR$_0$) for overcritical sampling case – rectangular pulse $L_g = N = 32$.

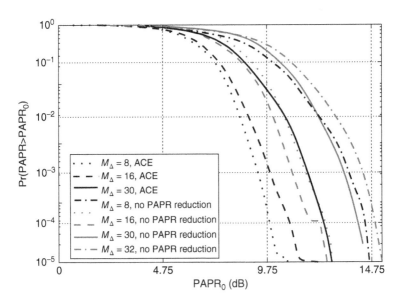

Figure 4.15 CCDF(PAPR$_0$) for overcritical sampling case – Gaussian pulse $L_g = N = 32$.

For this purpose, the ACE method with only one iteration has been considered. The Clipping and Filtering (C–F) method has been tested in two versions – with one iteration and with five consecutive iterations. In the Reference Signal Subtraction (RSS) method, the raised-cosine function multiplied by the *sinc* function has been applied. The amplitude of the subtracted reference function has been arbitrarily chosen to be equal to the half value of a certain peak. For the Peak Windowing (PW) method, the Gaussian pulse has been selected. In the Selective Mapping (SLM) case, four different representations (besides the original one) of the transmitted signal have been generated to select the GMC symbol with the lowest PAPR value.

In Figure 4.16–4.18, the simulation results are presented, that is, CCDF of PAPR for the overcritical sampling scenario and $L_g = N$ (Figure 4.16), for critical sampling and pulse of long duration scenario $L_g > N$ (Figure 4.17), and for critical sampling scenario $L_g = N$ and nonrectangular (Gaussian) pulse (Figure 4.18).

The results confirm that PW and C–F (with five iterations) seem to have the best performance in terms of PAPR and CM reduction. However, it must be remembered that these methods cause in-band signal distortions and additional out-of-band radiation. The PAPR reduction gain achieved for ACE method applied to GMC transmitters with only one iteration is comparable to the results known from the literature for OFDM. More iterations within the ACE algorithm would improve its PAPR reduction gain.

Moreover, one should also consider the BER performance of all these methods. Let us note that only ACE and SLM do not introduce distortion to the transmitted signal. Therefore, it is important to characterize the effectiveness of the PAPR reduction methods also in terms of resultant BER versus IBO. These results are shown in Figure 4.19. They have been obtained for 16QAM transmission with the Rapp model of SSPA and hard-decision receiver (Figure 4.19). To observe the influence of PA on the BER characteristic, SNR has been set to be at least 30 dB. One can conclude that the modified ACE method not only reduces PAPR and CM described earlier but also improves the BER performance. Obviously, it is the result of the slight power increase when some symbols are moved in the constellation. Moreover, the most effective PAPR reduction methods (C–F and PW) seem to degrade BER strongly.

Let us note that if one was to compare BER obtained for various PAPR reduction methods for a fixed SNR, one would actually compare systems with various minimum distances between constellation points. This is because in the ACE method, the outer symbols in the constellation are occasionally up-scaled, and thus, to maintain the average signal energy, the whole constellation should be scaled down, which reduces the minimum distance. If one compares the systems with the same minimum distance between constellation points, the ACE method would outperform other methods even further due to the slight increase of power.

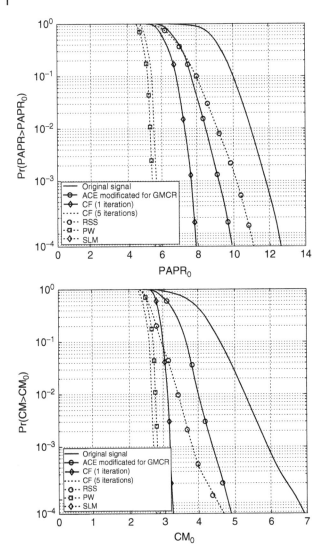

Figure 4.16 CCDF(PAPR$_0$) and CCDF(CM$_0$) for overcritical sampling and rectangular pulse ($L_g = N = 128, M_\Delta = 64$).

The ACE method together with its modifications and refinements can be applied to any pulse-shaped nonorthogonal multicarrier signal, with various pulse shapes. With the proposed improvements of ACE, the GMC transmission becomes attractive for future modern communication systems, such as opportunistic radio, because its major disadvantage (high PAPR) is diminished.

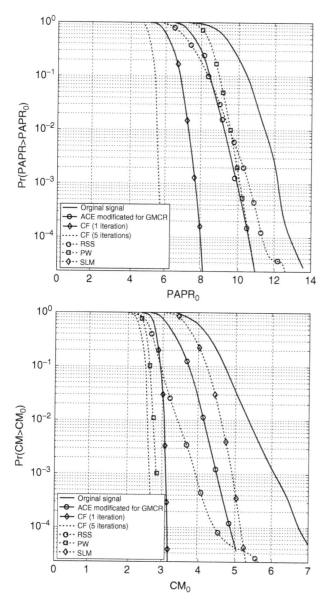

Figure 4.17 CCDF($PAPR_0$) and CCDF(CM_0) for rectangular pulse of long duration ($N = M_\Delta = 128, L_g = 192$).

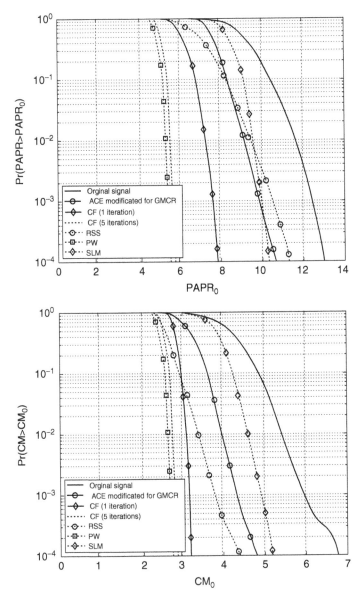

Figure 4.18 CCDF(PAPR$_0$) and CCDF(CM$_0$) for Gaussian pulse ($L_g = N = M_\Delta = 128$).

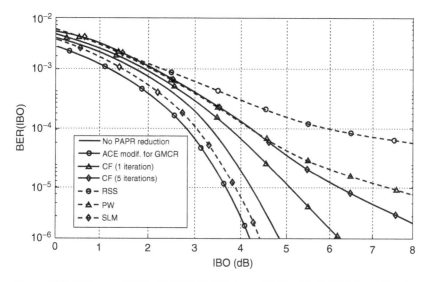

Figure 4.19 BER versus IBO for 16 QAM, AWGN channel SNR = 30 dB, SSPA, hard-decision receiver.

4.3 Link Adaptation in GMC Systems

As discussed in Section 2.4, link adaptation, adaptive modulation, and coding are advantageous techniques to maximize the bit rate. These techniques are applied successfully in many contemporary wireless multicarrier systems, particularly in OFDM transmission. In case of GMC transmission and the TF representation of signals, the problem of adaptive Power Loading (PL) becomes two-dimensional. The accurate representation of the power distribution of signal $s(t)$ on the TF plane is of the highest importance. Here, the application of the water-filling principle in two-dimensional case is examined based on various ways of calculation of the time–frequency power distribution of the signal $s(t)$.

4.3.1 Two-Dimensional Water-Filling

Let us assume the perfect knowledge of the channel gains in time and frequency, or if these values change within the frame duration, an appropriate channel prediction and estimation have been applied. We start with dividing the channel frequency band $F_{GMC} = NF$ into infinitesimally small subbands df in such a way that the channel characteristic could be considered flat in frequency band df and invariable within the time instant dt. The channel capacity can be calculated using the Shannon formula [66]:

$$C = \frac{1}{T_{\text{GMC}}} \int_{F_{\text{GMC}}} \int_{T_{\text{GMC}}} \log_2 \left[1 + \frac{P(f,t)|H(f,t)|^2}{\mathcal{N}(f,t)} \right] dt df, \tag{4.42}$$

where $\mathcal{N}(f,t), H(f,t)$ denote the TF noise power density and the channel characteristic, respectively, at the time instant t and frequency f, $P(f,t)$ denotes the power assigned to the signal localized at frequency f and the time instant t. Moreover, T_{GMC} and F_{GMC} denote the frame time duration and the bandwidth of the frequency band occupied by the GMC signal, respectively. (According to our notation, $T_{\text{GMC}} = MT$ and $F_{\text{GMC}} = NF$.)

As a result of the maximization of C using the Lagrange multipliers, function $P(f,t)$ that maximizes formula (4.42) has the following form:

$$P(f,t) = \begin{cases} W_{\text{level}} - \dfrac{\mathcal{N}(f,t)}{|H(f,t)|^2} & \text{for} \quad \dfrac{\mathcal{N}(f,t)}{|H(f,t)|^2} \leq W_{\text{level}} \\[3mm] 0 & \text{for} \quad \dfrac{\mathcal{N}(f,t)}{|H(f,t)|^2} > W_{\text{level}}. \end{cases} \tag{4.43}$$

The 2D *water-surface* W_{level} (the notion of *water-surface* is used here instead of *water-line* to distinguish between one- and two-dimensional water-filling scheme; in both cases, the term *water-level* can be used) can be computed using the power constraint, which results in

$$W_{\text{level}} \cdot F_{\text{GMC}} \cdot T_{\text{GMC}} = P_{\text{tot}} + \int_{F_{\text{GMC}}} \int_{T_{\text{GMC}}} \frac{\mathcal{N}(f,t)}{|H(f,t)|^2} dt df. \tag{4.44}$$

Let us now consider how this 2D water-filling principle relates to the GMC signal and to the channel model. As mentioned earlier, the subsequent pulses of the GMC signal can overlap the neighboring pulses both in the time and in the frequency domain. Thus, any change of the power assigned to one pulse has its repercussions on the power of the neighboring pulses. This can be understood as time–frequency *self-interference*. To include the pulse shape and this overlapping phenomena into the calculation of the optimal power allocation, let us represent the transmit signal power as a function of the power assigned to the time- and frequency-shifted synthesis pulse $g_{n,m}(t)$ and the power assigned to all respective data symbols $d_{n,m}$. Let us consider the STFT-based time–frequency distribution $S(f,t)$ of the transmit signal $s(t)$ defined by (4.4):

$$S(f,t) = \int_{-\infty}^{\infty} \sum_{n=0}^{N-1} \sum_{m=0}^{M-1} d_{n,m} g_{n,m}(\tau) \gamma^*(\tau - t) \exp(-j2\pi f \tau) d\tau$$

$$= \int_{-\infty}^{\infty} \sum_{n=0}^{N-1} \sum_{m=0}^{M-1} d_{n,m} g(\tau - mT) e^{j2\pi nF\tau} \gamma^*(\tau - t) e^{-j2\pi f\tau} d\tau$$

$$= \sum_{n=0}^{N-1} \sum_{m=0}^{M-1} d_{n,m} \int_{-\infty}^{\infty} g(\tau - mT) e^{j2\pi nF\tau} \gamma^*(\tau - t) e^{-j2\pi f\tau} d\tau$$

$$= \sum_{n=0}^{N-1} \sum_{m=0}^{M-1} d_{n,m} \text{STFT}[g_{n,m}(t)] = \sum_{n=0}^{N-1} \sum_{m=0}^{M-1} d_{n,m} G_{n,m}(f,t). \qquad (4.45)$$

In the aforementioned equations, $(\cdot)^*$ denotes complex conjugate (as before), γ denotes the analysis function used to represent the signal $s(t)$ on the TF plane, and STFT denotes Short-Time Fourier Transform operator. One can observe that the TF representation of the signal $s(t)$ is equivalent to the sum of weighted TF representations of the original synthesis pulse $g(t)$ denoted as $G(f,t)$, that is, $G(f,t) = \text{STFT}(g(t))$, and consequently, $G_{n,m}(f,t) = \text{STFT}(g_{n,m}(t))$. Now, the power distribution of signal $s(t)$ on the time–frequency plane can be calculated in the following way:

$$P(f,t) = \mathbb{E}\{|S(f,t)|^2\} = \mathbb{E}\left\{ \left| \sum_{n=0}^{N-1} \sum_{m=0}^{M-1} d_{n,m} \text{STFT}(g_{n,m}(t)) \right|^2 \right\}$$

$$= \sum_{n=0}^{N-1} \sum_{m=0}^{M-1} \mathbb{E}\{|d_{n,m}|^2 |\text{STFT}(g_{n,m}(t))|^2\}$$

$$= \sum_{n=0}^{N-1} \sum_{m=0}^{M-1} \mathbb{E}\{|d_{n,m}|^2\} |G_{n,m}(f,t)|^2, \qquad (4.46)$$

where $\mathbb{E}(\cdot)$ is the expected value. To obtain the final expression of (4.46), we used the fact that although pulses $g_{n,m}(t)$ and their STFT are not orthogonal, the data symbols $d_{n,m}$ are mutually independent random variables and their expected value is zero. Moreover, the aforementioned relation shows that based on the statistical independence of the user's data $d_{n,m}$, the power distribution on the TF plane can be computed as the weighted sum of the spectrograms of the synthesis pulse $g(t)$, shifted in time and in frequency. In other words, formula (4.46) can be rewritten as

$$P(f,t) = \sum_{n=0}^{N-1} \sum_{m=0}^{M-1} P_{d_{n,m}} \cdot P_{g_{n,m}}(f,t), \qquad (4.47)$$

where $P_{d_{n,m}}$ is the power assigned to the data symbol $d_{n,m}$, and $P_{g_{n,m}}(f,t) = |G_{n,m}(f,t)|^2$ is the STFT-based power density of pulse $g_{n,m}(t)$ ($g(t)$ shifted to the (n, m) location in the time–frequency plane). Let us note that

$$
G_{n,m}(f, t) = \text{STFT}(g_{n,m}(f, t))
$$

$$
= \int_{-\infty}^{\infty} g(\tau - mT)e^{j2\pi nF\tau}\gamma^*(\tau - t)e^{-j2\pi f\tau}d\tau
$$

$$
= \int_{-\infty}^{\infty} g(\tau - mT)\gamma^*(\tau - t)e^{-j2\pi(f-nF)\tau}\ d\tau. \tag{4.48}
$$

By substituting $\tau' = \tau - mT$, we obtain

$$
G_{n,m}(f, t) = \int_{-\infty}^{\infty} g(\tau')\gamma^*(\tau' - (t - mT))e^{-j2\pi(f-nF)(\tau'+mT)}\ d\tau'
$$

$$
= e^{-j2\pi(f-nF)mT}G_{0,0}(f - nF, t - mT). \tag{4.49}
$$

Thus,

$$
|G_{n,m}(f, t)|^2 = |G_{0,0}(f - nF, t - mT)|^2. \tag{4.50}
$$

Finally,

$$
P(f, t) = \sum_{n=0}^{N-1}\sum_{m=0}^{M-1} P_{d_{n,m}} \cdot P_{g_{n,m}}(f, t)
$$

$$
= \sum_{n=0}^{N-1}\sum_{m=0}^{M-1} P_{d_{n,m}} \cdot P_{g_{0,0}}(f - nF, t - mT). \tag{4.51}
$$

Let us now change the indices n and m to n' and m', so that

$$
P(f, t) = \sum_{n'=0}^{N-1}\sum_{m'=0}^{M-1} P_{d_{n',m'}} \cdot P_{g_{0,0}}(f - n'F, t - m'T), \tag{4.52}
$$

and substitute the time variable: $t = mT + \tau$ and the frequency variable: $f = nF + \varphi$, where $\varphi \in \left(-\frac{F}{2}, \frac{F}{2}\right)$ and $\tau \in \left(-\frac{T}{2}, \frac{T}{2}\right)$. In such a case,

$$
P(f, t)\Big|_{\substack{t = mT + \tau \\ f = nF + \varphi}} = P_{n,m}(\varphi, \tau)
$$

$$
= \sum_{n'=0}^{N-1}\sum_{m'=0}^{M-1} P_{d_{n',m'}}P_{g_{0,0}}(nF + \varphi - n'F, mT + \tau - m'T)
$$

$$
= \sum_{n'=0}^{N-1}\sum_{m'=0}^{M-1} P_{d_{n',m'}}P_{g_{0,0}}(\varphi + (n - n')F, \tau + (m - m')T).
$$

$$
\tag{4.53}
$$

The aforementioned formula can be substituted to (4.43):

$$P_{n,m}(\varphi, \tau) = \sum_{n'=0}^{N-1} \sum_{m'=0}^{M-1} P_{d_{n',m'}} P_{g_{0,0}} (\varphi + (n - n')F, \tau + (m - m')T)$$

$$= \begin{cases} W_{\text{level}} - \dfrac{\mathcal{N}_{n,m}(\varphi, \tau)}{|H_{n,m}(\varphi, \tau)|^2} & \text{for } \dfrac{\mathcal{N}_{n,m}(\varphi, \tau)}{|H_{n,m}(\varphi, \tau)|^2} \leq W_{\text{level}} \\[4mm] 0 & \text{for } \dfrac{\mathcal{N}_{n,m}(\varphi, \tau)}{|H_{n,m}(\varphi, \tau)|^2} > W_{\text{level}} \end{cases} \quad (4.54)$$

where $\mathcal{N}_{n,m}(\varphi, \tau)$ and $H_{n,m}(\varphi, \tau)$ denote the TF representation of the noise power density and the TF channel characteristic, respectively, in the $(mT + \tau)$-th time domain and the $(nF + \varphi)$-th frequency location. Let us assume that in the rectangular areas determined by (φ, τ), that is, areas of size F-by-T around atoms, whose location on the TF grid is defined by (n, m)-th atom position, the channel TF characteristic and the noise power density are constant, that is, $\mathcal{N}_{n,m}(\varphi, \tau) = \mathcal{N}_{n,m}$ and $H_{n,m}(\varphi, \tau) = H_{n,m}$. (Similarly, in case of OFDM, we assume flat fading in the subchannel defined around each subcarrier.)

By integrating function (4.54) over the mentioned rectangular areas (over possible values of φ and τ) in the GMC frame, we obtain the following result:

$$\frac{1}{TF} \int_{-\frac{F}{2}}^{\frac{F}{2}} \int_{-\frac{T}{2}}^{\frac{T}{2}} P_{n,m}(\varphi, \tau) d\tau d\varphi = \sum_{n'=0}^{N-1} \sum_{m'=0}^{M-1} P_{d_{n',m'}} \cdot$$

$$\frac{1}{TF} \int_{-\frac{F}{2}}^{\frac{F}{2}} \int_{-\frac{T}{2}}^{\frac{T}{2}} P_{g_{0,0}} (\varphi + (n - n')F, \tau + (m - m')T) d\tau d\varphi. \quad (4.55)$$

Let us denote

$$\psi_{g_{n,m}} = \frac{1}{TF} \int_{-\frac{F}{2}}^{\frac{F}{2}} \int_{-\frac{T}{2}}^{\frac{T}{2}} P_{g_{0,0}} (\varphi + nF, \tau + mT) d\tau d\varphi, \quad (4.56)$$

which can be interpreted as the part of the power of impulse $g_{0,0}$ observed in the F-by-T rectangular area around (n, m)-th atom location on the TF grid in a GMC frame. Thus,

$$\sum_{n'=0}^{N-1} \sum_{m'=0}^{M-1} P_{d_{n',m'}} \cdot \psi_{g_{n-n',m-m'}}$$

$$= \begin{cases} W_{\text{level}} - \dfrac{\mathcal{N}_{n,m}}{|H_{n,m}|^2} & \text{for } \dfrac{\mathcal{N}_{n,m}}{|H_{n,m}|^2} \leq W_{\text{level}} \\[4mm] 0 & \text{for } \dfrac{\mathcal{N}_{n,m}}{|H_{n,m}|^2} > W_{\text{level}} , \end{cases} \quad (4.57)$$

and finally, the closed-form formula for water filling in the time- and frequency-dispersive (fading) channels (power loading in a GMC transmitter) is the following:

$$
P_{d_{n,m}} = \begin{cases} \dfrac{W_{\text{level}}}{\psi_{g_{0,0}}} - \dfrac{\Upsilon_{n,m}}{\psi_{g_{0,0}}} & \text{for } \Upsilon_{n,m} \leq W_{\text{level}} \\[2ex] 0 & \text{for } \Upsilon_{n,m} > W_{\text{level}}, \end{cases}
\tag{4.58}
$$

where

$$
\Upsilon_{n,m} = \frac{\mathcal{N}_{n,m} + |H_{n,m}|^2 \sum_{\substack{(n',m') \in \mathbb{Z}' \\ (n',m') \neq (n,m)}} P_{d_{n',m'}} \cdot \psi_{g_{n-n',m-m'}}}{|H_{n,m}|^2},
\tag{4.59}
$$

$\mathbb{Z}' = \{(n,m) \in \langle 0,1,\ldots,N-1\rangle \times \langle 0,1,\ldots,M-1\rangle\}$, and here, $\times$ denotes the Cartesian product. It is worth mentioning that $\psi_{g_{0,0}} P_{d_{n,m}}/\Upsilon_{n,m}$ can be interpreted as the Signal-to-Interference-plus-Noise Ratio (SINR) for the considered pulse location (n,m). Moreover, *interference* here means self-interference, that is, the one originating from other shifted pulses representing the same signal:

$$
\mathcal{I}_{n,m} = |H_{n,m}|^2 \sum_{\substack{(n',m') \in \mathbb{Z}' \\ (n',m') \neq (n,m)}} P_{d_{n',m'}} \cdot \psi_{g_{n-n',m-m'}}.
\tag{4.60}
$$

Assuming that the useful power of an atom is in the rectangular F-by-T area around its location, while the rest of its power constitutes interference to other atoms, and that the values of $H_{n,m}$ and $\mathcal{N}_{n,m}$ are constant in these areas, we could actually use a different initial formula than (4.42):

$$
C = \sum_{n=0}^{N-1} \sum_{m=0}^{M-1} \log_2 \left[1 + \frac{P_{d_{n,m}} \psi_{g_{0,0}} |H_{n,m}|^2}{\mathcal{N}_{n,m} + \mathcal{I}_{n,m}} \right],
\tag{4.61}
$$

and obtain the same water-filling results.

Formulas (4.58) and (4.59) show that the optimal power allocation to a particular pulse representing the $d_{n,m}$ data symbol depends on the power allocated to all other pulses. To find a joint solution to this problem, one needs to solve a set of NM equations with NM unknown variables, with the powers allocated to all of the NM pulses, and a set of inequalities representing constraint for non-negative power assignments. For this purpose, (4.58) can be rewritten in the matrix form:

$$
\mathbf{P_g} \cdot \mathbf{P_d} = \mathbf{X}
\tag{4.62}
$$

where

$$\mathbf{P_g} = \begin{pmatrix} \boldsymbol{\Psi}_{g_{0,0}} & \boldsymbol{\Psi}_{g_{0,1}} & \cdots & \boldsymbol{\Psi}_{g_{0,N-1}} \\ \boldsymbol{\Psi}_{g_{1,0}} & \boldsymbol{\Psi}_{g_{1,1}} & \cdots & \boldsymbol{\Psi}_{g_{1,N-1}} \\ \vdots & \vdots & \vdots & \vdots \\ \boldsymbol{\Psi}_{g_{N-1,0}} & \boldsymbol{\Psi}_{g_{N-1,1}} & \cdots & \boldsymbol{\Psi}_{g_{N-1,N-1}} \end{pmatrix} \tag{4.63}$$

and

$$\boldsymbol{\Psi}_{g_{n,n'}} = \begin{pmatrix} \psi_{g_{n-n',0-0}} & \psi_{g_{n-n',0-1}} & \cdots & \psi_{g_{n-n',0-M+1}} \\ \psi_{g_{n-n',1-0}} & \psi_{g_{n-n',1-1}} & \cdots & \psi_{g_{n-n',1-M+1}} \\ \vdots & \vdots & \ddots & \vdots \\ \psi_{g_{n-n',M-1-0}} & \psi_{g_{n-n',M-1-1}} & \cdots & \psi_{g_{n-n',M-1-M+1}} \end{pmatrix}, \tag{4.64}$$

$$\mathbf{P_d} = (P_{d_{0,0}} \ P_{d_{0,1}} \ \cdots \ P_{d_{0,M-1}} \ \cdots \ P_{d_{N-1,0}} \ P_{d_{N-1,1}} \ \cdots \ P_{d_{N-1,M-1}})^{\mathrm{T}}, \tag{4.65}$$

$$\mathbf{X} = (X_{0,0} \ X_{0,1} \ \cdots \ X_{0,M-1} \ \cdots \ X_{N-1,0} \ X_{N-1,1} \ \cdots \ X_{N-1,M-1})^{\mathrm{T}}. \tag{4.66}$$

In the $\mathbf{X}$ vector, the elements are defined as $X_{n,m} = W_{\text{level}} - \frac{\mathcal{N}_{n,m}}{|H_{n,m}|^2}$. The solution of the TF power allocation problem in GMC transmission can be obtained by solving matrix equation (4.62). However, the calculated power values cannot be negative, and thus, apart from solving the set of equations, the following set of inequalities (constituting constraints to the power-optimization problem) has to be solved:

$$\forall d_{n,m} : P_{d_{n,m}} \geq 0 \ \wedge \ W_{\text{level}} \geq \Upsilon_{n,m}. \tag{4.67}$$

The parameter W_{level} has to be calculated from the initial condition on the total power P_{tot} constraint:

$$W_{\text{level}} \cdot N \cdot M = P_{\text{tot}} + \sum_{n=0}^{N-1} \sum_{m=0}^{M-1} \frac{\mathcal{N}_{n,m}}{|H_{n,m}|^2}. \tag{4.68}$$

4.3.2 Adaptive Modulation in GMC Transmitters

For the bit- and power-loading purpose, the function from (4.42) has to be complemented by parameter ρ accounting for a target BEP and defined in Section 2.4. The goal is now to maximize the link throughput – as the following objective function:

$$C = \frac{1}{T_{\text{GMC}}} \int_{F_{\text{GMC}}} \int_{T_{\text{GMC}}} \log_2 \left[1 + \frac{\rho \cdot P(f,t)|H(f,t)|^2}{\mathcal{N}(f,t)} \right] dt df. \tag{4.69}$$

Hence, one can follow the similar path as described earlier. However, the obtained formula should take parameter ρ into account. In such a case, the power assigned to a particular pulse on time–frequency plane can be calculated as

$$
P_{d_{n,m}} = \begin{cases} \dfrac{W_{\text{level}}}{\psi_{g_{0,0}}} - \dfrac{\tilde{\Upsilon}_{n,m}}{\psi_{g_{0,0}}} & \text{for } \tilde{\Upsilon}_{n,m} \leq W_{\text{level}} \\ 0 & \text{for } \tilde{\Upsilon}_{n,m} > W_{\text{level}}, \end{cases}
\tag{4.70}
$$

where $\tilde{\Upsilon}_{n,m} = \Upsilon_{n,m}/\rho$. The water-surface W_{level} (or $W_{\text{level}} : \psi_{g_{0,0}}$) has to be calculated from the power constraints as in the power-loading approach. Yet, integrating both sides of formula (4.70) over time and frequency leads to the following relation, slightly different from the one obtained for the water filling:

$$
\begin{aligned}
&W_{\text{level}} \cdot N \cdot M \\
&= P_{\text{tot}} + \sum_{n=0}^{N-1} \sum_{m=0}^{M-1} \frac{\mathcal{N}_{n,m} + (1-\rho)|H_{n,m}|^2 \sum_{\substack{(n',m') \in \mathbb{Z}' \\ (n',m') \neq (n,m)}} P_{d_{n',m'}} \psi_{g_{n-n',m-m'}}}{\rho|H_{n,m}|^2}.
\end{aligned}
\tag{4.71}
$$

One can observe that now, the water-surface W_{level} depends on the power levels $P_{d_{n',m'}}$ assigned to all data symbols $d_{n',m'}$. In such a situation, the power values $P_{d_{n',m'}}$ have to be calculated jointly with the water-surface value W_{level}, that is, the set of $N \cdot M + 1$ equations has to be solved. Moreover, the set of inequalities constituting nonnegative power constraints have to be satisfied:

$$
\tilde{\Upsilon}_{n,m} = \frac{\mathcal{N}_{n,m} + |H_{n,m}|^2 \sum_{\substack{(n',m') \in \mathbb{Z}' \\ (n',m') \neq (n,m)}} P_{d_{n',m'}} \cdot \psi_{g_{n-n',m-m'}}}{\rho \cdot |H_{n,m}|^2} \leq W_{\text{level}}
\tag{4.72}
$$

for all (n, m) pairs. The set of equations can also be represented in the matrix form as it was in the case of the power loading for the GMC transmitter. The matrices defined for the power-loading problem have to account for the changed water-surface W_{level} due to determined BEP and associated parameter ρ. Therefore, one may be able to solve the problem of joint water-surface determination and optimal power allocation to the pulse locations.

Finally, once we determine the power allocation for a given target BER, the assignment of bits to the respective TF locations can be done according to the formula similar to (2.21):

$$
\mathcal{M}_{n,m} = 1 + \frac{\rho P_{d_{n,m}} \psi_{g_{0,0}} |H_{n,m}|^2}{\mathcal{N}_{n,m} + \mathcal{I}_{n,m}}.
\tag{4.73}
$$

Note that the result of (4.73) is purely theoretical, where $\mathcal{M}_{n,m}$ can be non-integer. The theoretical approach presented earlier requires solving of the large matrix equation and the set of inequalities, which is impractical from the real-time implementation point of view. Practical and possibly suboptimal solutions have to be found to reduce its computational complexity. They are discussed in the next paragraphs.

4.3.3 Application of the Modified Hughes–Hartogs Algorithm in GMC Systems

One of the well-known bit- and power-loading algorithm of practical meaning is the one proposed by Hughes–Hartogs [260]. Its main idea bases on iteratively increasing number of bits assigned to each subcarrier, which depends on the amount of power required to transmit one additional bit on this subcarrier. The consecutive bits are assigned to the subcarrier, for which the smallest amount of power is required to send one additional bit. Next, the modification of the Hughes–Hartogs algorithm for the TF-represented signals is briefly presented [67, 261]. The first modification extends the conventional algorithm to the two-dimensional signals (defined in time and frequency), but still orthogonal and not interfering in both dimensions. Thus, in order to directly extend the Hughes–Hartogs algorithm to two dimensions, we assume that the atoms' energies are concentrated in the F-by-T areas around their locations on the TF grid, or that they are orthogonal, so that $\forall (n, m) : \mathcal{I}_{n,m} = 0$. This 2D Hughes–Hartogs algorithm will serve for further comparisons of the bit- and power-loading performance when possible overlapping of the transmitted pulses (atoms) in time and frequency is considered and also when this overlapping and the lack of orthogonality of the pulses are neglected.

First, in the initialization phase, the incremental matrix for each time and frequency location is defined. In this matrix, the rows relate to the possible numbers of bits per symbol (constellation sizes), and the columns relate to all possible pulses TF-locations in a considered frame (see Figure 4.20). Each element of this matrix denotes the amount of power required to transmit the minimum number (q) of additional bits to go from one allowable constellation to the next one (e.g., $q = 2$ when M-QAM constellations are considered): $\Delta P_{i,n,m} = P_{i,n,m} - P_{i-1,n,m}$, where $P_{i,n,m}$ is the transmit power needed at the pulse

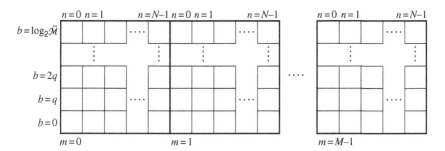

Figure 4.20 The incremental matrix for the two-dimensional Hughes–Hartogs algorithm; b – number of bits.

location (n, m) to transfer $b = iq$ bits per symbol at a required (and predefined) BEP, and i is the row index. The column index j is defined by n and m: $j = m \cdot N + n + 1$. Note that $\forall (n, m) \in \mathbb{Z}' : P_{0,n,m} = 0$. Values $P_{i,n,m}$ can be calculated from (4.73), assuming $\mathcal{I}_{n,m} = 0$:

$$P_{i,n,m} = (\mathcal{M}_{i,n,m} - 1) \frac{\mathcal{N}_{n,m}}{\rho \psi_{g_{0,0}} |H_{n,m}|^2}, \qquad (4.74)$$

where $\mathcal{M}_{i,n,m}$ is the constellation order assigned to the pulse localized at (n, m) point on the TF plane, and $\mathcal{M}_{i,n,m} = 2^{iq}$. The value of $\mathcal{M}_{i,n,m}$ is upper-bounded by a maximum constellation order considered in the algorithm ($\mathcal{M}_{i,n,m} \leq \tilde{\mathcal{M}}$).

After the initialization phase, the main loop of the algorithm is executed. The modified two-dimensional algorithm is presented in Table 4.1. The major difference between this algorithm and the conventional one is in the size of the incremental matrix.

The general assumption in the case Hughes–Hartogs algorithm is the orthogonality of signals baring the information bits. This is not the case for the GMC transmission with overlapping nonorthogonal pulses. Assume that in the initial phase of the algorithm, the $P_{i,n,m}$ values are calculated by taking the self-interference $\mathcal{I}_{n,m}$ observed at any (n, m) TF location into account,

Table 4.1 The two-dimensional Hughes–Hartogs algorithm.

Initialization phase:

Fill in the incremental-power matrix, assign $b = 0$ to each pulse that is, fill in the vector $B_{\mathrm{HH}} = \{b_j\}, j = 0, 1, \dots, N \cdot M - 1$ with zeros, and set: $P_{\mathrm{act}} := 0, B_{\mathrm{act}} = 0$

Main loop:

1) Find the smallest value $\Delta P_{i,n,m}$, that is, the indices of the selected cell: row $i^\star$ and column $j^\star = m^\star \cdot N + n^\star + 1$

2) Assign q more bits (e.g., $q = 2$ for QAM) to the pulse localized at $(n^\star, m^\star)$ point on the TF plane;

3) Increment $b_{j^\star}$ by q in vector B_{HH}

4) Move all terms of column j one place up (in terms of indices): $\Delta P_{i^\star,n,m} := \Delta P_{i^\star+1,n,m}$

5) Update the actual number of bits B_{act} and power P_{act} to be transmitted: $B_{\mathrm{act}} = B_{\mathrm{act}} + q, P_{\mathrm{act}} = P_{\mathrm{act}} + \Delta P_{i^\star,n^\star,m^\star}$

6) If $B_{\mathrm{act}} \leq B_{\mathrm{tot}} \wedge P_{\mathrm{act}} \leq P_{\mathrm{tot}}$ go to the step 1; otherwise, finish the loop.

Result:

The number of bits assigned to each pulse is in vector B_{HH}.

---

that is,

$$
\begin{aligned}
P_{i,n,m} &= (\mathcal{M}_{i,n,m} - 1)\frac{\mathcal{N}_{n,m} + \mathcal{I}_{n,m}}{\rho\psi_{g_{0,0}}|H_{n,m}|^2} \\[2mm]
&= (\mathcal{M}_{i,n,m} - 1)\frac{\mathcal{N}_{n,m} + |H_{n,m}|^2 \sum\limits_{\substack{(n',m')\,\in\,\mathbb{Z}'\\ (n',m')\,\neq\,(n,m)}} P_{i,n',m'}\cdot\psi_{g_{n-n',m-m'}}}{\rho\psi_{g_{0,0}}\cdot|H_{n,m}|^2}.
\end{aligned}
$$

$$(4.75)$$

Note that for a given number of bits $b = iq$ and for all (n, m) pairs, (4.75) defines the set of NM linear equations that can be easily solved. This way, every row in the incremental matrix can be filled with the elements obtained from the solution of this set of equations. However, the main loop of the Hughes–Hartogs algorithm cannot be executed as in Table 4.1. This is because assigning more power and increased number of bits to one atom creates higher interference to other neighboring atoms, so that the incremental powers calculated in the former phase for these atoms are not correct anymore. One should recalculate all incremental powers for all positions in the incremental power matrix in each iteration of the Hughes–Hartogs algorithm; however, this would be highly computationally complex solution.

Example Results

Computer simulations have been carried out to look at the performance of the Hughes–Hartogs algorithm for bit- and power loading in the GMC transmission. For this purpose, $N = 16$ and $M = 32$ have been assumed. Consequently, the TF plane has the dimensions 32-by-16, and the number of pulses in a GMC frame equals $NM = 512$. The constraint of the total transmit power has been set to the value M times higher than the normalized power of the signal in every time instant. The synthesis function $g(t)$ has been the Hanning window (pulse). To obtain the TF power distribution of the synthesis pulse $g(t)$, the STFT spectrogram has been calculated using the Gaussian pulse as the analysis function. The example fading channel model with $L_s = 12$ paths of exponentially decaying power has been selected. The assumed target BER for an uncoded system has been equal to BER $= 10^{-3}$, while the maximum number of bits per the constellation point has been equal to 10 ($\tilde{\mathcal{M}} = 2^{10}$), and the minimal increment of bits transmitted by a single atom has been $q = 1$.

The 2D Hughes–Hartogs algorithm has been tested in its two versions: the one that assumes the orthogonality of pulses (although in the simulation setup, they are not orthogonal), and the complex one that takes all the self-interference into account, as discussed in the previous paragraphs. A number of channel instantiations have been generated to average the results. An example TF channel characteristic is presented in Figure 4.21. The shades of gray in the rectangular areas denote the absolute value of the channel characteristic in the atoms locations. In Figure 4.22, one can see the results of the two versions of

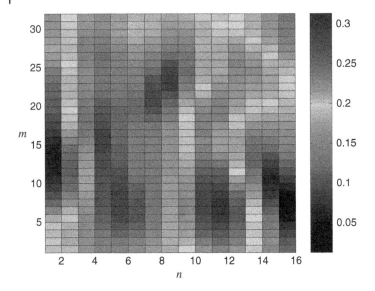

Figure 4.21 Example TF channel characteristic. (Kliks and Bogucka, 2009 [67]. Reproduced with the permission of EUSIPCO 2009.)

the Hughes–Hartogs algorithm for this particular example. Note that the bit allocation resulting from the modified version, taking the self-interference into account seems to better reflect the adjustment to the channel TF characteristic.

Let us now consider the achievable channel capacity versus an SNR for the selected methods. Two variants of pulses overlapping have been considered – strong and weak overlapping. In the case of strong overlapping, the pulse duration $L_g \Delta t$ has been more than two times longer than the time distance between consecutive pulses T, whereas for the weak-overlapping case – the pulse duration has only been around 1.2 times longer than T. The results are presented in Figure 4.23. One can observe that the channel capacity versus the average SNR is higher for the GMC-system-modified Hughes–Hartogs algorithm than for the simple (original, assuming orthogonality of pulses) 2D Hughes–Hartogs algorithm, regardless of the type of overlapping.

4.3.4 Remarks on Link Adaptation in the GMC Transmission

Apart from the Hughes–Hartogs algorithm, a number of other practical algorithms for bit- and power loading have been considered for the application in the GMC transmission together with their necessary modifications to account for two-dimensional signaling and the lack of orthogonality of the applied pulses. For example, the well-known Campello algorithm proposed in [262] has been considered in [263], and some simplifications to calculate the bit- and power allocations to the transmitted atoms have been suggested.

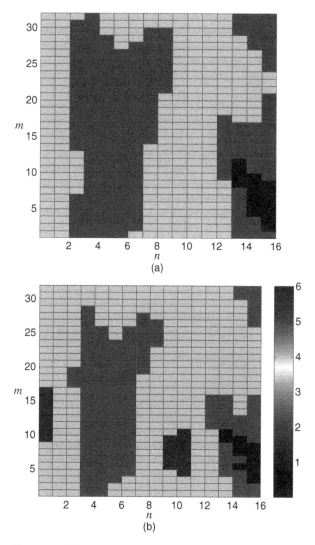

Figure 4.22 Bit assignment using: (a) original two-dimensional Hughes–Hartogs algorithm, (b) Hughes–Hartogs algorithm modified for GMC transmission. (Kliks and Bogucka, 2009 [67]. Reproduced with the permission of EUSIPCO 2009.)

The aim of the algorithms proposed by Hughes–Hartogs and by Campello is to optimize the usage of the transmit power. Quite different approach has been proposed by Fischer and Huber [264]. The authors have stated that there is no need to reach the system capacity in practice. Instead of maximizing the throughput, they find a way to transmit fixed data rate with fixed power at the lowest error rate possible. Based on these assumptions, the authors have

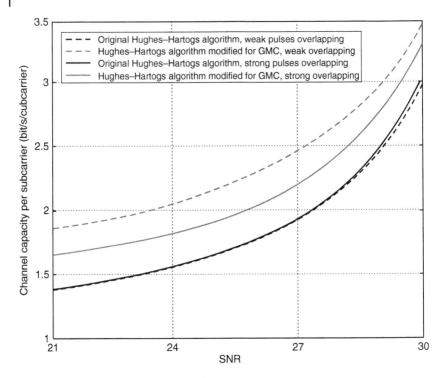

Figure 4.23 Channel capacity versus SNR obtained for original and modified Hughes–Hartogs algorithm for two cases: strong and weak overlapping (defined earlier). (Kliks and Bogucka, 2009 [67]. Reproduced with the permission of EUSIPCO 2009.)

derived the formula that allows to find the number of bits to be assigned to each subcarrier with the assumed bound on the highest probability of the erroneous decision at the receiver. In the next step, the number of assigned bits is rounded to the closest allowable value, and the final adjustment of the bit and power allocation is realized.

The main idea of the algorithm proposed by Chow *et al.* [265] is to allocate the assumed total number of transmit bits among the subcarriers in much less complex way as it is done in the Hughes–Hartogs algorithm [260]. It is assumed that the initial power distribution among subcarriers is uniform and transforms to a number of bits that can be transmitted by taking the SNR value at each subcarrier into account. Then, subsequently, in every iteration, the number of assigned bits and power levels are decreased, until the rate and power requirements are met. The modification of the original Chow algorithm to the GMC transmission is straightforward. In particular, instead of SNR for each subcarrier, the SINR value for each pulse in the TF plane should be calculated. Recall that SINR in the GMC case can be calculated as $\text{SINR}_{n,m} = P_{d_{n,m}} \psi_{g_{0,0}} / \Upsilon_{n,m}$. Similarly, as in the case of the Hughes–Hartogs algorithm, the change of the power at a single atom location (n, m) changes the values of SINR at all other locations, so that these values have to be recalculated in each iteration of the Chow algorithm.

We shall conclude that adaptive bit- and power loading for the GMC transmitters is relatively simple when the transmitted pulses are orthogonal or well-localized in the time or in the frequency domain. In such a case, the standard adaptive modulation algorithms developed for OFDM can be easily extended to the 2D signals. However, when the pulses overlap in time and frequency in a way that they create significant self-interference in the transmitted signal, the standard algorithms cannot be applied, and they must be altered to account for the mutual dependence of these pulses. This in turn increases their computational complexity significantly.

When discussing the aspects of link adaptation, one should also consider adaptive modulation and coding, that is, when a pair of the modulation constellation and the coding scheme is selected adaptively. For practical reasons, the coding scheme is usually selected for a block of resources (for a block of subcarriers or for the so-called basic resource block in an LTE system). This modulation and coding adaptation is usually based on reading the Look-up Table (LUT), where these values are stored for various SNR.

4.4 GMC Receiver Issues

One of the advantages of OFDM when compared with a single-carrier transmission is that the complexity of channel estimation and equalization can be significantly reduced [53, 96, 244]. There, subchannels defined in the vicinity of each subcarrier frequency (i.e., the fragment of frequency band corresponding to this subcarrier) can be treated as flat fading and constant in a certain period of time. This is because (i) subcarriers are orthogonal in the defined orthogonality period, (ii) the coherence time is longer than the OFDM symbol duration, and (iii) the coherence bandwidth is much higher than the OFDM symbol rate. Moreover, in order to avoid Inter-Symbol Interference (ISI), a cyclic prefix is placed before the transmitted symbol. Such specific features of OFDM signals enable the channel coefficients can be easily calculated, and the influence of the channel can be equalized in the frequency domain [53, 266–268]. Consequently, an OFDM receiver can be practically realized by means of serial-to-parallel converter and FFT module, followed by a simple equalizer consisting of single-tap adaptive filters operating in the frequency domain.

However, in general, when overlapping between neighboring pulses in the TF plane is allowed, and the application of the cyclic prefix is not obligatory, the receiver structure has to be modified. It is well known that the filter-bank-based structure, widely used in the context of transmultiplexers, can be treated as a form of generic receiver structure (see Figure (4.4) as the reference) [49, 64,

84]. In such a case, the received signal is first filtered in the bank of properly designed filters followed by the FFT module and the equalizer. As discussed earlier in this chapter, the transmitted pulses adjacent in TF plane may overlap each other, may not be orthogonal, and the self-interference phenomenon (i.e., interference between neighboring pulses in the time and in the frequency domain) may occur in the GMC signal already at the transmitter output. Thus, signal reception methods have to account for this self-interference, for example, the iterative interference cancellation methods, widely used for multiuser interference cancellation [269, 270], can be applied at the GMC receivers. It has to be mentioned that some GMC-like signal reception proposals have been presented in the state-of-the art literature [51, 52, 80, 84, 85, 240]. Next, we examine some interference cancellation methods for the GMC signals. We consider the transmit pulses constituting an overcomplete set when it is impossible, in general, to create a unique transmit–receive pair of pulses.

4.4.1 Received Signal Analysis

As it has been stated earlier in this chapter, the GMC signal can be represented as

$$s[k] = \sum_{m \in \mathbb{Z}} \sum_{n=0}^{N-1} d_{n,m} \cdot g_{n,m}[k], \tag{4.76}$$

where $s[k]$ ($k \in \mathbb{Z}$, $\mathbb{Z}$ being the set of integers) is the discrete-time signal belonging to the complex Hilbert space of square-summable sequences l_2, $\{d_{n,m}\}$ are the so-called frame coefficients, N is the number of subcarriers, and $\{\gamma_{n,m}[k]\}$ is a sequence of basis functions, defined as [51, 82, 240]:

$$g_{n,m}[k] = g[k - mM_\Delta] \cdot \exp\{j2\pi n(k - mM_\Delta)/N\}. \tag{4.77}$$

As it has already been mentioned, M_Δ is the time spacing between N parallel symbols (in samples) modulating N subcarriers, and $\gamma[k]$ is the shape of the pulse. In general, frame coefficients $\{d_{n,m}\}$ can be found by applying the Gabor transform or STFT [82, 240], described as

$$d_{n,m} = \sum_{k \in \mathbb{Z}} s[k] \cdot \gamma_{n,m}^*[k], \tag{4.78}$$

where

$$\gamma_{n,m}[k] = \gamma[k - mM_\Delta] \cdot \exp\{j2\pi n(k - mM_\Delta)/N\} \tag{4.79}$$

is referred to as the analysis window, and $(\cdot)^*$ denotes the complex conjugate.

The existence of the dual Gabor frame allows for restoration of the data symbols at a GMC receiver. Based on the Balian–Low theorem [239], good TF localization of atoms (concentrated around the (n, m) coordinates on the discrete TF plane) results in reduced ISI and Inter-carrier Interference (ICI) and allows for the omission of the guard periods necessary in the case of OFDM [51, 82, 240, 271].

Let us focus on the transmission schemes using nonorthogonal pulses, where the basis functions are linearly dependent and create an overcomplete set (the overcritical sampling case). Such assumptions lead to the situation when typical reception methods, known for the OFDM-based systems, cannot be applied in order to recover the user data correctly [80, 85]. The reason for such a situation is that the arbitrarily selected two-dimensional set of complex values (corresponding to the user data) may not be a *valid* result of the application of STFT for any signal [237], that is, there is no real time-domain signal for which the considered two-dimensional set of complex values constitutes its time-domain representation that can be obtained by the application of STFT or Gabor transform. On the other hand, for an overcomplete set of basis functions, dual functions (pulses, windows) are not unique. However, it has been proved that in the case where the set of input values is strictly defined, that is, the number of possible values that can be chosen is limited (e.g., the set of all constellation points from the M-QAM set), it is possible to define a unique pair of transmit–receive pulses that allow for perfect reconstruction of the transmitted data [80].

In order to extract the desired signal and the interference term, let us analyze the received signal, in case of the ideal (nondistorting) channel, that is,

$$\tilde{d}_{n,m} = \sum_{k \in \mathbb{Z}} r[k] \cdot \gamma_{n,m}^*[k]$$

$$= d_{n,m} \overbrace{\sum_{k \in \mathbb{Z}} g_{n,m}[k] \cdot \gamma_{n,m}^*[k]}^{\text{Desired signal}} + \overbrace{\sum_{\substack{(n',m') \in \mathbb{Z}' \\ (n',m') \neq (n,m)}} \sum_{k \in \mathbb{Z}} d_{n',m'} g_{n',m'}[k] \cdot \gamma_{n,m}^*[k]}^{\text{Self-interference}}$$

$$+ \overbrace{\sum_{k \in \mathbb{Z}} w[k] \cdot \gamma_{n,m}^*[k]}^{\text{Colored Noise}}, \tag{4.80}$$

where $\mathbb{Z}'$ is the Cartesian product of the sets of indexes n and m ($\mathbb{Z}' = \{(n,m) \in \langle 0, 1, \ldots, N-1 \rangle \times \langle 0, 1, \ldots, M-1 \rangle\}$), $w[k]$ is the noise sample at the k-th sampling moment. The received and demodulated signal, even in the case of an ideal channel, suffers from the self-interference coming from the neighboring (in time and frequency) pulses carrying the user data $d_{n,m}$. Even when the synthesis and analysis pulses are unique, and although it is theoretically possible to recover the user data perfectly (i.e., $\hat{d}_{n,m} = d_{n,m}$), the self-interference between neighboring pulses cannot be canceled ideally in the practical implementation of such systems. This is because in practice, the pulses have to be finite (e.g., in the time domain), and in theory, very often at least one of the dual pulse pair is of infinite duration. Thus, the residual interference exists causing degradation of SINR. It gets even worse when the overcomplete basis is used (the overcritical sampling is applied, that is, $M_\Delta \leq N$), because then, the dual window for a given synthesis function is not unique and standard processing cannot be applied [80, 85]. Furthermore, due to the limitations of practical implementation of the (even unique) transmit and receive pulse pair, the biorthogonality property is not fulfilled. This is the reason why the self-interference term in formula (4.80) exists. Moreover, the lack of biorthogonality causes degradation of the desired part of the received signal, since the expression $\sum_{k\in\mathbb{Z}} g_{n,m}[k] \cdot \gamma_{n,m}^*[k]$ is not equal to 1.

It is worth mentioning that due to the filtering on the receiver side, the observed noise is no longer white. It can be observed in formula (4.80), where the last term refers to the colored noise [272–274]. The presence of the colored noise implies that specific reception methods (including, for example, the Cholesky factorization [275] or application of the noise-plus-interference whitening matched filters [269, 276, 277]) could be considered.

In case of the multipath fading channel, formula (4.80) can be rewritten as follows:

$$
\begin{aligned}
\tilde{d}_{n,m} &= \sum_{k\in\mathbb{Z}} r[k] \cdot \gamma_{n,m}^*[k] = H_{n,m} d_{n,m} \sum_{k\in\mathbb{Z}} g_{n,m}[k] \cdot \gamma_{n,m}^*[k] \\
&+ \sum_{\substack{(n',m')\in\mathbb{Z}' \\ (n',m')\neq(n,m)}} \sum_{k\in\mathbb{Z}} H_{n',m'} d_{n',m'} g_{n',m'}[k] \cdot \gamma_{n,m}^*[k] + \sum_{k\in\mathbb{Z}} w[k] \cdot \gamma_{n,m}^*[k] \\
&= H_{n,m} d_{n,m} \sum_{k\in\mathbb{Z}} g_{n,m} \cdot \gamma_{n,m}^*[k] + \tilde{I}_{n,m} + \tilde{\mathcal{N}}_{n,m},
\end{aligned}
\tag{4.81}
$$

where $H_{n,m}$ represents the channel coefficient corresponding to the user data localized at time–frequency point (n, m). Moreover, $\tilde{I}_{n,m}$ and $\tilde{\mathcal{N}}_{n,m}$ denote the residual interference and the colored noise, respectively, observed in the time–frequency symbol location (n, m) at the output of the receiver analysis filter. It is assumed that the channel gain is constant in the vicinity of each pulse, that is, in the rectangular area of size F-by-T around the pulse center location, so that $H_{n,m}(\varphi, \tau) = H_{n,m}$ for $\varphi \in \left(-\frac{F}{2}, \frac{F}{2}\right)$ and $\tau \in \left(-\frac{T}{2}, \frac{T}{2}\right)$. After received signal demodulation and channel equalization, the decision is taken on the transmitted data: $\hat{d}_{n,m}^{(q)} = D\{\tilde{d}_{n,m}^{(q)}\}$. (For simplicity, we omit other arguments of the decision operator D that may include the set of estimated channel coefficients $\{\hat{H}_{n,m}\}$ and the chosen equalization method.)

One way to cope with the problem of residual self-interference is the application of the successive or parallel interference cancellation procedures. Interference cancellation techniques have been widely used in CDMA and Multiple Input, Multiple Output (MIMO) systems and allow for removal of the interference components from the received signal [270, 278–282]. For CDMA systems, interference cancellation techniques are used to eliminate the multiuser interference from the received signal, for example, in the uplink transmission, when the base station observes a great number of interfering signals coming from mobile terminals [280]. Similarly, in the MIMO systems, there is the need of reduction of the interference between antennas, and the well-known BLAST algorithm (or its modifications) [283, 284] is used to cope with this problem [282]. In general, however, two most popular classes of iterative interference cancellation techniques can be recognized, with their presentation as follows.

4.4.2 Successive Interference Cancellation (SIC)

The main idea behind the Successive Interference Cancellation (SIC) is to decode only one particular signal (corresponding to one user in the CDMA case or to one antenna in the MIMO case) at a time (in a single iteration), which is of the best reception quality (usually in terms of SINR) among all other received signals. When this signal is decoded, it is encoded again and subtracted from the sum of received signals, thus eliminating the interference caused by it. The procedure is repeated iteratively, until all of the received signals are decoded. The advantage of such an approach is its effectiveness of interference cancellation. Through this step-by-step algorithm, it is possible to cancel one interferer in each iteration and improve the quality of the remaining part of the received signals. Unfortunately, the SIC method has some disadvantages. One of them is its long processing time and, in consequence, high delay in the decoding algorithm due to elimination of the interference in a serial manner. Moreover, the algorithm is prone to the error propagation, that is, if one signal is incorrectly decoded, the error propagates in the next successive steps of the SIC algorithm.

Let us discuss adaptation of SIC in the GMC receiver. The goal is to remove the self-interference between the neighboring pulses (atoms) so that data transmitted by these pulses are decoded subsequently starting from the one with the highest SINR. The generic block diagram of the SIC receiver adapted to the GMC signaling is presented in Figure 4.24.

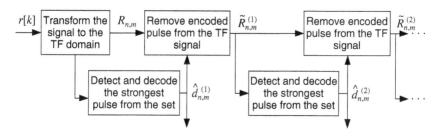

Figure 4.24 Generic structure of the GMC-SIC receiver.

In the first step, the received signal $r[k]$ is transformed to its time–frequency representation $R_{n,m}$ (using the Gabor transform or STFT). Next, the strongest pulse (having the highest SINR) is selected, and the data symbol transmitted using this pulse is detected, producing the estimate of this symbol $\hat{d}_{n,m}^{(1)}$. This pulse is then modulated one more time and subtracted from the originally received signal $R_{n,m}$, resulting in an output signal $\tilde{R}_{n,m}^{(1)}$. The procedure is repeated $N \cdot M$ times, that is, as many times as there are symbols in the GMC frame. At a q-th step ($q = 1 \ldots NM$) of interference cancellation, the demodulated symbol $\tilde{d}_{n,m}^{(q)}$ that has the highest SINR is processed, that is,

$$\tilde{d}_{n,m}^{(q)} = H_{n,m} d_{n,m} \sum_{k \in \mathbb{Z}} g_{n,m}[k] \gamma_{n,m}^*[k]$$

$$+ \sum_{\substack{(n',m') \in \mathbb{Z}'^{(q)} \\ (n',m') \neq (n,m)}} \sum_{k \in \mathbb{Z}} H_{n',m'} d_{n',m'} g_{n',m'}[k] \gamma_{n,m}^*[k] + \sum_{k \in \mathbb{Z}} w_{n,m}[k] \gamma_{n,m}^*[k]$$

$$= H_{n,m} d_{n,m} \sum_{k \in \mathbb{Z}} g_{n,m}[k] \gamma_{n,m}^*[k] + \tilde{I}_{n,m}^{(q)} + \tilde{\mathcal{N}}_{n,m}, \qquad (4.82)$$

where $\mathbb{Z}'^{(q)} \subset \mathbb{Z}'$ is the set of indexes pairs of all symbols that have not been detected in the previous $(q - 1)$ iterations, and the term $\tilde{I}_{n,m}^{(q)}$ describes the interference observed at the time–frequency location (n, m) in the q-th step. Moreover, $\tilde{d}_{n,m}^{(q)} = d_{n_q, m_q} : (n_q, m_q) = \arg \max_{(n,m) \in \mathbb{Z}'^{(q)}} \text{SINR}_{n,m}^{(q)}$, where

$$\text{SINR}_{n,m}^{(q)} = \frac{H_{n,m} d_{n,m} \sum_{k \in \mathbb{Z}} g_{n,m}[k] \gamma_{n,m}^*[k]}{\tilde{I}_{n,m}^{(q)} + \tilde{\mathcal{N}}_{n,m}}. \qquad (4.83)$$

At each iteration, the estimate of the transmitted symbol is calculated based on some decision rule $\hat{d}_{n,m}^{(q)} = \mathcal{D}\{\tilde{d}_{n,m}^{(q)}\}$ that possibly includes channel equalization. Assuming that $\text{SINR}_{n,m}^{(q)}$ is sufficiently high, there is a high probability that the decision is correct, so that $\hat{d}_{n,m}^{(q)} = d_{n_q, m_q}$, where n_q and m_q are indexes of the actual transmitted pulse. Then, it should be possible to remove the pulse modulated by this data symbol and distorted by the channel from the received signal, thus, lowering the interference to other pulses and increasing $\text{SINR}_{n,m}^{(q+1)}$ at the next step.

Thus, the GMC-SIC receiver can have the structure as presented in Figure 4.25. In the first step, the time-domain signal $r[k]$ is transformed into its time–frequency representation $R_{n,m} = R_{n,m}^{(0)}$, where $R_{n,m}^{(q)}$ denotes the time–frequency representation of the received signal in the q-th step of the algorithm. Then, from the actual set of indexes $\mathbb{Z}'^{(q)}$, the pulse with the highest SINR value is selected and detected using the channel estimate to obtain $\hat{d}_{n,m}^{(q)}$. (Again, in case of perfect detection, $\hat{d}_{n,m}^{(q)} = d_{n_q, m_q}$.) This symbol is

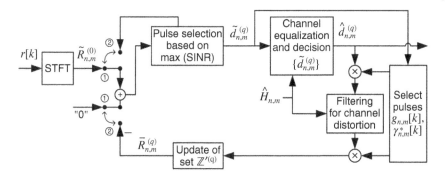

Figure 4.25 Practical GMC-SIC receiver.

then modulated by the corresponding transmit pulse, filtered using channel coefficients estimates, and demodulated by the dual receive pulse, which yields $\overline{R}_{n,m}^{(q)}$. Finally, the set of indexes $\mathbb{Z}'^{(q)}$ of the pulses to be detected is updated.

4.4.3 Parallel Interference Cancellation (PIC)

In contrast to the successive approach, the Parallel Interference Cancellation (PIC) method is characterized by a short processing time obtained at the cost of high complexity of the receiver expressed in terms of the required number of correlators, which grows linearly with the number of users (in case of CDMA) or antennas (in the MIMO case) [279, 282]. The idea behind PIC is to treat all of the received signals individually but simultaneously. After the matched filter block [53], the received signals (corresponding to the number of users or number of antennas) are quantized to obtain the first decision value. Then, in the next step, from each received signal (i.e., not quantized), the interference coming from the remaining signals is subtracted, and the new representation of the received signal is obtained, resulting in a more accurate decision. This process is repeated Q times. Obviously, the higher the number of repetitions, the more efficient interference cancellation at the cost of higher complexity.

When adapting the PIC receiver to the GMC receiver, the goal is to detect all NM data symbols simultaneously. In the first step, all pulses in time–frequency domain are decoded. Then, for each separate pulse, the interference coming from all other pulses is subtracted, creating new time–frequency representation of the received signal. Such an operation can be repeated Q times, increasing the receiver efficiency expressed in terms of BER versus SINR. The generic block diagram of PIC receiver is presented in Figure 4.26.

For the purpose of analyzing the PIC receiver, let us consider another representation of the demodulated symbol in the q-th iteration $\tilde{d}_{n,m}^{(q)}$

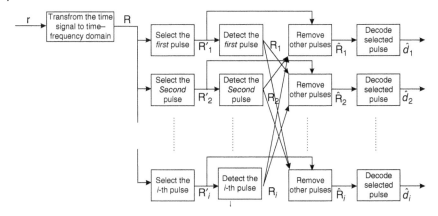

Figure 4.26 Generic structure of the GMC-PIC receiver.

$(q = 1 \ldots Q)$:

$$\tilde{d}_{n,m}^{(q)} \approx H_{n,m} d_{n,m} \sum_{k \in \mathbb{Z}} g_{n,m}[k] \gamma_{n,m}^*[k]$$

$$+ \sum_{\substack{(n',m') \in \mathbb{Z}' \\ (n',m') \neq (n,m)}} \sum_{k \in \mathbb{Z}} \beta_{n',n,m',m}^{(q)} H_{n',m'} \hat{d}_{n',m'}^{(q-1)} g_{n',m'}[k] \gamma_{n,m}^*[k]$$

$$+ \sum_{k \in \mathbb{Z}} w_{n,m}[k] \gamma_{n,m}^*[k] = H_{n,m} d_{n,m} \sum_{k \in \mathbb{Z}} g_{n,m}[k] \gamma_{n,m}^*[k]$$

$$+ \tilde{I}_{n,m}^{(q)} + \tilde{\mathcal{N}}_{n,m}, \tag{4.84}$$

where $\beta_{n',n,m',m}^{(q)}$ defines the scaling factor (depending on the correlation between the impulses localized on time–frequency points (n', m') and (n, m)). Similarly, as earlier, terms $\hat{d}_{n,m}^{(q-1)}$, $\tilde{I}_{n,m}^{(q)}$ and $\tilde{\mathcal{N}}_{n,m}^{(q)}$ describe the decoded symbol in the previous ($q - 1$-th) step, interference observed in the q-th iteration, and noise observed at the (n, m)-th location on the TF grid, respectively. Note that in case of perfect detection, $\hat{d}_{n,m}^{(q)} = d_{n,m}$ for any q, and $\beta_{n',n,m',m}^{(q)} = 1$, formula (4.84) becomes the same equation (not approximation) as (4.81). Moreover, approximation of $d_{n,m}$ for all n and m allows for estimation of $\tilde{I}_{n,m}^{(q)}$ and its removal at all atom locations. Regarding the scaling factor, an observation can be made that it should be possible to derive its optimal value, since it depends on the correlation between neighboring pulses.

The structure of the GMC-PIC receiver is presented in Figure 4.27. After the received signal transforms from the time to time–frequency domain, all symbols are detected resulting in the estimates $\hat{d}_{n,m}^{(q)}$, for all $(n, m) \in \mathbb{Z}'$. Then, all of the detected symbols are modulated by corresponding transmit (synthesis) pulses, filtered using the channel coefficients estimates in order

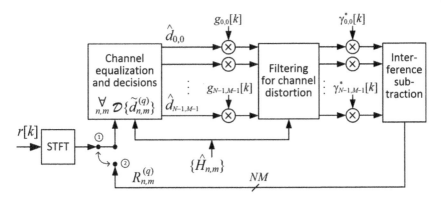

Figure 4.27 Practical GMC-PIC receiver.

to reflect channel distortions, and demodulated using the corresponding receive (analysis) pulses. Finally, the interference is eliminated, creating new time–frequency representation $R_{n,m}^{(q)}$ of the received signal for all $(n, m) \in \mathbb{Z}'$.

4.4.4 Hybrid Interference Cancellation (HIC)

To overcome the problems of high complexity in parallel interference cancellation and high delay in serial interference cancellation, some hybrid solutions have been presented in the literature that merge the characteristic features of both kinds of interference cancellation techniques [285, 286]. The main idea behind the Hybrid Interference Cancellation (HIC) technique is to divide the received set of signals into smaller groups. Next, all groups are decoded simultaneously (in parallel), whereas for all signals within one group, the interference is canceled in a serial manner. Clearly, such a procedure can be reversed, that is, parallel approach can be applied in each block, and the interference between the blocks of symbols is eliminated one by one.

Example Results

Let us look at the performance of the iterative interference cancellation methods applied in the GMC systems. In the considered scenarios, the frame is defined by transmission of $M = 4$ symbol periods in the time domain and $N = 32$ subcarriers in the frequency domain, that is, one GMC-frame consists of $NM = 128$ pulses that carry the data symbols (usually from the QPSK constellation). Perfect knowledge of the channel characteristics has been assumed, and the equalization method based on the MMSE criterion has been implemented. The channel model has been the exponentially decaying path power model. Moreover, various types of pulse shapes, that is, rectangular, Hanning, and Gaussian, have been considered.

Prior to presenting the simulation results, it should be highlighted that the results have been obtained for two cases:

- Case A, when the critical sampling is considered, that is, $N = M_\Delta$, but the duration of the transmit (synthesis) pulse in samples is higher than the IFFT size, that is, $L_g \geq M$; the pulses overlap in time.
- Case B, when the overcomplete set of synthesis pulses is considered, that is, when $N \geq M_\Delta$, and for a given synthesis pulse, the dual pulse is not unique. In this case, conventional reception methods cannot be applied in order to recover the transmit data perfectly [64, 80, 81, 85]. It has to be mentioned that for a given set from which the input data can be selected (e.g., a set of M-QAM symbols), the transmit–receive pair of pulses can be designed in such a way that the transmission for an oversampling rate is feasible [80].

In both cases, for a given synthesis pulse shape, the dual one has been derived based on the Prinz algorithm, described in [255]. Moreover, all of the proposed interference cancellation methods have been compared with the standard ZF or MMSE-based reception methods, that is, when the self-interference is treated as noise, no interference cancellation algorithm is applied, and the ZF or MMSE criterion is applied for the data symbol detection.

Results for the SIC Algorithm

First, the results of the SIC algorithm in Case A as the obtained BER versus SINR are shown in Figure 4.28, where the AWGN channel has been assumed, the pulse duration was set to $L_g = 128$ samples, and the time distance between the corresponding samples of the consecutive pulses was equal to $M_\Delta = 32$. It means that one transmit pulse overlaps three neighboring pulses, thus causing strong ISI. Since the Gaussian pulse has been applied, the ICI between the adjacent pulses in frequency domain is also present. The shapes of the transmit (synthesis) and receive (analysis) pulses are known; therefore, one can expect that self-interference shall be canceled to a good degree.

One can state that due to the application of the SIC algorithm, BER at the receiver has been significantly reduced (up to the level of 10^{-5} at SNR= 20 dB). Since such strong interference both in the time and in the frequency domain has been eliminated, it is worth verifying what is the effectiveness of the algorithm (in terms of BER vs. SNR) in the multipath channel, where nonideal estimates have been used during the equalization process (i.e., the MSE has been equal to $\mathbb{E}\{(H_{n,m} - \hat{H}_{n,m})^2\} = 0.0032$, where $\mathbb{E}(x)$ represents the mean value of x. The results are shown in Figure 4.29. One can observe that the decrease of BER is higher for lower overlapping of pulses, but in both cases, the influence of the residual self-interference is limited, that is, the BER level 10^{-4} has been achieved for SNR$\approx$ 15 dB.

Finally, let us investigate the influence of the pulse shape on the BER. The rectangular and the Gaussian pulses have been considered, whose duration has

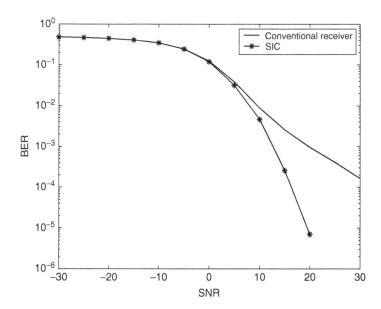

Figure 4.28 BER versus SNR for the conventional MMSE and SIC receiver; AWGN channel, $N = M_\Delta = 32, L_g = 128$ (strong overlapping of pulses).

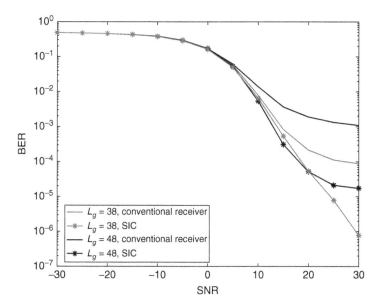

Figure 4.29 BER versus SNR for conventional and SIC receiver; various pulse duration, multipath channel with nonideal channel estimates, $N = M_\Delta = 32$.

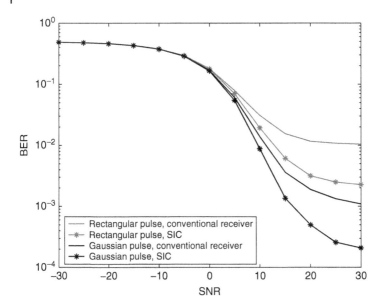

Figure 4.30 BER versus SNR for the conventional MMSE and SIC receiver; various pulse shapes, $N = M_\Delta = 32$, $L_g = 48$.

been set to $L_g = 48$ sampling periods. The results are presented in Figure 4.30. Note that the results obtained for the Gaussian pulse, which has better energy concentration compared to the rectangular one, are much better, that is, the achieved BER is much lower for the same SNR.

The results presenting the obtained BER versus SNR for the successive interference cancellation method in Case B are shown in Figure 4.31. Now, the following conclusions can be drawn:

- the SIC algorithm proposed for self-interference cancellation in GMC systems reduces BER with respect to the conventional MMSE receiver regardless of the pulse shape for SNR > 0 dB;
- the performance of SIC algorithm depends on the shape of the transmit and receive pulse pair; the best results have been obtained for the rectangular and Gaussian pulses. One can observe that BER in an uncoded system where $N > M_\Delta$, that is, when the pulses overlap, can be significantly reduced;
- unfortunately, the SIC algorithm seems not to work efficiently enough in the case of strong overlapping when $N > M_\Delta$ and when the pulse shape is not specifically designed for the arbitrarily selected set of data symbols (such as the set of M-QAM symbols). (The relevant results are not presented here.) Although the so-called error floor is reduced even more significantly (e.g., for low overlapping in time when $\frac{M_\Delta}{N} = \frac{30}{32} = 0.9375$), the obtained BER is rather not sufficiently low for practical application. Clearly, some coding methods

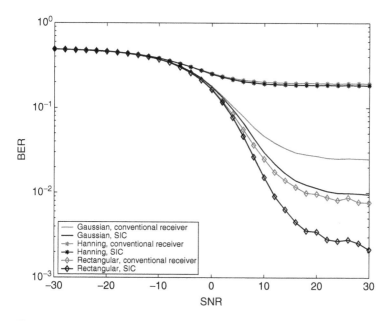

Figure 4.31 BER versus SNR for the conventional MMSE and SIC receiver; $N = 32$, $M_\Delta = 30$, $L_g = 32$.

shall be applied to reduce BER; however, the design of the optimal transmit and receive pulses is also of particular importance. This efficient design of the pulses pair (dedicated for a specific set of symbols) is still a bottleneck of the GMC transmission when $N > M_\Delta$. For low overlapping of atoms (i.e., $\frac{M_\Delta}{N} = 0.9375$), one can obtain BER at the level of 10^{-3} in the uncoded system and without cyclic prefix. These results are in opposition to the claims that in such a case, transmission is even impossible, because it is not possible to recover the user symbols correctly. Moreover, as it has been repeatedly mentioned in this chapter, practical implementation of the transmit–receive pulses assumes the finite duration of the pulses. Such limitations can lead to a situation when even the pulses designed in the optimal way cannot be practically implemented and the suboptimal solutions have to be used. In such a case, it seems that the residual self-interference between the pulses can be efficiently canceled.

Finally, the results for the combined Cases A and B are presented in Figure 4.32. There, the presence of AWGN has been assumed, the time distance between two neighboring pulses is set to $M_\Delta = 30$, and the pulse duration is equal to $L_g = 48$. The number of subcarriers is $N = 32$, as before. The results obtained for $M_\Delta = 32$ and $L_g = 48$ have been presented for reference. In both cases, the Gaussian pulse has been used.

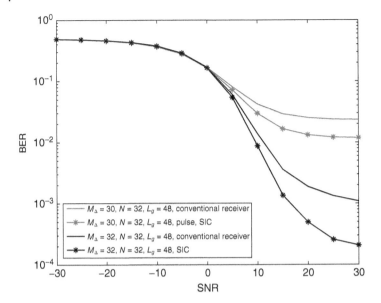

Figure 4.32 BER versus SNR for the conventional MMSE and SIC receiver; Gaussian pulse, $N = 32, L_g = 48$.

It can be concluded that if the distance between the pulses in the time domain is at least equal to the size of FFT, the SIC algorithm reduces the interference significantly. On the other hand, if the pulses are not properly designed for the transmission scheme, that is, when $N > M_\Delta$, it results in the BER performance degradation.

Results for the PIC Algorithm

The results obtained for the PIC algorithm mostly outperform the results presented earlier. This is mainly due to the fact that in the classic approach, the PIC procedure can be iteratively repeated to improve the quality of the receiver performance, however, at the cost of computational complexity. Let us first look at the results of the PIC receiver with 10 repetitions for Case A. In Figure 4.33, BER versus SNR in the AWGN channel is shown in the AWGN channel, when the duration of the Gaussian pulse equals $L_g = 128$. Similarly to the case when the successive interference cancellation is applied, BER is been reduced significantly (to the value lower than 10^{-5} at SNR = 20 dB).

Moreover, for the PIC receiver, the influence of imperfect channel estimation (assuming $\mathbb{E}\{(H_{n,m} - \hat{H}_{n,m})^2\} = 0.0032$) on the system performance has been tested (Figure 4.34). Two pulse durations have been taken into account, that is, $L_g = 38$ and $L_g = 48$. The results are presented in Figure 4.35. Note that the PIC algorithm limits the impact of the imperfect channel estimates on the overall

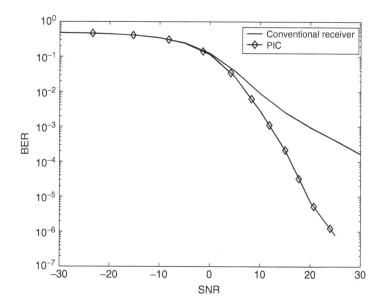

Figure 4.33 BER versus SNR for the conventional MMSE and PIC receiver; AWGN channel, $M_\Delta = 32, L_g = 128$.

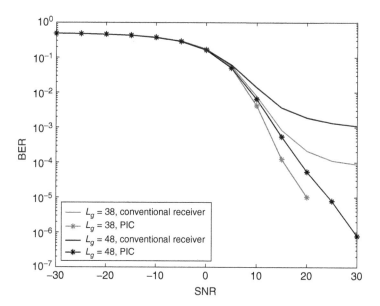

Figure 4.34 BER versus SNR for the conventional MMSE and PIC receiver; multipath channel with nonideal channel estimates, various pulse durations, $M_\Delta = 32$.

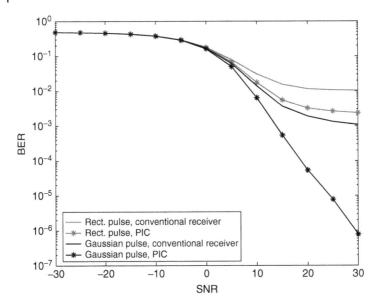

Figure 4.35 BER versus SNR for the conventional MMSE and PIC receiver; various pulse shapes, $N = M_\Delta = 32$, $L_g = 48$.

system performance. Moreover, it outperforms the results obtained for the SIC receiver, in particular for stronger pulse overlapping.

Finally, the impact of the shape of the pulse has been investigated. Again, rectangular and Gaussian pulses of time duration equal to $L_g = 48$ sampling periods have been applied. Analyzing the obtained results (see Figure 4.35), it can be stated that similarly to the case of the SIC reception method, the Gaussian shape is much more suitable for the transmission (lower BER is achieved for this pulse).

In Figure 4.36, BER versus SNR is presented for Case B, that is, when $N > M_\Delta = 30$. The number of iterations for the PIC algorithm has been fixed to 20. It can be concluded that the performance of the PIC method depends on the pulse shape, but on the other hand, the overall efficiency of this algorithm is significantly better than for the SIC receiver (see Figure 4.31 as the reference).

Finally, combination of Cases A and B has been tested, when the number of subcarriers equals $N = 32$, the time distance between the consecutive pulses is equal to $M_\Delta = 30$, and the duration of the considered Gaussian pulse is $L_g = 48$. For this purpose, the PIC procedure with 10 iterations has been implemented. The results are presented in Figure 4.37. The curves of BER versus SNR for the critical sampling case ($N = M$) are also printed for reference. One can observe that if the distance between the pulses in the time domain is at least equal to the size of (I)FFT, the algorithm reduces the interference noticeably. If the shape of

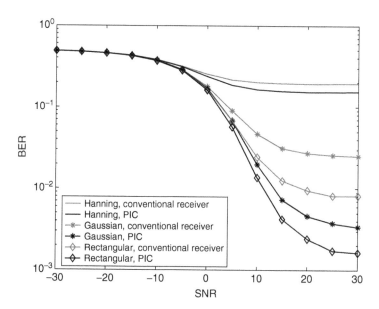

Figure 4.36 BER versus SNR for the conventional MMSE and PIC receiver; the number of iterations = 20, $M_\Delta = 30$, $L_g = 32$.

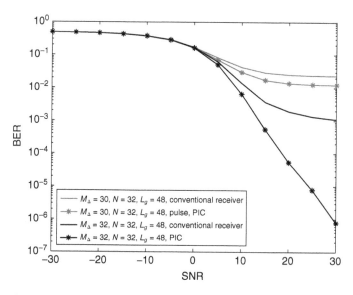

Figure 4.37 BER versus SNR for the conventional MMSE and PIC receiver for the Gaussian pulse; $M_\Delta = 30$ or $M_\Delta = 32$, $L_g = 48$.

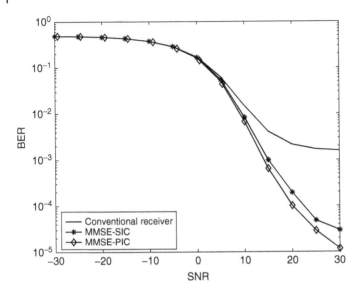

Figure 4.38 BER versus SNR for the conventional MMSE, SIC and PIC receivers; the application of the Gaussian pulse, $N = M_\Delta = 32, L_g = 48$.

the pulse is not matched to the assumed transmission scheme, the obtained BER is degraded.

Finally, the effectiveness of both SIC and PIC algorithms in the multipath fading channel has been investigated and compared. Example results for the mean normalized attenuation equal to $\overline{H} = 0.8$ are presented in Figure 4.38. Again, we can observe that the PIC algorithm outperforms the SIC method, which has already been highlighted in the previous paragraphs.

4.5 Summary

So far, we presented the key aspects of the transmission and reception of the GMC signals. We can conclude these considerations as follows.

It has been recognized that spectral efficiency can be increased by means of proper shaping of the synthesis and analysis pulses. The shape of these pulses can be adapted to ensure the fulfillment of the given system parameters such as the distance between the transmit pulses either in the time or in the frequency domain, the presence of the cyclic prefix, and the frequency bandwidth. In other words, appropriate application of GMC representation can lead to better utilization of the available resources (e.g., frequency band or available power). Moreover, it has been observed that all multicarrier transmission schemes, using orthogonal or nonorthogonal subcarriers, can be treated as a special form of the broad class of GMC signals. As a consequence, all of the contemporary transmission and reception methods can be modified and applied in the future multicarrier systems. The analysis of the problem of high PAPR in GMC signals shows that the price paid for the possibility of appropriate pulse shaping (signal filtering) can be the high variation of the envelope of the time-domain signal. Thus, the value of PAPR in the case of GMC signal can be higher than the values of this metric observed in the multicarrier signals with orthogonal subcarriers. This implies that efficient methods of PAPR reduction have to be applied in order to limit the level of out-of-band emission when the nonlinear elements (such as power amplifiers) are placed inside the transmission chain. Various PAPR reduction methods have been analyzed earlier with particular focus on the ACE method. Appropriate modifications and refinements of the ACE method have been proposed to improve the overall efficiency of this method, applied in GMC transmitters, in terms of the Complementary Cumulative Distribution Function (CCDF) of PAPR and in terms of the resulting BER versus SNR. It can be concluded that, although the GMC signals suffer from high PAPR, it is possible to cope with this problem efficiently. Hence, the properties of the GMC representation of signals can be utilized while satisfying the requirements described by the spectrum masks, that is, the power of the out-of-band radiation can be maintained under the acceptable level.

Various aspects of link adaptation in the GMC-based systems have been considered in this chapter, exploiting the possibility of signal parameters adaptation according to the actual channel characteristic, thus optimizing the defined figures of merit. Some example experimental results that have been presented show that it is possible to increase the system performance (in terms of reduction of power consumption, increase of the overall throughput) and optimize the transmit parameters. It means that one of the main features, or even paradigms of the standard OFDM transmission scheme, is also valid in GMC systems, provided that the modifications of the bit- and power-loading methods are applied.

As highlighted earlier, in some cases of the GMC signal representation, it is impossible to design ideally biorthogonal pair of transmit–receive pulses in practical realizations. In these cases, the effect of self-interference exists between the pulses at the transmitter output, which is impossible to reject even when the channel state information is perfectly known at the receiver. Therefore, reception methods that base on the iterative interference cancellation methods (widely used in CDMA systems to cope with the problem of multiuser interference) can be applied to cope with this kind of interference observed at the receiver at each pulse position in the TF plane. It is worth mentioning that the SIC and PIC algorithms significantly reduce the self-interference in the received GMC signal and improve the BER performance also in the most demanding case, that is, when the set of transmit pulses create the overcomplete set of functions. It can also be concluded that since it is possible to apply the iterative reception method to deal with the problem of self-interference between adjacent pulses, the same algorithm can be used simultaneously to cancel the multiuser interference.

Finally, it can be stated that it is possible to adapt multiple signal parameters to the actual overall requirements obliged to the designed system. Such adaptation of transmit signal allows for better spectrum utilization, for the increase of the user-data throughput, and for the improvement of the quality of offered services. Consequently, the GMC transmission schemes can be suitable for future multistandard mobile radio applications.

5

Filter-Bank-Based Multicarrier Technologies

Orthogonal Frequency-Division Multiplexing (OFDM) and noncontiguous OFDM (NC-OFDM), discussed in Chapters 2 and 3, provide several advantages for system designers. They are characterized by high spectral agility (i.e., individual subcarriers can be independently activated or deactivated), high robustness to the fading channels (i.e., subcarriers that are particularly distorted can be excised from the set of usable ones, and one-tap equalizers can be applied as long the channel is treated as flat within one subcarrier), and simplicity of hardware implementation due to availability of the Fast Fourier Transform (FFT) and Inverse Fast Fourier Transform (IFFT) Integrated Circuit (IC). However, Orthogonal Frequency-Division Multiplexing (OFDM) has some drawbacks, and the presence of high-power Out-of-Band (OOB) spectrum components is treated as one of the key issues. As a consequence, high OOB-power emission is observed, which may induce unwanted interference to other coexisting systems. Moreover, the need for addition of the Cyclic Prefix (CP) results in the OFDM spectral efficiency degradation. Therefore, some advanced filtering of OFDM transmitted waveforms is considered for future radio-communication systems to eliminate the unwanted OOB power leakage to the neighboring frequency bands. One of the competitors in a Fifth Generation (5G) air-interface contest is the so-called Filter-Bank Multi-Carrier (FBMC) waveform, which does not require the use of CP, thus, improving the spectral efficiency of the transmission based on this type of waveform. In principle, FBMC scheme is based on filtering each subcarrier separately in order to achieve low OOB power of this subcarrier, that is, the spectrum of a pulse transmitted on a single subcarrier is shaped using the transmit (synthesis) filter, which requires application of the properly selected receiver (analysis) filter. While in the OFDM system, both (synthesis and analysis) pulses are known to be rectangular, FBMC-based systems can utilize pulse shapes, such as those described in Chapter 4. The FBMC systems have their drawbacks as well, such as intrinsic (self-)interference, which creates a serious problem for Multiple Input, Multiple Output (MIMO) system implementation. Moreover,

Advanced Multicarrier Technologies for Future Radio Communication: 5G and Beyond, First Edition.
Hanna Bogucka, Adrian Kliks, and Paweł Kryszkiewicz.
© 2017 John Wiley & Sons, Inc. Published 2017 by John Wiley & Sons, Inc.

it results in higher complexity of algorithms applied at the transmitter and at the receiver as compared to OFDM.

In fact, FBMC waveforms can be considered as an important class of the Generalized Multicarrier (GMC) signals. Since it is a very strong candidate for the future 5G radio-communication waveforms, let us discuss its principles and key aspects related to its future applications. The literature on FBMC signals and systems is vast, and an interested reader is encouraged to delve further into the overview and research articles (e.g., Ref. [287–295]).

5.1 The Principles of FBMC Transmission

Historically, the concept of FBMC systems is dated to the mid-1960s, when the application of filter banks has been proposed for Pulse Amplitude Modulation (PAM) symbol delivery within the given frequency band (maximizing bandwidth efficiency). Next, vestigial sideband (VSB) signaling [296] and double-sided modulated formats (DSB) [297] have been used for PAM and QAM symbol delivery, respectively. Polyphase filter banks have been applied in that context for the first time in [298] and later in [299]. Furthermore, similar solutions to FBMC waveforms have been discussed from the perspective of their application in wired systems (such as digital subscriber lines), in particular the Discrete Wavelet Multi-Tone (DWMT) and Filtered Multi-Tone (FMT) schemes have been addressed in, for example, [300, 301]. In principle, FBMC-based systems rely on separate filtering of each subcarrier by means of dedicated filter pairs, $g(t)$ and $\gamma(t)$, applied at the transmitter and at the receiver, respectively. These sets of filters constitute the structures known as synthesis filter bank at the transmitter side and analysis filter bank at the receiver side (see Figure 5.1).

For the sake of clarity, let us recall the mathematical formula describing the transmit signal in its continuous, $s(t)$ (5.1), and discrete $s[k]$ (5.2) forms, as discussed in Chapter 4.

$$s(t) = \sum_{n=0}^{N-1} \sum_{m=0}^{M-1} d_{n,m} g_{n,m}(t)$$

$$= \sum_{n=0}^{N-1} \sum_{m=0}^{M-1} d_{n,m} g(t - mT) \exp(j2\pi nF(t - mT)), \tag{5.1}$$

$$s[k] = \sum_{n=0}^{N-1} \sum_{m=0}^{M-1} d_{n,m} g_{n,m}[k]$$

$$= \sum_{n=0}^{N-1} \sum_{m=0}^{M-1} d_{n,m} g[k - mM_\Delta] \exp[j2\pi nN_\Delta(k - mM_\Delta)]. \tag{5.2}$$

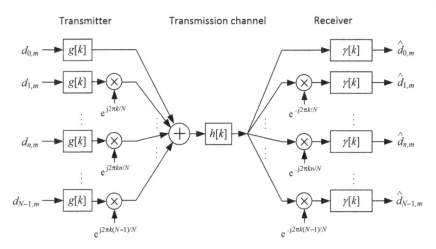

Figure 5.1 The structure of the FBMC transmultiplexer.

One may observe that the mathematical description of an FBMC signal is the same as GMC signal provided in Chapter 4 and very similar to the one of OFDM. The key differences between FBMC and OFDM transmissions are in the definition of the time duration of the transmit pulse applied in both systems as well as of their shapes. In case of OFDM, rectangular pulse of amplitude 1 is chosen for the transmission (and reception) chain. Moreover, the transmit pulse duration is $T > T_{\mathrm{FFT}} = \frac{1}{F}$, and at the receiver, the receive pulse duration is set to $\frac{1}{F}$. In OFDM systems, CP is added to each transmitted symbol in order to minimize the influence of time-dispersive channel; this prefix is typically discarded at the receiver side. Contrarily, in FBMC systems, the pulses are usually well localized in the frequency domain (high attenuation of sidelobes is achieved by advanced filtering), but their duration is typically an integer multiplication of $\frac{1}{F}$, leading to overlapping of consecutive signals. Moreover, no CP is introduced in the system, which increases the spectrum efficiency when compared to the OFDM systems. One may observe that, in general, good TF localization of the spectrally shaped and modulated data (called *atoms* as in Chapter 4) allows for minimization of Inter-Carrier Interference (ICI) and Inter-Symbol Interference (ISI) between adjacent pulses [302–304].

The FBMC systems are generally divided into two classes, related to the types of data carried by these systems. In the first class, the complex-valued symbols (more precisely, QAM symbols) are transferred, whereas in the second case, only real-valued data are considered (i.e., PAM symbols). When the complex symbols are transmitted, the formulas (5.1) and (5.2) do not have to be specifically adjusted, as the rectangular time–frequency lattice is applied, that is, each pulse carrying complex data is separated from its neighbor by T in the time domain and by F in the frequency domain. The key issue in this case is to design a pair of transmit and reception pulses in such a way to allow for the distortion-free symbol recovery at the receiver side.

Another approach widely discussed in the literature is to consider only the PAM symbols, thus allowing for transmission of real-only data symbols. In such a case, it is required to narrow the distance between the pulses either in the time domain or in the frequency domain in order to maintain the rate at least at the same level as in the OFDM case. Various approaches to this aspect have been proposed, such as the so-called Staggered Multitone (SMT) [305] or Cosine-Modulated Multitone (CMT) [306]. In general, however, this scheme is widely referred to as FBMC with Offset Quadrature Amplitude Modulation (OQAM) (FBMC/OQAM), where the time-domain signal is represented as

$$s[k] = \sum_{n=0}^{N-1} \sum_{m=0}^{M-1} d_{n,m} g_{n,m}[k] \tag{5.3}$$

$$= \sum_{n=0}^{N-1} \sum_{m=0}^{M-1} \tilde{d}_{n,m} g \left[k - m \frac{M_\Delta}{2} \right] \exp \left[j(n+m) \frac{\pi}{2} \right] \cdot$$

$$\exp \left[j 2 \pi n N_\Delta \left(k - m \frac{M_\Delta}{2} \right) \right],$$

where $\tilde{d}_{n,m} = d_{n,m} \exp[j(n+m) \frac{\pi}{2}]$ represents the OQAM symbol localized at (n, m) coordinates on the TF plane. Note that component $\exp[j(n+m) \frac{\pi}{2}]$ takes the following values: 1 for even values of $n + m$, and j for odd values of $n + m$.

5.2 FBMC Transceiver Design

In the FBMC transmitter, the impulse modulating each subcarrier is filtered individually, and the key challenge is to design the filter impulse response efficiently, so that various criteria are fulfilled (e.g., good TF localization, duration of the filter impulse response). At the receiver, the dual impulse has to be applied to reconstruct the transmitted data symbols. The existence of the dual pulse and the perfect-reconstruction filter bank have been discussed in Chapter 4. There, also the polyphase decomposition of the original low-pass filter to pass-band bank of filters has been briefly analyzed. All these considerations and the structures of the transceivers presented in Figures 4.5 and 4.6 are valid for FBMC transceivers. In the FBMC transmitter, the set of N user data symbols (either complex or real) are filtered in the low-pass filter $g[k]$. These spectrally shaped symbols modulate the N subcarriers (e.g., by the means of IFFT), so that they are transmitted in the combined N parallel streams. After passing through the channel, the received signal is subject to demodulation and low-pass filtering in the filter of the impulse response $\gamma[k]$.

The generic transceiver architecture can be modified in order to meet various optimization criteria, for example, to minimize the complexity burden. It can be achieved by application of either the so-called polyphase filters, which can be efficiently realized by means of IFFT, or one of the other filter structures discussed in the literature, for example, in [64]. In the former case, the original prototype filter $g[k]$ is decomposed into a bank of polyphase transmit filters $\tilde{g}_n[k]$ (where $n = 1, \dots, N-1$) at the modulator (IFFT) output. The N signal components at the filters outputs are combined, constituting the transmit signal $s[k]$. At the receiver, a bank of polyphase reception filters $\tilde{\gamma}_n[k]$ is applied at the input of the demodulator (FFT). The set of transmit filters is widely called the synthesis filter bank, whereas the set of reception filters is identified as analysis filter bank. This filter-bank structure of the FBMC transceiver can be treated as the low-complexity solution, particularly in a situation when the duration of the transmit pulse is not long. For example, if the pulse duration (in samples) is equal to the number of subcarriers N, the per-subchannel filtering reduces to the simple multiplication by one (complex) scalar. More details on polyphase filtering can be found in Chapter 4. The transceiver architecture that applies the bank of polyphase filters is shown in Figure 4.6.

Regarding the FBMC waveform generation, the input data can be of either real or complex type. However, FBMC system is widely referred to as the filter-bank-based multicarrier system with OQAM data symbols. Following the literature, for example, [307, 308], the commonly understood FBMC transceiver can be implemented by following the approach shown in Figure 5.2. Clearly, the computationally efficient structure realized by means of IFFT/FFT blocks and polyphase filters, as discussed earlier, can also be applied. In the mentioned figure, one may observe the preprocessing of the QAM symbols called *staggering* as in [309], which is a simple complex-to-real conversion, where the real and imaginary parts of a complex-valued symbol are separated to form two new symbols. Note that this complex-to-real conversion increases the sampling rate twice. Moreover, the up-sampling and down-sampling by $\frac{N}{2}$ (as opposed to up- and down-sampling by M_Δ shown in Figure 4.5) are necessary to reflect the signal modification due to the application of OQAM system and the distance between the symbols in the time domain (in samples). The staggering operation (the time delay between the real and imaginary parts of the symbol) shown in Figure 5.2 is explained by Figure 5.3. Naturally, at the receiver, the opposite *destaggering* operation is performed [309] to form the complex-valued symbols for further QAM detection.

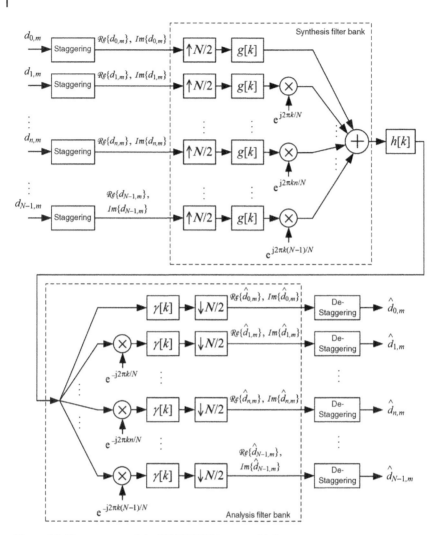

Figure 5.2 The structure of the FBMC/OQAM transmultiplexer.

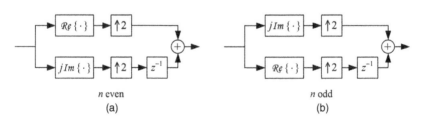

Figure 5.3 The staggering block for even (a) and odd (b) subcarriers.

5.3   Pulse Design

One of the key issues in designing the FBMC-based system is the definition of a pair of transmit and reception pulses (prototype filter impulse response and its dual, perfect-reconstruction filter). They should be characterized by good time- and frequency-localization properties and meet other design criteria, such as energy minimization, low OOB power, hardware limitations, or spectrum nulling [288]. As mentioned already in Chapter 4, various types of pulse shapes can be considered. For the specific application in the FBMC systems, as they are commonly understood, the following ones are being proposed: the Isotropic Orthogonal Transform Algorithm (IOTA) function, Extended Gaussian Function (EGF), prolate functions, and the one proposed by M. Bellanger [310, 311]. A comprehensive discussion on the pulse shapes (prototype filters) and corresponding lattice structures as well as on the implementation aspects associated with a particular pulse selection can be found in one of the recent surveys [288, 312]. Let us now briefly describe the selected solutions to the problem of proper pulse-shaping filter design in the context of the defined FBMC waveform generation.

5.3.1   Nyquist Filters and Ambiguity Function

One of the prototype filter design criteria is zero-crossing of the generated signal at interval T, which guarantees ISI avoidance. The prototype filter can be designed to meet perfect-reconstruction conditions, that is, the condition of the existence of the (dual) receiving filter able to perfectly reconstruct the original data in case of the perfect channel. In this case, the designed prototype filter is typically a spectral factor of an Nth-band Nyquist filter. If $g(t)$ is the impulse response of the root-Nyquist prototype filter, $g_n(t) = g(t) \exp(j2\pi t f_n)$ is the modulated synthesis filter at the n-th subcarrier of frequency f_n, $\gamma(t)$ is the perfect-reconstruction receiver filter, and $\gamma_l(t) = \gamma(t) \exp(j2\pi t f_l)$, the following orthogonality condition holds:

$$\langle g_n(t - aT), \gamma_l(t - bT) \rangle = \int_{-\infty}^{\infty} g_n(t - aT)\gamma_l^*(t - bT)dt = \delta_{n,l}\delta_{ab}, \quad (5.4)$$

where $(\cdot)^*$ denotes complex conjugate, and δ_{ab} is the Kronecker delta function. The inner product $\langle \cdot, \cdot \rangle$ defines the similarity between the elements and can be evaluated as a specific case of an ambiguity function defined later in this section. The discussion on the orthogonality criterion and the ways of design of pulse pair ($g(t)$ and $\gamma(t)$) is presented in Chapter 4; some discussion is also provided in [288, 312].

Usually, the requirement of the perfect reconstruction is not essential because it is sufficient that the designed filter bank creates self-interference between the transmitted pulses that is much lower than the one resulting from the transmission channel time and frequency dispersion.

Moreover, near-perfect reconstruction designs can be more efficient then perfect-reconstruction filters, resulting in lower OOB-power components for a given prototype filter order. An interesting discussion on the pulse shape design for the FBMC systems is presented in [313].

One of the key metrics used in the assessment of the time–frequency properties of the designed pulses is the so-called *ambiguity function* $A_g(\tau, v)$, which identifies the relation between the particular points on the TF lattice. Mathematically, it is defined as

$$A_g(\tau, v) = \int_{-\infty}^{\infty} g_n \left(t + \frac{\tau}{2} \right) g_l^* \left(t - \frac{\tau}{2} \right) e^{-j2\pi v t} dt. \tag{5.5}$$

In the aforementioned formula, tuple (τ, v) defines the time delay and frequency shift, respectively. Note that (5.5) is proportional to the inner product $\langle g_n(t - aT), \gamma_l(t - bT) \rangle$ for the case when $g(t) = \gamma(t)$, $aT = bT = \frac{\tau}{2}$ and $v = f_n - f_l$. In fact, in many applications, the synthesis and analysis filers are the same. The ambiguity function can be presented on the TF plane, showing the quality of the used pulses in terms of their TF localization, created self-interference between the pulses (atoms), and the possibility of perfect or near-perfect signal reconstruction at the receiver. It is the real-valued function in case when $g(t)$ is an impulse of even symmetry. In the ideal case, the ambiguity function should be close to the Kronecker impulse (on the 2D time–frequency plane). In Figures 5.4 and 5.5, the ambiguity function of the rectangular pulse (as in OFDM signal but without the addition of CP) is presented.

5.3.2 IOTA Function

The IOTA function is the result of the orthogonalization process applied to the Gaussian function, defined as

$$\rho_\alpha(t) = (2\alpha)^{0.25} \exp(-\pi \alpha t^2), \tag{5.6}$$

where α denotes the Gaussian function parameter. Let us define the orthogonalization operator $\mathcal{O}_a[x(t)]$ with parameter a as

$$\mathcal{O}_a[x(t)] = \frac{x(t)}{\sqrt{a \sum_k |x(t - ka)|^2}}, a > 0. \tag{5.7}$$

One can state that the effect of operator $\mathcal{O}_a$ is the orthogonalization of function $x(t)$. Now, the IOTA function (based on the Gaussian function (5.6)) can be defined as [304]:

$$g_a(t) = g_\alpha(t) = \mathcal{F}^{-1}\{\mathcal{O}_{M_\Delta}[\mathcal{F}\{\mathcal{O}_N[\rho_\alpha(t)]\}]\}, \tag{5.8}$$

where $\mathcal{F}$ denotes the Fourier transform operator, and $\mathcal{F}^{-1}$ is its inverse.

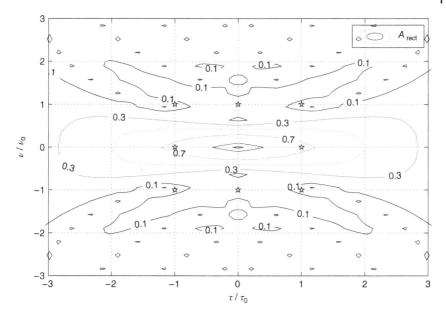

Figure 5.4 Ambiguity function of the rectangular function—surface plot. (Jackowski, 2014 [314]. Reproduced with the permission of Faculty of Electronics and Telecommunication.)

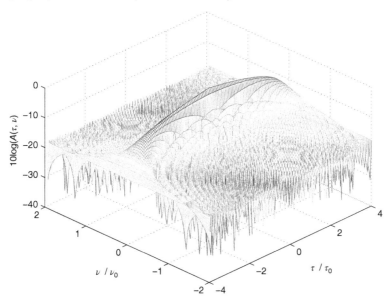

Figure 5.5 Ambiguity function of the rectangular function – logarithmic plot. (Jackowski, 2014 [314]. Reproduced with the permission of Faculty of Electronics and Telecommunication.)

In [303], the idea of orthogonalization has also been applied to the EGF defined as

$$z_{\alpha,T,F} = \frac{1}{2}\left(\sum_{k=0}^{\infty} d_{k,\alpha,T}\left(\rho_\alpha\left(t+\frac{k}{T}\right) + \rho_\alpha\left(t-\frac{k}{T}\right)\right)\right). \tag{5.9}$$

$$\cdot \sum_{l=0}^{\infty} d_{l,1/\alpha,F} \cos\left(2\pi l \frac{t}{F}\right).$$

There, it is assumed that $TF = 0.5$ (the density of the rectangular lattice defined in Chapter 4 by (4.17) is equal to $\delta(\Lambda) = \frac{1}{TF} = 2$), $\alpha \in (0.528 M_\Delta^2, 7.568 M_\Delta^2)$, and the coefficients $d_{x,y,z}$ can be found as proposed in [63]. The function ρ_α is the Gaussian function defined by (5.6). In this light, the IOTA function can be interpreted as the special case of EGF, where $T = F = \frac{1}{\sqrt{2}}$. (More details on EGF can be found in [315].)

The IOTA function is well characterized by the ambiguity function. An illustration of the $A_p(\tau, v)$ for the IOTA pulse is shown in Figure 5.6 (surface plot) and in Figure 5.7 (logarithmic plot). One may observe the characteristic nulls on the time–frequency plane. This property is utilized in the FBMC systems in order to minimize the intersymbol interference. (Compare the ambiguity

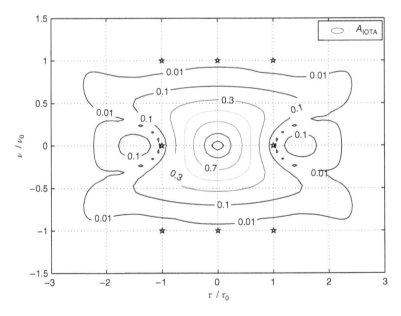

Figure 5.6 Ambiguity function for the IOTA function—surface plot; star markers indicate the center locations of the surrounding pulses. (Jackowski, 2014 [314]. Reproduced with the permission of Faculty of Electronics and Telecommunication.)

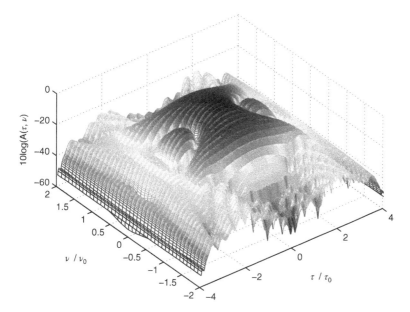

Figure 5.7 Ambiguity function for the IOTA function—logarithmic plot. (Jackowski, 2014 [314]. Reproduced with the permission of Faculty of Electronics and Telecommunication.

functions of the IOTA pulse with the rectangular pulse from Figures 5.4 and 5.5.)

Finally, the IOTAfilter impulse responses and the amplitude characteristic are presented in Figures 5.8 and 5.9 (solid lines) respectively.

5.3.3 PHYDYAS Pulse

One of the most popular solutions in the context of FBMC/OQAM systems is the so-called *Bellanger* filter (known also as PHYDYAS filter), named after its designer Bellanger [310, 311]. The key idea here is to apply the frequency-sampling technique. Typically, in digital systems, the impulse response of the transmit filter crosses zeros at integer multiples of the pulse duration, which corresponds to the cutoff frequency of the design filter. As discussed in [310], the idea is to design the Nyquist filter while considering the appropriate frequency coefficients, apply the symmetry condition, and then calculate the time-domain pulse shape by performing the Fourier transform. In [311], it has been proposed that the $KN - 1$ coefficients in the frequency domain, denoted as $\overline{P}_k$, of the pulse $g(t)$ shall fulfill the following criteria:

$$\begin{cases} \overline{P}_0 = 1, \\ \overline{P}_i^2 + \overline{P}_{K-i}^2 = 1 \quad \text{and} \quad \overline{P}_i = P_{K-i}, \quad \text{for } 1 \leq i \leq K - 1 \\ \overline{P}_i = 0, \qquad\qquad\qquad\qquad \text{for } K \leq i \leq KN - K, \end{cases} \quad (5.10)$$

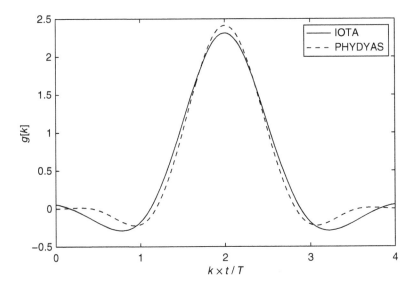

Figure 5.8 IOTA and PHYDYAS filters' impulse responses; the prototype filter order $L_g = 64$, the overlapping factor $K = 4$, the IFFT order $N = 16$. (Jackowski, 2014 [314]. Reproduced with the permission of Faculty of Electronics and Telecommunication.)

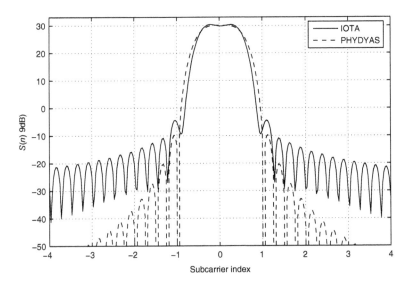

Figure 5.9 IOTA and PHYDYAS filters' amplitude characteristics; the prototype filter order $L_g = 64$, the overlapping factor $K = 4$, the IFFT order $N = 16$. (Jackowski, 2014 [314]. Reproduced with the permission of Faculty of Electronics and Telecommunication.)

where K is the so-called *overlapping factor*, the integer value, typically set to $K = 2, 3, 4,....$[1] The overlapping factor is the prototype filter length (order) over the number of subchannels (subcarriers). In case of the polyphase decomposition, it is also the single polyphase filter order. The contiguous frequency response of the filter, achieved by means of interpolation:

$$G(f) = \sum_{i=-K+1}^{K-1} \overline{P}_i \cdot \frac{\sin\left(\pi\left(f - \frac{i}{NK}\right)NK\right)}{NK\sin\left(\pi\left(f - \frac{i}{NK}\right)\right)} \tag{5.11}$$

Table 5.1 lists the values of the calculated coefficients $\overline{P}_i$ for various values of K. These can be used for definition of the time-domain pulse:

$$g(t) = 1 + 2\sum_{i=1}^{K-1} \overline{P}_i \cdot \cos\left(\frac{it}{KT}\right). \tag{5.12}$$

The exemplary pulse shapes of the Bellanger (PHYDYAS) filter with the arbitrarily selected duration of 128 samples are shown in Figure 5.10. Moreover, one can compare the IOTA and the PHYDYAS filters' impulse responses in Figure 5.8, as well as their amplitude characteristics in Figure 5.9.

Finally, the ambiguity function of the PHYDYAS function is presented in Figures 5.11 and 5.12.

5.3.4 Other Pulse-Shape Proposals for FBMC

Apart from the typical IOTA and PHYDYAS pulse-shaping filters for FBMC/OQAM, there are other designs offered to implement this type of waveform. The usage of a specific discrete Zak transform [239] for pulse design is presented in [302]. In [252], the Hermite function is proposed instead of the IOTA function (as a candidate for OFDM/OQAM systems) because it allows better spectral efficiency. Following these results, the authors of [313] propose an isotropic filter that is robust to the time and frequency spread of the channel.

Table 5.1 Calculated coefficients of the Bellanger filter.

K	$\overline{P}_0$	$\overline{P}_1$	$\overline{P}_2$	$\overline{P}_3$
2	1	$\dfrac{\sqrt{2}}{2}$	0	0
3	1	0.911438	0.411438	0
4	1	0.971960	$\dfrac{\sqrt{2}}{2}$	0.235147

---

1 Although in Chapter 4, variable K has been used to denote the number of samples in the GMC frame (see Figure 4.2), in this chapter, we use it to denote the overlapping factor. This is to be consistent with the most popular notation in the literature.

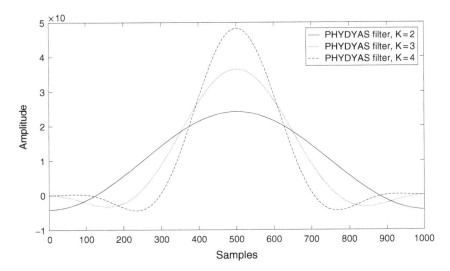

Figure 5.10 Bellanger (PHYDYAS) pulse for $K = 2, 3, 4$.

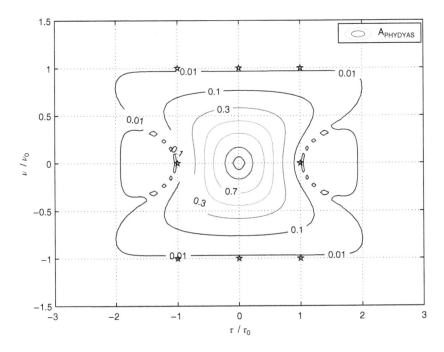

Figure 5.11 Ambiguity function for the Bellanger (PHYDYAS) function—surface plot; star markers indicate the center locations of the surrounding pulses. (Jackowski, 2014 [314]. Reproduced with the permission of Faculty of Electronics and Telecommunication.)

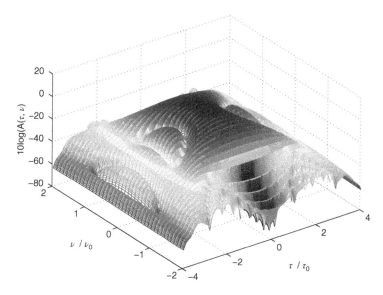

Figure 5.12 Ambiguity function for the Bellanger (PHYDYAS) function—logarithmic plot. (Jackowski, 2014 [314]. Reproduced with the permission of Faculty of Electronics and Telecommunication.)

Another interesting approach toward the minimization of the OOB power emission has been proposed in [316]. As a result of the optimization (minimization) procedure, the proposed pulse shape $g(t)$ is obtained as a linear combination of the truncated (finite in time domain) Prolate Spheroidal Wave Functions [317–319]. These functions are characterized by some excellent features: very good joint time–frequency localization (thus, good energy concentration), the fulfillment of the orthogonality and completeness properties [320], and maximization of the ratio of the main-lobe to sidelobe energy. The in-depth discussion on the filter design is presented in [288, 312].

5.4 Practical FBMC System Design Issues

Let us now briefly discuss the selected practical issues related to the FBMC system design. The identified issues, self-interference problem, computational complexity, the implementation issues of burst transmission, and FBMC MIMO systems are inherent in the filter-bank-based transmitters and receivers and result from the transceiver structure and applied system parameters.

5.4.1 Self-Interference Problem in the FBMC Systems

In the generic FBMC transceiver structure, regardless of whether it is realized by means of polyphase filters or not, the data symbols can be either complex (when $d_{n,m}$ are the QAM symbols) or real-valued (when $d_{n,m}$ are the PAM symbols). It is mainly the filter impulse-response shape that has a direct impact on the potential data format used for the transmission. A typical assumption in the context of the FBMC systems is to significantly reduce the amount of energy emitted out of the nominal frequency band. In consequence, the slopes of the frequency response of the transmit filter $g(t)$ are often steep, leading to long time duration of the impulse response. One may conclude that the selection of the pulse shape (analyzed in both time and frequency domains) entails the definition of the pulse location on TF plane (i.e., distances between the neighboring pulses). Improper selection of the TF grid may result in severe self-interference, that is, overlapping of the neighboring atoms both in the time domain and in the frequency domain. This is already observed at the output of the transmitter. Now, let us analyze the coefficients of the impulse and frequency response of two prototype filters based on the exemplary PHYDYAS and IOTA pulses, which are presented in Tables 5.2 and 5.3, respectively (see Refs [310, 321] for more details). In both cases, we assume that these pulses are located at the specified location (n_0, m_0), so we analyze the TF distribution of the pulse on the n_0-th subcarrier in the m_0-th symbol. Note that these coefficients are normalized.

One may observe that in both cases, the highest value is indeed observed at location (n_0, m_0). Moreover, the IOTA and PHYDYAS pulses have significant

Table 5.2 PHYDYAS-pulse time- and frequency-response coefficients.

	$m_0 - 3$	$m_0 - 2$	$m_0 - 1$	m_0	$m_0 + 1$	$m_0 + 2$	$m_0 + 3$
$n_0 - 2$	0	0	0	0	0	0	0
$n_0 - 1$	0.043j	−0.125	−0.206j	0.239	0.206j	−0.125	−0.043j
n_0	−0.067	0	0.564	1	0.564	0	−0.067
$n_0 + 1$	−0.043j	−0.125	0.206j	0.239	−0.206j	−0.125	0.043j
$n_0 + 2$	0	0	0	0	0	0	0

Based on [321].

Table 5.3 IOTA-pulse time- and frequency-response coefficients.

	$m_0 - 3$	$m_0 - 2$	$m_0 - 1$	m_0	$m_0 + 1$	$m_0 + 2$	$m_0 + 3$
$n_0 - 2$	0.0016	0	−0.0381	0	−0.0381	0	0.0016
$n_0 - 1$	−0.0103j	−0.0381	0.228j	0.4411	0.228j	−0.0381	−0.0103j
n_0	−0.0182	0	0.4411	1	0.4411	0	−0.0182
$n_0 + 1$	−0.0103j	−0.0381	0.228j	0.4411	0.228j	−0.0381	−0.0103j
$n_0 + 2$	0.0016	0	−0.0381	0	−0.0381	0	0.0016

Based on [321].

impact on the preceding and succeeding pulses in the time domain and also on the neighboring pulses in the frequency domain. The PHYDYAS impulse has steeper slopes of the frequency response as compared to the IOTA function. (One may see the value 0.239 for the table entry at location $n_0 \pm 1; m_0$ in Table 5.2 and 0.441 for the same locations in Table 5.3; refer also to the ambiguity plots to analyze this effect in detail.) This means that the pulses transmitted using odd-index subcarriers only or even-index subcarriers only do not overlap. Moreover, the points where the coefficients are equal to zero create a grid of time–frequency points ($n_0 \pm 2i; m_0 \pm 2j$) for both i and $j \in \{0, \pm1, \pm2, \cdots\}$. However, one may observe that the self-interference observed at the surrounding TF points ($n_0 \pm 1; m_0 \pm 1$) is only present in the imaginary part of the complex symbol. Thus, by analyzing only the real part of the observed symbol at the receiver, the interference can be fully eliminated. This observation has originally led to the actual introduction of an dedicated offset (as in OQAM), where complex data $d_{n,m}$ are split into their real and imaginary parts and are transmitted on two shifted time–frequency grids, respectively. In particular, the real part of the data symbol is transmitted using the pulses located at points ($n_0 \pm 2i; m_0 \pm 2j$) for both i and $j \in \{0, \pm1, \pm2, \cdots\}$, whereas the imaginary part is carried on the points creating the following defined grid: ($n_0 \pm (2i + 1); m_0 \pm (2j + 1)$) (also for both i and $j \in \{0, \pm1, \pm2, \cdots\}$). In other words, the real and the imaginary parts of the symbol are not transmitted simultaneously as in OFDM, but the imaginary part is delayed by half the symbol duration. Please note that by splitting the complex data symbols into their real and imaginary parts, we reduce twice the transmission rate as compared to OFDM. However, application of the time offset by half of the pulse duration compensates this drawback. This is possible due to the symmetry of the transmitter and receiver filters, which in fact are also identical.

5.4.2 Computational Complexity

While considering the FBMC techniques, the question concerning the computational complexity and the feasibility for practical implementation arises. The price for high spectral efficiency, for the subcarrier OOB-power attenuation and the adjacent-channel interference mitigation of these techniques is usually the increase of the transceiver complexity. This is due to the fact that additional filtering operations are required when compared with the OFDM transmission.

In [3], the number of required operations (the number of real multiplications and additions) employed at the transmitter (specifically at the modulator) and at the receiver (demodulator) per sampling period versus the number of subcarriers is presented for the case of GMC synthesis and analysis filter polyphase decomposition and with a number of overlapping-factor values. Moreover, no simplifications in the IFFT and FFT implementation has been considered as possible, apart from the standard split-radix-2 algorithms. (Note that in practice, various techniques can be used to reduce the number of operations needed for modulation and demodulation implemented using Discrete Fourier Transforms (DFTs), with one example being the pruning of the IFFT and FFT operations.) There, it can be observed that GMC-system transceiver is more demanding than OFDM transceiver in terms of computational complexity due to the application of filter banks.

In [309], the total number of real multiplications for the synthesis filter bank (SFB) is derived as the sum of multiplications in each processing (including preprocessing) block, in the transmitter with N-point IFFT, and N-branch polyphase filter section. It is given by the following formula:

$$C_{\text{SFB}} = 2 \cdot \{2N + [N(\log_2 N - 3) + 4] + 2KN\} \tag{5.13}$$

where K is the overlapping factor. The preprocessing section is considered to be multiplication-free, while the IFFT is again assumed to be implemented using the split-radix-2 algorithm. The complexity of the analysis filter bank is equal to the complexity of the synthesis filter bank due to similar processing blocks at the receiver. Thus, in [309], the total complexity of an FBMC modulator and demodulator is assumed to be twice the complexity described by the formula (5.13). Moreover, in the mentioned work, the complexities (in terms of the number of real multiplications) of the FBMC and the OFDM systems (actually modulators and demodulators) are compared. Naturally, the FBMC system is more complex than the conventional OFDM (approximately 10 times for the system parameters assumed in [309]), and its computational complexity is proportional to the overlapping factor K.

Although the complexity of the FBMC modulator and demodulator appears to be higher when compared with OFDM, it should be noted that the complexity of several associated FBMC signal processing algorithms, such as channel equalization or interference cancellation, can be significantly more complex relative to the same algorithms employed by the OFDM-based system. The complexity increase associated with these signal processing modules also depends on the employed pulse-shaping filters and other FBMC parameters [3]. On the other hand, if the transmitted atoms are well localized on the TF plane, particularly in the frequency domain, the frequency offset in the received signal, resulting from the Doppler effect or the Local Oscillator (LO) instability and causing ICI may not be that severe, as in the case of OFDM. The algorithms necessary to eliminate this offset may be less computationally complex in FBMC receivers. Moreover, if properly modified, the synthesis filters at the receiver can equalize the channel distortions, thus eliminating the additional operations needed for channel equalization. An interested reader is encouraged to refer to the complexity analysis derived for GMC transmission in [51] and ponder on the simplifications that can be performed for the subclass of GMC systems, namely FBMC transmissions.

5.4.3 Limitations of FBMC in Burst Transmission Schemes

One of the key advantages of the multicarrier transmission based on filter banks is its low OOB emission due to per-subcarrier filtering with relatively low excess bandwidth. As mentioned earlier, this results in the long prototype-filter impulse response and in the consecutive pulses overlapping in the time domain, that is, in the ISI. For example, in the PHYDYAS case, the typical values of the overlapping factor K are set to 2, 3, or 4 and also indicate the number of overlapped pulses. Due to the proper pulse design, however, this problem can be minimized, for example, when the pulse at a given subcarrier frequency is located exactly at the zero-crossings of the other pulses in the time domain. However, if we consider a burst of M symbols consisting of N-subcarriers modulating parallel pulses, and creating the FBMC frame, the next bursts can be distorted by ISI (or to be more precise, interburst or interframe interference) [310]. Note that such an effect is boosted by the influence of the multipath propagation channel. In practice, the duration of the selected pulse has a direct impact on the time-critical transmission schemes, for example, in Time-Division Duplex (TDD) transmission, or fast burst transmission regardless of the duplexing scheme. When the FBMC is employed in the burst transmission, the length of the burst must be extended in time, for example, by the application of some guard period to allow for transitions due to the filter impulse response. These transitions may be shortened if some temporary signal leakage in the frequency domain is allowed. Some interesting discussions on FBMC transmission in burst scenarios can be found in [310].

5.4.4 MIMO technique for FBMC Transmission

Let us now consider the problem of application of FBMC concept in the MIMO transmission schemes. As clearly indicated in [310], two key cases are possible, which correspond to the two approaches to data transmission in FBMC systems. In the first case, the self-interference between subcarriers (pulses) is fully eliminated or is negligible, and complex QAM data are transmitted on active subcarriers. In this situation, the subcarriers used for carrying data are selected to avoid interference or to keep it below the acceptable level. If this criterion is guaranteed, the classical MIMO techniques originally applied to OFDM-based systems can also be used for FBMC transmission. However, the situation is more complicated in the other case, when self-interference exists between the transmitted pulses localized closely on the time–frequency plane and when the OQAM scheme is applied. Let us discuss it briefly as follows.

Due to the aforementioned self-interference issue and possible lack of orthogonality between the neighboring subcarriers (i.e., the subcarriers with indexes n_0 and $n_0 \pm 1$, carrying either the real or the imaginary parts of data), the application of the MIMO techniques, known from the OFDM-based systems, is not straightforward. This is because in the MIMO transmission and coding techniques, the assumption that the MIMO channels are statistically independent is made. If the specificity of the FBMC/OQAM signal is not taken into account, the well-known and mature MIMO techniques cannot be implemented without any modification. For example, the simple and well-known Alamouti coding scheme for 2×2 MIMO cannot be applied, as the self-interference influences the original decoding schemes applied in OFDM [322].

A number of papers have been published on the methods designed to overcome the problem if FBMC signals self-interference effect in the MIMO transmission. In [323], the authors propose an interesting FBMC transceiver architecture called FFT-FBMC, where IDFT at the transmitter and DFT at the receiver are performed on each subcarrier separately, in order to get rid of the interference. This new transceiver scheme allows for the application of the solutions developed for MIMO-OFDM schemes, at the expense of significantly increased computational complexity. In [295], the authors studied the joint design of the transmit- and receive beamformers for frequency-selective MIMO channels. This is done on per-subcarrier basis and applies multiple-tap filtering structure. Two approaches are proposed. The first approach aims at maximization of the signal to leakage-and-noise power ratio and applies the precoder based on multitap filtering and single-tap equalizers. The second approach, in which the goal is to maximize the signal to interference-and-noise power ratio, applies one-tap filtering at the transmitter and the multitap equalizers at the receiver. In [324], the authors propose a solution to reduce the channel dispersion based on applied linear precoding, in the context of the MIMO transmission over the broadband frequency-selective fading channels. The precoder and the corresponding equalizer are defined based on the polynomial singular-value decomposition. Another approach for the per-subcarrier precoder design has been proposed in [325], where the idea of frequency sampling has been utilized.

The application of MIMO-FBMC systems for multiuser scenario has been discussed in [326], where the goal was to overcome the influence of the channel frequency selectivity by the proper iterative design of beamformers. In particular, the precoder is designed in such a way that, at each user terminal, only a real-valued single-tap spatial filter is applied, and the effect of multiuser, inter-symbol, and intercarrier interference is mitigated. In [327], the multiuser case has been discussed in the context of the appropriate design of Zero Forcing (ZF)-based precoder and equalizer. The problem of effective equalization in the MIMO-FBMC systems has also been analyzed by the authors of [328]. In particular, the single-tap per-subcarrier precoders and equalizers have been proposed for the case, in which the channel is considered as *not-flat-fading* at the subcarrier level. Finally, an interesting early comparison of OFDM- and FBMC-based MIMO is provided in [310] and [329], whereas recent approaches and achievements in designing the FBMC-MIMO systems can be found in an excellent survey [307].

5.5 Filter-bank-Based Multicarrier Systems Revisited

We have mentioned earlier that FBMC signals can be understood as a specific case (a subclass) of the GMC waveforms. This is because the GMC transceiver structure includes filter banks for the signal synthesis and analysis, as in FBMC. In fact, the generalized description of multicarrier signals allows for inclusion of any signal waveform that has been mentioned in our classification in Chapter 2. The FBMC modulation format, and in particular its version known as OFDM/OQAM, has been widely studied. In parallel to this process, however, other waveform candidates have been proposed, which try to solve the problems identified during investigations carried on FBMC. Therefore, let us now revisit the multicarrier signal classification and indicate some other than FBMC important multicarrier transmission methods incorporating filtering that are recently being proposed for their important advantages in case of application in the future 5G radio-communication systems.

OFDM and DMT

For the sake of classification, let us briefly address the issue on how the GMC system parameters describe the standard multicarrier waveforms, namely OFDM and Discrete Mutli-Tone(DMT) signals, although subcarrier filtering is not actually applied for their generation.When projecting the OFDM waveform onto GMC signals, the frequency spacing between the adjacent pulses is equal to the inverse of the pulse duration, function $g(t)$ is rectangular, and the duration of the pulse L_g in samples is equal to the IFFT order and equal to the time distance between consecutive pulses, that is, $L_g = M_\Delta = N$. Since the main difference between OFDM and Biorthogonal Frequency Division Multiplexing (BFDM) or Non-Orthogonal Frequency Division Multiplexing (NOFDM) signals lies in the orthogonality of the used pulses, one can state that OFDM is a subclass of the *orthogonal multicarrier* systems that constitutes a subclass of GMC signal set, where a set of basis functions creates a linearly independent system. In case of OFDM, $TF = 1$ before a cyclic prefix is added, and $TF < 1$ if the cyclic prefix is used to avoid ISI.

The DMT waveforms constitute a subclass of the OFDM signals. Thus, DMT can be interpreted in the context of GMC in the same way as OFDM. The main differences between DMT and OFDM are the following:

- the spectrum of the DMT signal is symmetric referring to subcarrier of frequency equal to zero. Thus, the time-domain signal is real-valued.

- since the DMT modulation is usually used in the wired systems (e.g., Asymmetric Digital Subscriber Line (ADSL) [60]), the channel is usually assumed to be known (quasi-static) at the transmitter, which is not the case in the wireless OFDM transmission. This feature is, however, not strictly related with DMT modulation itself.

Filtered OFDM

In principle, filtered OFDM [287] differs only slightly from the classical OFDM. In the latter case, the rectangular pulse shape is applied for data transmission, which, in practice, in the context of digital systems, is equivalent to the lack of additional filtering. As a consequence, due to the well-known time–frequency relation between the rectangular shape of the pulse in OFDM and its *Sinc*-shaped frequency spectrum, high OOB-power emission is observed. The highest sidelobe of the pulse spectrum is just 13 dB lower than the main lobe, which results in high power leakage. In the contemporary applied communication systems, additional filtering is applied to shape the transmit signal spectrum, for example, instead of rectangular pulse, the square-root-raised-cosine pulse is used. If we generalize this approach, one can assume, as in [287], that in filtered OFDM, the dedicated filter is applied to the transmit signal at the output of the modulator with the purpose of the whole-signal spectrum shaping and of minimization of the OOB emission. What is important here is that unlike in FBMC systems described earlier in this chapter, the subcarriers spectra are not shaped individually, but the whole signal spectrum is shaped by the transmit filter.

FMT

In the so-called FMT systems, the transmit signal can be defined in a very similar way as it is done for GMC signal [271, 301, 330]:

$$s[k] = \sum_{m} \sum_{n=0}^{N-1} d_{n,m} g[k - mM_{\Delta}] \exp\left\{ \frac{j2\pi nk(1 + \alpha)}{M_{\Delta}} \right\}, \qquad (5.14)$$

where α is a modulation parameter. The pulses used for shaping the data symbols $d_{n,m}$ are defined in such a way that the subcarriers are quasi-orthogonal, that is, although the orthogonality condition is not met, the subcarrier spectra overlap to a negligible degree on the frequency axis. Furthermore, functions $g[k - mM_{\Delta}]$ (or in continuous form $g(t - mM_{\Delta}\Delta t)$) may overlap in the time domain. Usually, the square-root-raised-cosine Nyquist filter with some *roll-of* parameter α is used. The utilization of various pulses in FMT transmission has been discussed in [271]. It is worth mentioning that the usage of cyclic prefix is not necessary in the case of FMT systems [301]. In the context of the GMC set of waveforms, signals with the FMT modulation can be treated as quasi-orthogonal.

Cosine-Modulated Multitone Signaling

In the Cosine Modulated Multi-Tone (CMT), the parallel streams of the PAM data symbols are transmitted through a set of Vestigial Side-Band (VSB) subcarrier channels. It is assumed that the subcarriers are minimally spaced to maximize the bandwidth efficiency of the system [89, 331]. The CMT modulator can be efficiently realized by means of the filter banks [332].

Lattice OFDM

The idea of changing the lattice structure in order to increase the system capacity has been proposed in [244]. By changing the lattice structure, one is able to adapt the pulse shape to the actual parameters of the scattering functions (defining the transmission channel). Moreover, it is also possible to adapt (by means of specific precoding) the hexagonal lattice to the rectangular one while still achieving the same spectral efficiency.

Universal Filtered Multicarrier (UFMC)

The Universal Filtered Multicarrier (UFMC) transmission scheme discussed in [333, 334] is one of the natural solutions to the problem of complexity of per-subcarrier filtering applied in FBMC modulators. As discussed earlier, in FBMC, each subcarrier is filtered separately by the means of dedicated filter (e.g., PHYDYAS, IOTA, etc.).such an approach results in increased computational burden when compared to the traditional OFDM scheme or its filtered version (filtered-OFDM, [287]). In the UFMC transmitter, subcarriers are grouped in subsets (e.g., of a size of one physical Basic Resource Block (BRB) as used in the LTE system), and these subsets of subcarriers are subject to filtering. This can lead to simplification of the transceiver structure because a lower number of filters is required, that is, equal to the number of the considered subcarrier subsets (or BRB) instead of the total number of subcarriers N. In terms of computational complexity, the UFMC approach can be then treated as a mid-solution between the two extrema, either filtered OFDM or FBMC.

The natural consequence of the per-subcarrier filtering in FBMC is the requirement placed on the high selectivity of the frequency response of the applied filters. As the subcarrier spectra cannot overlap too much in the frequency domain (these are rather narrow and are comparable to the subcarrier bandwidth), the impulse responses of the filters are naturally long. Recall that in practice, for example, in case of the PHYDYAS filter, the overlapping factor is set to 2 or even up to 4, which means that the applied pulse overlaps some of the (2–4) consecutive pulses. This may be highly problematic in burst transmission, where long ramp-up or ramp-down periods are required [335]. The UFMC waveforms are well-tailored to the communication systems with small bursts, as in the machine-type communication [46], as per-subband filtering scheme results in shorter time-domain impulse responses of the applied filters.

If we assume that the whole considered band is divided into N_S subsets consisting of N' carriers, and for each subband, we apply the IFFT operation of size N' and the subband filtering using filters of the length L_B, the UFMC transmit signal can be represented as

$$s(t) = \sum_{i=0}^{N_S-1} s_i(t) = \sum_{i=0}^{N_S-1} \sum_{n=0}^{N'-1} \sum_m d_{i,n,m} g_i(t - mT) e^{\frac{j2\pi(t-mT)n}{N'}} e^{\frac{j2\pi(t-mT)i}{N_S}}, \qquad (5.15)$$

where $d_{i,n,m}$ is the data symbol transmitted in the i-th subcarrier subset (or BRB) on the n-th subcarrier in the m-th symbol period. The high-level structure of the UFMC transmitter and receiver is shown in Figure 5.13.

Generalized Frequency-Division Multiplexing, GFDM

Another extension of the standard OFDM transmission has been called the Generalized Frequency Division Multiplexing (GFDM)[44–48] and is assumed

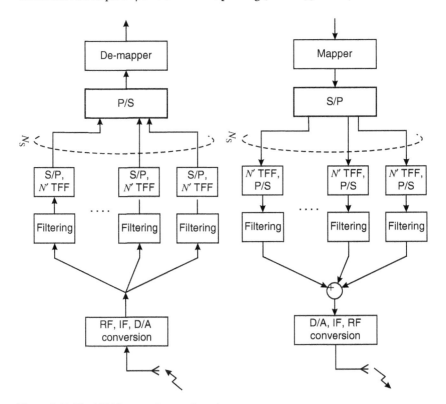

Figure 5.13 The UFMC transmitter and receiver structure.

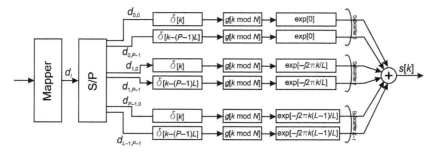

Figure 5.14 GFDM transmitter structure.

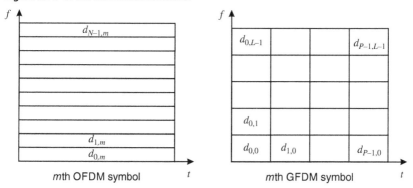

mth OFDM symbol t

mth GFDM symbol t

Figure 5.15 Concept illustration—OFDM and GFDM symbols on time–frequency plane.

to be a flexible version of OFDM, where the subcarriers are not orthogonal to each other. In principle, GFDM bases on modulation of a set of LP independent blocks, each consisting of L subcarriers and P subsymbols. Each subcarrier is filtered by the circularly shifted time and frequency responses of the prototype filter, allowing for the so-called tail-biting operation [44]. On the one hand, the subcarrier filtering results in reduction of the OOB power emission, and on the other, it results in the loss of the orthogonality between the subcarriers and in both ISI and ICI increase. Following [45], the block diagram of the GFDM transmitter is shown in Figure 5.14. One may observe that the block of $N = LP$ data symbols is split into L subcarriers and P subsymbols, meaning that the data symbol $d_{p,l}$ is carried on the l-th subcarrier and p-th subsymbol. The data transmitted on each subcarrier are then subject to filtering with the prototype filter $g[k]$. The output signal from the modulator is the summation over all subcarriers, and the output frame of LP samples is extended with the cyclic prefix. Following the notation used in the mentioned figure, the time-domain signal can be represented as

$$s[k] = \sum_{p=0}^{P-1} \sum_{l=0}^{L-1} d_{l,p} \cdot g_{l,p}[(k - pL) \mod N] e^{-j2\pi \frac{lk}{L}}. \tag{5.16}$$

In order to better understand the difference between the OFDM and GFDM signals, the time–frequency representation of one transmitted symbol has been presented in Figure 5.15. One may observe the set of N parallel subcarriers in the m-th OFDM symbol covering the whole available transmission bandwidth. At the same time, the time–frequency plane has been split into smaller areas.

5.6 Summary

The importance of filter-bank-based multicarrier systems has significantly increased in the recent years. The application of per-subcarrier or per subband filtering allows for substantial reduction of unwanted emission out of the nominal band of the transmit signal, at the expense of increased computational complexity. This computational complexity, however, is usually scalable with the overlapping factor K. As a consequence, the FBMC systems are well-tailored for application in future wireless communication systems where the coexistence of various types of systems (and their waveform spectra in the frequency domain) is envisioned. This observation is particularly important in the context of advanced spectrum-sharing strategies, and introduction of noncontiguous transmission schemes, where stringent and rigorous requirements on the interference induced to outer frequency bands need to be fulfilled. The issues of spectrum aggregation and sharing using this kind of multicarrier waveforms are addressed in the next chapter. Of course, besides their advantages, FBMC systems have their drawbacks, such as long tails influencing burst transmission or the signal self-interference, which makes it difficult to be applied in the MIMO systems. However, a lot of research effort has been put to overcome these shortcomings and to utilize the FBMC waveforms in the future radio communication.

6

Multicarrier Technologies for Flexible Spectrum Usage

6.1 Cognitive Radio

The Cognitive Radio (CR) concept has been introduced by Mitola and Maguire [336] and Mitola [337] and further described in [338]. Since the publication of his PhD thesis in the year 2000, this concept has been intensively researched as an attractive technology for radio communication systems. The list of comprehensive publications and *best readings* on CR include the collection of books [338–343] and overview articles [74, 336, 344–346]. Together with the development of the programmable radio platforms, novel algorithms, and regulations, the CR technologies are being introduced in practical systems to address the goals of modern radio communication, such as environmental awareness, flexibility in adapting to the radio environment, and operational effectiveness.

The CR operation bases on the so-called cognitive cycle presented in Figure 6.1, defined in [337], which includes its basic functions and actions, namely to *observe* (the environment), *orient, plan, decide, act*, and *learn*. The radio environment is observed by CR using various sensors and measurements of the external stimuli in order to obtain enough data of the operational context. Then, these gathered data are analyzed in order to extract useful information of the context. For example, CR may analyze the Global Positioning System (GPS) coordinates, mobility parameters, temperature, and light to determine indoor or outdoor location and then use this information for choosing the most favorable radio interface and waveform to communicate with other terminals. Another typical example of this phase of the cognitive cycle is the frequency analysis for the purpose of available frequency band detection. The result of this analysis and related conclusions is context-understanding (*orientation* in the context). In the orientation phase, priorities of further operation are set, which trigger normal, urgent, or immediate type of action.

In the normal mode, the cognitive cycle includes planning phase, in which alternatives for operation are being generated and evaluated. These alternatives may include possible radio interfaces, operational frequency bands, radio access techniques, power levels, signal reception methods, and so on.

Advanced Multicarrier Technologies for Future Radio Communication: 5G and Beyond, First Edition.
Hanna Bogucka, Adrian Kliks, and Paweł Kryszkiewicz.

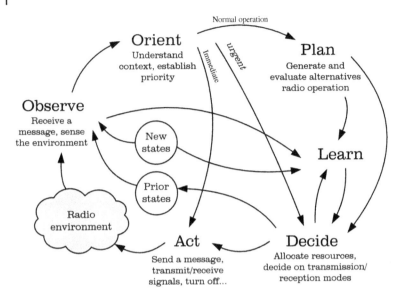

Figure 6.1 Cognitive cycle. (Based on [337].)

Evaluation of the alternatives impacts the states (of knowledge) in the learning process, which also takes former decisions and their observed results into account. The last process in the cycle is to *act*, that is, to activate appropriate hardware and software resources to implement the actual communication: generation, transmission, and reception of signals in a given frequency band, time period, and with a given power level.

The general idea of CR is commonly narrowed and referred to the radio network (and nodes) capable of dynamic and flexible use of the spectrum resources. Cognition of CR here means operational-context awareness and taking intelligent decisions (optimized in some sense) based on learning capability on accessing the spectrum and satisfying dynamically changing Spectrum Emission Masks (SEMs). Thus, CR is considered as the potential solution to increase the efficiency of spectrum utilization, as it enables opportunistic access of temporarily unused frequency bands once the presence of the incumbent (licensed or primary) system transmissions in these bands is excluded. In order to determine, with high probability, whether the considered frequency band is occupied or available, the CR nodes should either have the access to the appropriate data-base or implement real-time measurement of the Primary Users (PUs) activity, that is, to sense the spectrum resources. The spectrum-sensing methods have been intensively studied in the past decade. Some excellent examples of overview papers regarding these methods are [347–350]. Reliability of the spectrum sensing is of particular importance. This is measured by the probability of detection and the probability of false alarm.

In general, it is beneficial to maximize the detection probability and minimize the false alarm probability. However, the optimizations of these two metrics are contradictory goals, that is, usually an increase of the detection probability leads to a higher number of false alarms and consequently to an increase of the probability of false alarm. In order to improve both, cooperative spectrum sensing is considered, instead of often unreliable, autonomous sensing. Still, there is a number of other trade-offs in spectrum sensing, such as energy consumption versus reliability, accuracy versus granularity of reported sensing information, sensing time versus transmission time, and so on. These trade-offs are summarized in [351]. There, cooperative spectrum-sensing methods are surveyed, and the energy consumption of these methods is discussed.

Intelligent decision-making is another key paradigm of CR. It usually refers to the selection of one of the possible actions taking all the constraints into account, which relate to either internal state or to the external conditions. In the case of the considered dynamic and flexible use of radio resources, these external conditions encompass regulations related to the Dynamic Spectrum Access (DSA), licensed systems protection, rules of spectrum sharing, and so on. Moreover, users Quality of Service (QoS) and Quality of Experience (QoE) expectations, parameters of the propagation environment, coexisting-systems standards, characteristics of the networks, and the nodes constitute mentioned decision-making constraints. Thus, one group of the methods in CR is constraint optimization. However, optimization often requires some a priori knowledge, which due to the nature of the radio environment and limitations of the control traffic may be not complete or not accurate or outdated [352].

Optimization theory deals with the problem of finding the best solution given the constraints. In case, when the goals and the constraints are not fixed and the decision outcome depends on decisions of the other decision-makers (the players), such as other CR nodes in the network, it may be desirable to consider methods of game theory for the efficient decision-making. In game theory, each player when taking his or her decision must take possible choices (decisions in the so-called strategy space) of the other players into account. The key assumption here is that the players are rational (care for their own benefit only) and that they are intelligent enough to analyze the situation in the game. Analysis of the players strategies may lead to finding the equilibrium (if it exists), which is a set of choices that all rational players have to take in order for them to not deteriorate their payoff (outcome of the game). The key difference in using the game theory as opposed to using the optimization theory is that the solutions found by the former may not be optimal. The Pareto-optimal outcome of a game is possible when cooperation between the players is allowed. The literature regarding game theory is vast. Example classic works containing game-theoretic analysis are We should also mention works by the pioneers of game theory: John von Neumann and John Nash, whose doctoral thesis on non-cooperative games and concept of Nash equilibrium revolutionized economics.

Application of game theory for decision-making in CR networks has been recently intensively researched. Examples of these applications include algorithms for power control, DSA methods, and radio-resource auctions. However, making the assumption on the complete information or cooperation possibility is not always possible; therefore, considerations on the allowable simplification of the game-theoretic model of the considered problem (e.g., partial-information exchange or limited cooperation) are very important direction in research.

The CR nodes are expected to operate in a network. For the network of CR nodes, we can consider a number of scenarios depending on the existence of the protected PU of the licensed systems and the network coordination. The classification of these scenarios is presented in Figure 6.2 [394].

Note that the two scenarios presented in the upper part (as opposed to the scenarios in the lower part of the figure) do not assume the occurrence of any licensed systems. The two scenarios on the left (as opposed to the scenarios

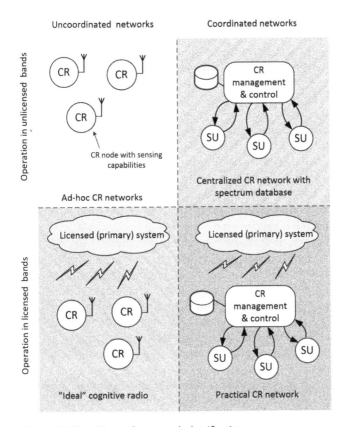

Figure 6.2 Cognitive radio network classification.

presented on the right) do not assume any coordination of the CR nodes. Thus, the network in the lower left corner may be viewed as the *ideal* CR network, in which the CR nodes are fully independent, that is, they sense the spectrum availability independently and use the radio resources in a distributed manner [394]. At the same time, they still are able to protect the licensed systems and handle interference between themselves. Naturally, such an idealistic vision on CR network operation is not yet introduced in practice. This is because of the shortcomings of the distributed Radio Resource Management (RRM) approaches and usually low quality of detection for the autonomous sensing schemes, when no information about the PU transmission is available.

From the standpoint of future Fifth Generation (5G) communication with cognitive capabilities, the most practical CR network scenario seems to be the one presented in the lower right corner of Figure 6.2. In this case, there is some coordination of the Secondary Users (SU) in the network, which may sense the radio environment but also have access (through the central control entity) to the database, for example, the knowledge base containing and continuously updating the information about radio resource availability. (These kind of knowledge bases are often called Radio Environment Map (REM).) Moreover, the centralized entity coordinates the spectrum sharing between PU and SU and between SU themselves. Thus, the quality and efficiency of the CR transmission, as well as protection of the licensed systems, in this network category, may be guaranteed.

6.2 Spectrum Sharing and Licensing Schemes

The DSA refers to making an opportunistic use of the spectrum resources. According to the principles of CR, the SU can use the radio resources (frequency, time, and power) at a certain geographical location and transmit their signals only if the PU transmission is protected from interference generated by SU. This calls for the application of some particular rules and policies of spectrum sharing, defining the ways and conditions on how the resources could be shared between PU and SU.

The theoretical spectrum-sharing options are presented in Figure 6.3 [394]. The first, most popular scheme (visible on the top of the figure), referred to as *interwave* spectrum sharing, allows SU to use the so-called *spectrum holes*, which are the frequency bands unused (sometimes temporarily) by the primary systems. The interwave spectrum-sharing approach is the one used in the context of secondary multicarrier systems, which typically use subcarriers located in the spectrum holes on the frequency axis. Another scheme, often considered in the literature, is the *underlay* spectrum-sharing scheme. There, standard spread-spectrum techniques can be used for this purpose, as well as the ultra-wideband transmission schemes, the techniques that distribute the

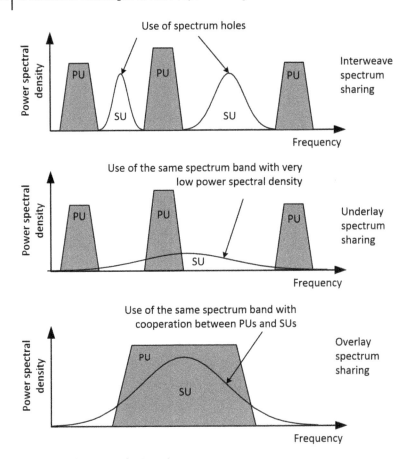

Figure 6.3 Spectrum-sharing schemes.

signal power over a wide frequency band, resulting in relatively low Power Spectral Density (PSD) level. This low PSD level is intended to be below the allowable interference level for the PU systems. The issue in both the interwave and underlay spectrum-sharing approaches is that the power of interference generated to the primary systems increases with the number of SU. For this reason, this number and the generated interference have to be coordinated. Finally, in the last approach presented at the bottom of Figure 6.3 and called *overlay* spectrum sharing, SU can use the same frequency band at the same time as PU. This, however, is not the technology-"agnostic" solution and requires some interference-cancellation schemes, possibly orthogonality of the PU and SU transmitted signals (e.g., through the use of coding) and cooperation between the PU and SU transceivers.

As mentioned earlier, spectrum sharing requires some policies and regulations that make the whole process effective for both the primary and the secondary systems. Let us briefly review these options and discuss their applicability to future CR multicarrier technologies.

6.2.1 Exclusive Use of Spectrum

In the classic, traditional case, the radio frequency bands are statically allocated to various players (such as network operators) for exempt use. Thus, a certain operator is authorized to utilize a given frequency band based on the individual licenses. Such a spectrum division is made usually at the national (country) level; however, international agreements (such as those made at the World Radiocommunication Conference (WRC) level) need to be fulfilled. Following these decisions, each country creates their own frequency allocation plan and defines the rules, how the particular frequency bands can be utilized (e.g., the maximum transmit power is defined, types of services that can be delivered to the end user), trying to guarantee high level of electromagnetic compatibility between the services and systems. On the one hand, such an approach can be treated as a way for efficient interference management between various systems, as each frequency spectrum band has its own strict transmission rules, the fulfillment of which must be ensured by the service provider. On the other hand, such static spectrum assignment can cause high spectrum underutilization, as the frequency bands that are not used at a certain time and geographical location cannot be used by other operators/service providers due to the lack of license. In this context, the application of various multicarrier schemes discussed in this book is possible as long as the transmission requirements defined by regulators are fulfilled (i.e., the level of the out-of-band emission).

6.2.2 License Exempt Rules

As opposed to the exclusive spectrum use, the approach known as *spectrum commons* or *unlicensed spectrum* is used in some frequency bands, which are exempted from any kind of licensing. In this case, each user is allowed to access a dedicated spectrum band with no additional permissions. The Industry–Science–Medicine (ISM) bands are the good examples illustrating such spectrum use opportunities, where various systems can coexist in the same geographical area (such as WiFi and Bluetooth operating in the 2.4 GHz band). The benefits of such an approach are evident – each interested player can start utilizing the spectrum resources as needed. However, as the main advantage of the licensed spectrum assignment lays in the strict interference management, the license-exempt spectrum bands are said to be "interference-limited". Indeed, when no interference control mechanisms are applied and anyone can start offering communication services without any agreements with the other users, system blockades may occur. Various solutions to this problem have been proposed. However, with the continuously increasing traffic, the fully unlicensed spectrum band would need to face the problem of the "tragedy of commons" described in [353]. In such a case, the vacant and unlicensed band will not be used at a particular moment due to the permanent collisions in accessing the spectrum or due to extreme levels of in-band interferences.

As in the previous case, also in this context, the application of the multi-carrier schemes discussed previously in this book is possible as long as the transmission requirements defined by regulators are fulfilled. In fact, a number of Wireless Local Area Network (WLAN) standards based on Orthogonal Frequency-Division Multiplexing (OFDM) operate in the ISM bands. However, the opportunities of adaptation of the transmit signal parameters (such as the number of occupied subcarriers in the multicarrier transmissions with noncontiguous subcarriers, pulse shape in Generalized Multicarrier (GMC),) create new degrees of freedom for advanced interference management in the scenarios where various system coexist.

6.2.3 Licensed Shared Access and Authorized Shared Access

In its principle, Licensed Shared Access (LSA) is the response to the industry interests in the new ways of utilization of the currently occupied band (assigned to the incumbent systems) by additional licensed users. The key idea lays in signing an individual license agreement between interested stakeholders (i.e., the incumbent spectrum "owner" and the new spectrum entrant) where the ways for spectrum sharing are defined. A special case of LSA is the so-called Authorized Shared Access (ASA) where the spectrum of interest covers bands allocated to the mobile services and identified for International Mobile Telecommunication (IMT) by the International Telecommunication Union Radiocommunication Sector (ITU-R) at WRC but currently encompassing other type of primary use [354].

Application of new and flexible multicarrier systems seems to be well tailored to this scenario, as the possibility of adaptation of the transmit parameters (such as Out-of-Band (OOB) power, or application of noncontiguous subcarriers) enables potentially easier fulfillment of signed coexistence requirements and obligations. These requirements will be defined in advance; thus, it will be possible to design the transmit signal in such a way to optimize certain parameters of the system, important for a given application.

6.2.4 Citizen Broadband Radio Service and Spectrum Access System

The Federal Communications Commission (FCC) has recently introduced new three-tier sharing model dedicated to bands in the range of 3550–3700 MHz, which is commonly referred to as CBRS sharing model [355, 356]. Three types of users have been introduced here, mainly, the tier of high-priority users (in fact, incumbents, whose transmissions have to be fully protected), the tier of additional licensed users called Priority Access Licenses (PALs) and the license-exempt General Authorized Access (GAA) users. The PAL users correspond to the LSA licensees in the LSA/ASA model, as the singed agreement has to be created. One may observe that in Citizen Broadband Radio Service (CBRS) spectrum sharing, the combination of licensed and license-exempt approaches coexist. Let us note that, in order to efficiently manage the interference between the users, a dedicated Spectrum Access System (SAS) has to be created to coordinate the spectrum usage. Within this system, the so-called Environmental Sensing Capability is envisaged to monitor the activity of the users in a certain area.

Similarly as in the LSA case, also in CBRS spectrum sharing, the application of flexible multicarrier systems seems to be an attractive solution due to the same reasons, that is, the wide opportunities in the definition of transmit signal format and parameters.

6.2.5 Pluralistic Licensing

The main idea of Pluralistic Licensing was proposed in [357] as an innovative approach to spectrum sharing between the primary and secondary players (operators). Let us quote the main definition: pluralistic licensing concept can be understood as "the award of licenses under the assumption that opportunistic secondary spectrum access will be allowed, and that interference may be caused to the primary with parameters and rules that are known to the primary at the point of obtaining the license" [357]. In this kind of spectrum-sharing policy, it is assumed that the primary operator will choose from a range of offered pluralistic licenses, each with a different fee structure and each specifying alternative opportunistic access rules, which can be mapped to associated interference characteristics [357, 358]. As the core license is still granted to the primary operator, the main control mechanism is kept by the primary operator. Thus, the primary operator may trade off the form and degree of opportunistic access for various licensing fees or another incentive.

6.2.6 Licensed Assisted Access

Recently, particular attention has been focused on the new ways of exploitation of unlicensed frequency band located around 5 GHz. The key industrial players consider the application of the Long Term Evolution (LTE) technology in the unlicensed bands, leading to the development of the so-called Long Term Evolution-Unlicensed (LTE-U) solution and establishment of the dedicated LTE-U forum [359]. Operation in an unlicensed band entails the need for fair coexistence of WiFi users with LTE-U. One of the approaches is to dynamically select *clear* channels in order to avoid the presence (and interference) of Wireless Fidelity (WiFi) transmission. However, as the presence of vacant WiFi channels cannot be guaranteed, one of the requirements related to the way of spectrum access is the well-known Clear Channel Assessment (CCA) technique or Listen-Before-Talk (LBT) approach, typical for unlicensed spectrum use. From the WiFi networks' perspective, the LTE-U transmission will look like another WiFi client.

The idea of Licensed Assisted Access (LAA), introduced in Rel. 13 of the LTE standard, assumes that the unlicensed band can be used alongside with the licensed one. As various scenarios can be considered here, one important example would be to transmit all control data using licensed band and support user data transmission with the aggregation of licensed and unlicensed spectrum. By assumption, if the traffic observed will be low enough to be managed by means of licensed band connections, the unlicensed bands will be released. Typically, LAA is considered mainly in the context of supporting the downlink traffic by means of utilization of unlicensed spectrum bands. A number of interesting papers have been published on this type of spectrum sharing, in the recent years, for example, [360–362]. Various modifications and improvements of this concept have also been proposed; let us just mention the idea of LTE-WiFi aggregation (known as LWA), where the operator decides to use already deployed WiFi network for data transmission, or the idea of enhanced LAA (eLAA), where spectrum aggregation will be considered also for the uplink connectivity. It is envisaged that the updated version of LAA will be included in the next LTE standard releases.

6.2.7 Co-Primary Shared Access

The concept of Co-primary Shared Access (CSA) bases on the assumption that multiple operators (with the same privileges) decide to jointly use a fragment of their licensed spectrum [363, 364]. Two CSA cases can be defined: mutual renting (MR), where operators keep their individual licenses to use spectrum, but they can mutually rent part of the spectrum based on prior requests; and limited spectrum pool (LSP), in which case, the dedicated fragment is commonly used by all operators based on the group licenses. The former case can be considered as a specific form of LSA between limited set of players, whereas in the latter, various forms of cooperation can be envisaged. In the so-called orthogonal sharing, common frequency resources can be shared in the time-, frequency-, or spatial-division multiple access modes (TDMA, FDMA, or SDMA), while the application of nonorthogonal approach entails simultaneous usage of resources causing some interoperator interference.

6.3 Dynamic Spectrum Access Based on Multicarrier Technologies

The DSA and flexible and efficient spectrum allocation procedures are considered as measures to increase the utilization of the spectrum resources in future wireless communication networks. Apart from the spectral efficiency, the QoE and the associated fairness in resources distribution are the focus of research on the cognitive DSA. The spectrum allocation procedures are often centralized and require either the detailed Channel State Information (CSI) (usually the link-channel coefficients) or at least Channel Quality Indicator (CQI) (a representative compact information on the channel quality) of all links in the network. Moreover, centralized methods involve the overhead traffic, which in turn occupies the radio resources. For the future communication concepts, the CR nodes are expected to take intelligent decisions at least partly in a distributed way, thus minimizing or eliminating the overhead traffic. An example of the multiple access technique that could be considered as the DSA method for multicarrier transmission is the well-known Orthogonal Frequency-Division Multiple Access (OFDMA). In the dynamic and opportunistic OFDMA, the network nodes are able to adopt a subset of accessible subcarriers individually, as well as the transmission rate and power allocated to these subcarriers [365].

Both centralized and distributed (cognitive) spectrum allocation procedures have been presented in the literature, where game theory is applied for the intelligent, rational decision-making in the competitive environment of the CR network. The centralized schemes allow for more efficient or more fair spectrum utilization; however, they require centralized management and possibly a considerable amount of control traffic. The CSI or CQI has to be exchanged or to be made available at a central unit every time the channels' qualities change for the nodes in the network area. Interesting solutions and results for centralized solutions based on cooperative complete information game models for opportunistic OFDMA have been presented in [366–368]. Distributed decision-making algorithms employ noncooperative game models with the dominant concept of *equilibrium* as the solution of the conflict between competitors. However, the complete-information games cannot be considered for the practical spectrum allocation and DSA, because the complete knowledge of all dynamically changing links (and related CSI or CQI) is not usually available for every player (CR node). Thus, noncooperative complete-information game models are only suitable for multicell environment, where the players are connected base stations, knowing the CSI of all links in their cell areas [369, 370].

6.3.1 DSA Based on Spectrum Pricing

The remedy for the incomplete information in the distributed spectrum allocation in the DSA is to incorporate mechanisms that require only the limited information, for example, the resource pricing mechanism. Resource pricing has been considered in the literature extensively for power allocation, also in OFDMA [371–373]. In these works, the resource that is subject to pricing is the power used by the network nodes, and the goal is to maximize the sum throughput given the total allowable transmission power in the network. However, the concept of the spectrum pricing for the purpose of DSA is different [374].

Distributed dynamic allocation of OFDMA resources (subcarriers) based on pricing in a multicell scenario, where the base stations act as players, has been considered in [375–377]. However, again, the pricing concepts developed for the multicell scenario cannot be considered for the DSA in decentralized networks. Another approach to the price-based DSA is based on iterative water filling, which allows all players to use the same frequency channels and adjust their power levels in these channels based on the pricing function. In this approach, definition of the pricing function for each player requires the information exchange between the neighboring nodes (players). In [378], the spectrum allocation model is the Bertrand-game-based oligopoly competition between PU and SU, which again requires the knowledge of the SU CSI by the PU players. Although the aforementioned works have contributed to significant advance in the game-theoretic price-based models, they all make an assumption on the complete information available for all players, or they narrow the considerations to the subcarrier power allocation.

In [379], a generalized framework for the pricing-based allocation of orthogonal frequency channels, as in OFDMA using noncontiguous subcarriers, is presented. The scheme requires only the limited knowledge of the channel quality (the effective Signal-to-Noise Ratio (SNR) as CQI). The *selfish* and *social behavior* of the players is considered, which is reflected in the utility functions. These utility functions include the linear-taxation (pricing) component dependent on the acquired spectrum resources (subcarriers). The nodes are able to sense the radio environment, detect the parameter called *tax rate* (or the pricing parameter) available in a given area, detect the available spectrum resources, and acquire a subset of these resources (subcarriers) usable for their intended transmission. The goal of each node is to make the best use of these resources, that is, to obtain high data rate at the lowest cost [379]. A selfish player aiming at throughput maximization would occupy the whole available spectrum; however, the cost arising from spectrum pricing prevents such a behavior. Too-low pricing parameter (taxation rate) may cause greedy behavior, that is, for the individual CR nodes to occupy maximal number of subcarriers, while too high pricing can be so restrictive, it does not pay for any node to access the spectrum. Thus, optimal pricing parameter should be determined to maximize the sum throughput or other defined goal of the CR network.

Example Results

In the considered scenario, the total transmission power is fixed. The average considered SNR has been equal to 30 dB, and an example channel model is the six-path channel, with paths having the same power, and delays uniformly spread between 0 and $1/B$, where B is the considered accessed bandwidth. (This is a test-channel model often used for the testing of equalizers that reflects particularly hostile environment with very small coherence bandwidth and very deep fading.) Moreover, the uncoded-system target Bit Error Probability (BEP) has been assumed to be equal to 10^{-3} for all links.

Example results of the spectrum-pricing game defined in [379] are shown in Figure 6.4. On the left-hand side of the figure, optimal pricing parameter (maximizing the sum throughput in the network) is plotted against the number of players. On the right, the resulting sum throughput is presented. For comparison, the throughput-per-node results are also presented for the centralized algorithms: the greedy maxSNR and the Round-Robin algorithm.

6.3.2 DSA Based on Coopetition

As mentioned earlier, DSA in multicarrier systems (mainly in OFDM networks) is often considered as fully centralized dynamic resource allocation of subcarriers. Distributed OFDMA-based CR networks using the idea of spectrum pricing have also been considered. Another distributed approach improving the results of the applied oligopoly models is based on a methodology called *coopetition*. Coopetition (cooperative competition) is a neologism reflecting a combination of competition and cooperation [380]. Coopetition bases on the creation of an added value in cooperation where its distribution is an element of competition. It has been successfully applied in economics, cybernetics, complex production, and logistic systems. In [381], a generic framework for DSA and resource allocation in distributed cognitive radio networks has been presented. The focus has been on OFDMA. There, the competition phase applying the Cournot oligopoly model is followed by the cooperation phase defined by the coalition game. Due to the flexibility in the definition of different modes at each stage of coopetition, the considered CR network nodes may access the spectrum dynamically under different policies, supporting hierarchical traffic, fairness, and efficiency of resource utilization.

6.4 Dynamic Spectrum Aggregation

Concerning the capacity and spectrum usage enhancement in the 5G systems, it will result from the so-called network densification, that is, a combination of increased number of nodes and antennas and spectrum aggregation [1]. Spectrum aggregation refers to making use of possibly discontinuous frequency bands and, thus, larger amounts of electromagnetic spectrum. A technique called Carrier Aggregation (CA) applied in the Long Term Evolution Advanced (LTE-A) standard is the practical instantiation of this idea to achieve the throughput of 1 Gbps in the downlink for Fourth Generation (4G) systems in a 20 MHz channel. It allows for aggregation of the so-called Basic Resource

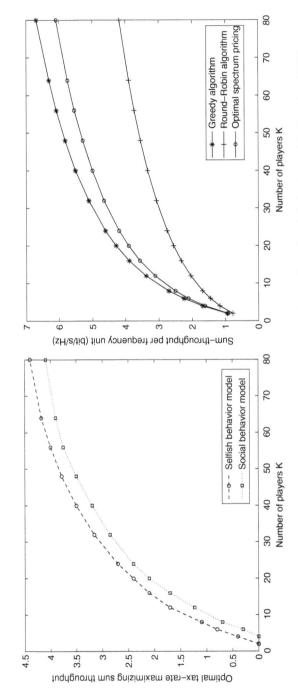

Figure 6.4 Optimal resource pricing for sum-throughput maximization in the selfish- and social node-behavior models. (Based on [379].)

Blocks (BRBs) even in different frequency bands [2]. This aggregation is handled at the Medium Access Control (MAC) layer, and each component carrier uses its own Physical Layer (PHY) protocol and the Hybrid Automatic Repeat Request (HARQ) mechanism. Following the idea of contiguous or noncontiguous OFDM, using adjacent or disjoint subcarriers, two types of CA approaches have been proposed [42]. In the contiguous CA approach, the used multiple carriers are adjacent to each other, which enables the usage of a single Inverse Fast Fourier Transform (IFFT) and a single Fast Fourier Transform (FFT) block at the transmitter and at the receiver, respectively. Moreover, a single Radio Frequency (RF) front end can be used at both sides of a transmission link. In case of the noncontiguous CA, the component carriers do not have to be adjacent, which has repercussions in the transceiver design, both for the Base-Band (BB) and for the RF front end.

The two methods of CA applied in LTE-A have limited flexibility in aggregating any kinds of available spectrum fragments, because only a limited and integer number of BRB can be used for a single-user data communication. Moreover, the proposed protocols do not allow for the dynamic spectrum access. To expand the LTE capacity, integration of unlicensed carriers (unlicensed spectrum) into an LTE system has been proposed, namely U-LTE. Both cases of contiguous and noncontiguous CA are presented in Figure 6.5 in the upper two example frequency arrangements. They are referred to as the *large-scale* spectrum aggregation [42, 382]. In this case, the spectrum fragments with a fixed granularity defined by the BRB are merged and can be assigned to a single user, as in an LTE-A system or in a multiple-carrier High Speed Packet Access (HSPA) system.

However, as shown in Figure 6.5 (the lower example frequency arrangement presented in that figure), spectrum aggregation can also be considered on a smaller scale and with higher flexibility [42]. In such a case, the aggregated fragments of spectrum are interleaved with the spectrum originating from the transmission of signals in another licensed system. Such a spectrum-sharing concept is typical for future cognitive systems coexistence, where one user transmitting signals within a given spectrum mask is geographically and spectrally neighboring the transmission and reception of another one. Naturally, interference between these two users depends on their OOB power emission, as well as on the selectiveness of their reception filters. This is presented in Figure 6.6. There, one can see that at the RX of one link or system (RX 1), interference can be observed originating from the TX of the other link or system (TX 2) due to the OOB power radiation of TX 2 (resulting from a given Spectrum Emission Mask (SEM) in this TX) as well as from the imperfect selectiveness of the RX 1 filter. Thus, appropriate measures have to be taken to keep the OOB power of both transmitters at the required level and to shape the signals spectra taking the coexisting-system reception-filter characteristics into account.

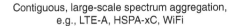

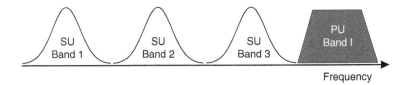

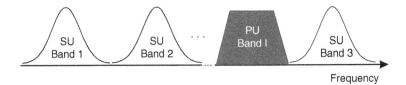

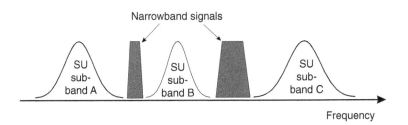

Figure 6.5 Cases of spectrum aggregation. (Based on [42].)

Let us consider the Non-contiguous Orthogonal Frequency-Division Multiplexing (NC-OFDM) and Non-contiguous Filter-Bank Multi-Carrier (NC-FBMC) techniques, presented in the previous chapters, for their potential to aggregate the spectrum in cognitive 5G communications. Both these techniques possess the ability to efficiently use fragmented spectrum opportunities and to suppress interference that may affect coexisting systems in the cognitive radio scenario and environment.

The enhanced NC-OFDM technique has been described in Chapter 3 in detail. It is characterized by relatively low computational complexity in making use of disjoint sets of OFDM subcarriers. The useful, dynamically selected subcarriers are modulated by information data symbols or by noninformation

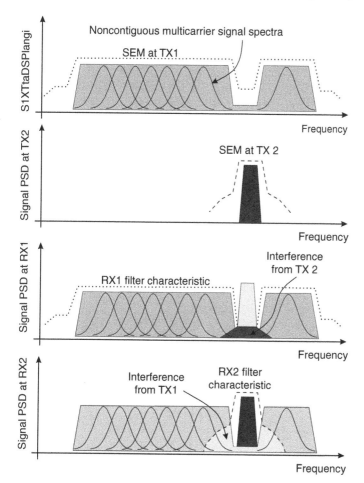

Figure 6.6 The spectral scenario of the coexisting systems and resulting interference. (Based on [42].)

(information-data dependent) values, applied to shape the NC-OFDM signal spectrum (in particular, to attenuate its OOB components). The other sub-carriers within the NC-OFDM signal band are *unused*, that is, modulated by zeroes. This flexible NC-OFDM signal spectrum shaping allows for accommodation of other transmissions in the adjacent bands or in the NC-OFDM spectrum notches. Note that these transmissions do not have to have the specific Radio Access Technology (RAT) defined and may use frequency carriers not orthogonal to the considered NC-OFDM signal, as long as their spectra do not overlap with the NC-OFDM system aggregated spectrum.

As discussed in Chapter 4, Filter-Bank Multi-Carrier (FBMC) technique is considered as potential RAT for next-generation wireless communication systems, enhancing many properties of OFDM [287]. Although various real-izations of this concept have been considered (such as complex exponentially modulated filter banks, cosine-modulated filter banks, transmultiplexers, perfect-reconstruction filter banks, oversampled filter banks, or modified Discrete Fourier Transform (DFT) filter banks), recently, the term FBMC refers to the schemes, where spectrally shaped offset-QAM (OQAM) symbols modulate orthogonal subcarriers [383]. Depending on the applied filters shaping the subband spectra, the FBMC scheme may offer high OOB-power reduction at the expense of higher complexity of the transceiver architecture. Among the pulse shapes proposed in the literature, the extended Gaussian function together with its special case – Isotropic Orthogonal Transform Algorithm (IOTA) – and the PHYDYAS pulse are the most promoted and promising [42, 333, 382]. Finally, FBMC is also considered for the use with noncontiguous sets of subcarriers and is called NC-FBMC.

The considered example NC-OFDM and NC-FBMC signal spectra are pre-sented in Figure 6.7. There, also an example coexisting PU system (signal) spec-trum is indicated to illustrate mutual interference between considered systems.

6.4.1 Complexity and Aggregation Dynamics

Energy efficiency is an important target of future cognitive 5G networks, and associated computational complexity becomes an important issue in

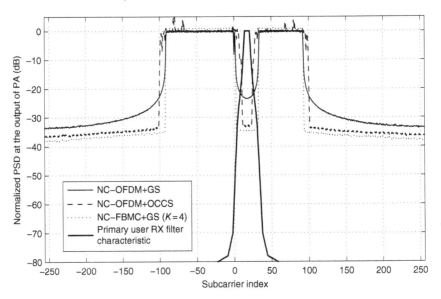

Figure 6.7 Example aggregated spectra of enhanced NC-OFDM with GS or OCCS, and of NC-FBMC system with the overlapping factor K; QPSK mapping, PA Rapp power model used, input back-off parameter IBO = 7 dB, and the smoothness factor p = 10). (Based on [42].)

the system design. The complexity of the enhanced NC-OFDM and of the NC-FBMC transceiver has been estimated in [42]. There, for the purpose of this complexity approximation, some assumptions have been made, that is, that a set of aggregated subcarriers is known both at the transmitter and at the receiver resulting from the connection parameters' configuration phase, the IFFT and FFT are implemented using the split-radix-2 algorithm, there are the Cancellation Carriers (CCs) applied in NC-OFDM, the number of coefficients of the polyphase-decomposed filters (applied in NC-FBMC) is relatively small, channel estimation is based on pilots, time and frequency synchronization is based on autocorrelation properties of the Cyclic Prefix (CP), and no additional self-interference cancellation algorithm is considered.

The computational complexity of the applied on-line spectrum aggregation methods impacts the spectrum aggregation dynamics offered to the system. Moreover, algorithms for aggregation of fragmented frequency bands, changing their positions and bandwidths on the frequency axis, and varying requirements concerning the spectrum masks (for the adjacent-channel interference protection) entail additional calculations for the design of the proper spectrum-shaping and aggregation algorithm. This calls for the estimation of the amount of off-line computations required to reshape the spectrum satisfying the upper limit associated with the dynamics of the required SEM changes. These estimates are also presented in [42]. In case of the enhanced NC-OFDM with OCCS, a redesign of the matrix used to calculate the values modulating CC requires some matrix operations and solving the scalar-based nonlinear equation (see derivations presented in Section 3.3). For this kind of spectrum shaping, the most computationally demanding operation is the matrix Singular Value Decomposition (SVD). For the case of NC-FBMC, the prototype-filter redesign may be required if the spectrum aggregation scenario demands change of the pulse shape, for example, for changed systems-coexistence parameters or allowable self-interference.

6.4.2 Transmitter Issues

As discussed in the previous chapters, the spectrum-shaping and aggregation algorithms are often designed by assuming ideal RF front ends. In practical transmitters' RF front ends however, the following effects are observable that distort the spectrum shape in the BB processing: the quantization noise in Digital-to-Analog Converters (DACs), the IQ signal components' imbalance, the LO leakage, or nonlinear distortions in a PA resulting from its nonlinear characteristics. Multicarrier techniques, such as enhanced NC-OFDM and NC-FBMC, are prone to nonlinear distortions due to high PAPR of the time-domain signals (see discussion in Section 2.2). High-amplitude peaks appearing in a signal at the output of the multicarrier modulator are clipped by nonlinear PA, resulting in energy leakage to the adjacent frequency bands, that is, the power of spectrum components in the OOB region is increased, which may naturally impact the coexisting systems. This source of the OOB-power emission is independent from the one observed in BB and resulting from the applied pulse shape. The presence of nonlinear distortions significantly limits the spectrum aggregation capabilities and calls for advanced OOB-power reduction algorithms (as in enhanced NC-OFDM) or per-subcarrier filtering (as in NC-FBMC). In Figure 6.8, the ACIR is presented versus the IBO parameter of PA for the considered noncontiguous multicarrier transmissions, whose spectra are presented in Figure 6.7. (The example victim system is narrowband, with the spectrum located in the NC-OFDM or NC-FBMC spectrum notch.) Note that ACIR is defined as the ratio of the interference power affecting the victim coexisting system receiver to the total power from the source reaching the victim receiver and is an indicator of the performance of the spectrum-shaping methods for systems coexistence using multiple adjacent frequency channels. In the mentioned figure, one can observe ACIR at the input of PA (or HPA) and at its output in case of the NC-OFDM and NC-FBMC signals. Moreover, these transmissions apply the known PAPR reduction method called C–F.

Note that in our example of systems coexisting scenario (from Figure 6.7), NC-FBMC outperforms enhanced NC-OFDM with OCCS in terms of ACIR, because apart from subcarrier filtering, the system uses some GS around the

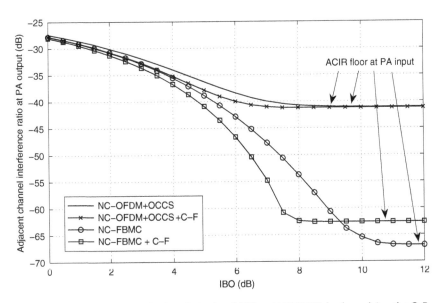

Figure 6.8 ACIR for enhanced NC-OFDM using OCCS and NC-FBMC, both applying the C–F method; the number of polyphase-filter coefficients (in NC-FBMC) $K = 4$, QPSK mapping, PA Rapp model parameters: $p = 10$, IBO = 10 dB. (Based on [42].)

coexisting (victim) system band. Moreover, it is worth noting that nonlinear effects and high ACIR are observed at low IBO and for lower-class PA. This means that there is no need for the application of sophisticated and computationally complex spectrum-shaping methods in the baseband, because their performance will be significantly reduced at the output of PA. In order to cope with this problem, joint spectrum-shaping and PAPR reduction methods should be applied, as well as algorithms for *linearization* of the input–output PA characteristics have been proposed. Their appropriate modifications for the use in NC-OFDM and NC-FBMC transmitter are also the subject of research.

6.4.3 Receiver Issues

The main receiver issues of the spectrum aggregating noncontiguous multicarrier transmission systems are as follows: synchronization in the presence of in-band interference (originating from the *in-notch* coexisting system) and reception quality. Time and frequency synchronization for noncontiguous multicarrier transmission is a challenge for the spectrum aggregating techniques as discussed in Section 3.5.1 in detail. Note that standard synchronization algorithms have been proposed for OFDM transmission; they may be not suitable for the operation of NC-OFDM or NC-FBMC systems. This is because a synchronization algorithm has to be flexible enough to adapt to the dynamically changing set of occupied fragments of spectrum (subcarriers) and because it has to be robust against possibly high interference power (originating from a coexisting system) inside the spectrum notch between the aggregated bands (called in-band or *in-notch* interference).

In [216], the in-notch interference-robust synchronization algorithm called LUISA has been proposed for NC-OFDM systems. The preamble-based algorithm estimates the time and frequency offset using cross-correlation between the received signal and the reference preamble, as well as multiple signal-path components. The new proposed algorithm has been compared against the standard ones: the S–C algorithm [148], the Ad-Hoc Detector no. 1 (AHD1) [384], and the Ziabari method [229]. Contrary to these existing algorithms, LUISA is robust against in-band interference including narrowband interference at the cost of increased complexity. In the interference-free system, the probability of frame synchronization error is also improved. Moreover, in [385], low-complexity algorithm for time and frequency synchronization in an NC-OFDM radio communication system in the presence of the same kind of interference has been proposed. The computational complexity of the proposed synchronization method is similar to the known S–C algorithm. The algorithm is partially blind, that is, it does not require information on the used and excluded subcarriers, nor on the center frequency of the interfering system. It has been shown that it performs best for the in-notch interference of constant carrier frequency, but it also provides good results for the frequency-modulated interfering signal.

Because in general, in FBMC systems, CP may not be needed, the CP-based synchronization algorithms are excluded from application in NC-FBMC receivers. The problems of reliable symbol timing and CFO calculation in the context of FBMC transmission, either blind or data-aided, have been intensively studied, resulting in the pilot-based or preamble-based synchronization algorithms [386–388]. However, for noncontiguous version of FBMC (for NC-FBMC), the issue of in-notch interference robustness of the proposed synchronization algorithms appears to be the same as for the NC-OFDM systems. It is a challenge to derive such algorithms of reasonable complexity for NC-FBMC receivers in the presence of interference from coexisting systems.

Another reception issue with the noncontiguous multicarrier schemes is the data-detection quality. The cost of the spectrum aggregation and shaping in enhanced NC-OFDM is in sacrificing some energy resources to the spectrum-shaping subcarriers (e.g., as in the CC or OCCS algorithms) and reducing the power for data subcarriers. This may result in the BER performance degradation. However, as discussed in Sections 3.1.1 and 3.3, values modulating CC can be treated as coded symbols correlated with the data symbols. Thus, they can be used to decode the data at the receiver and improve the detection quality. This requires the knowledge of the precoding matrix at the receiver and is associated with additional computations for decoding. Example results of such advanced reception based on the ZF and MMSE detection criteria are presented in the mentioned sections and in [42].

Regarding the reception performance of the other considered scheme (NC-FBMC), it is dependent on the prototype-filter characteristic. Additionally, for other filter-bank-based modulations using nonorthogonal pulses, self-interference among these pulses on the TF plane may deteriorate the performance. This interference has to be removed in order to achieve reasonable BER [387, 389]. As discussed in Section 4.4, Successive Interference Cancellation (SIC) and Parallel Interference Cancellation (PIC) receivers can be applied for this purpose. Moreover, some works propose a dedicated FBMC subcarrier model incorporating interference from the adjacent subbands, as well as advanced algorithms for self-interference management using MMSE or Maximum-Likelihood Sequence Estimator (MLSE) criteria. Application of an equalizer for each subcarrier and iterative decoding schemes have been proposed in the literature, as well as techniques for channel estimation using special preambles or scattered pilots. For example, in [386], auxiliary pilots are introduced to the transmitted signal for interference minimization. Advanced techniques for increased performance of the NC-FBMC signal reception accounting for self-interference cancellation, channel estimation, and equalization are still being an important research topic.

6.4.4 Throughput Maximization

In [382], the concept of spectrum aggregation using noncontiguous multicarrier waveforms in a future 5G system is discussed and evaluated in practical radio-communication setup. The analytical model of interference is derived, taking practical characteristics of the transmit and reception filters into account. Note that typically, the reception filter characteristic is not considered in papers dealing with systems coexistence and their mutually generated interference, although it noticeably impacts the interference power gathered by a wireless receiver. In the mentioned paper, both spectrum aggregating techniques are considered: NC-OFDM and NC-FBMC, which are also validated by real-world measurements. The considered scenario includes a single 5G-system link using available radio-frequency band opportunities, in the presence of the PU (incumbent) system interference, as shown in Figure 6.9.

Let us define the 5G-system rate optimization problem taking the reception filter characteristics into account (as discussed earlier in this chapter), using the interference model suggested in [382], and making it subject to efficient protection of the incumbents. The problem is defined so as to find the vector of the power levels $\mathbf{P} = \{P_n\}$ assigned to the noncontiguous subcarriers of the secondary system to maximize its Shannon data rate:

$$\mathbf{P}^* = \arg\max_{\mathbf{P}} \Delta f \kappa \sum_{n \in \mathbf{I}_{\text{DC}}} \log_2 \left(1 + \frac{P_n \left| H\left(\frac{nf_s}{N}\right) \right|^2}{FN_0 \Delta f + P_{1n}} \right) \tag{6.1}$$

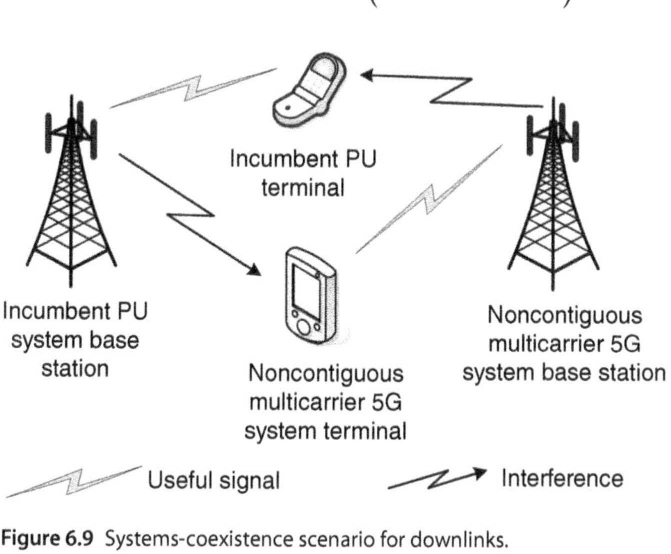

Figure 6.9 Systems-coexistence scenario for downlinks.

where Δf is the subcarrier spacing, $\mathbf{I}_{DC} = \{I_{DCj}\}$ for $j \in \{1, \dots, \alpha\}$ is the vector of occupied Data Carriers (DCs), as defined previously in Chapter 3, $H(f)$ is the considered link (5G using noncontiguous multicarrier waveforms) channel frequency response, f_s is the sampling frequency, N_0 is the white noise power spectral density, F is noise figure of the 5G system receiver, P_{In} is the power of interference observed at the n-th subcarrier of the victim 5G receiver at the output of the reception filter. Note that in [382], this interference power has been modeled taking this reception filter characteristic into account. Furthermore, κ is the rate-scaling factor accounting for the symbol duration extension (e.g., application of CP). It is equal to 1 in the ideal case (when no CP is used as in NC-FBMC transmission) and equal to $N/(N + N_{CP})$ in case of an NC-OFDM system, which typically uses CP. Moreover, in case of the time-domain windowing applied to NC-OFDM, each symbol is additionally extended by N_w windowed samples resulting in $\kappa = N/(N + N_{CP} + N_w)$. Finally, constraints to the formulated optimization include the minimum required Signal-to-Interference Ratio (SIR) for the incumbent system receiver (below which the PU transmission is not considered as protected), as well as the maximal level of the transmit power of the 5G system link.

Detailed analysis of the throughput-optimization problem in our considered systems-coexistence scenario is presented in [382]. There, also an efficient algorithm for finding the problem solution is provided. Next, we discuss some example results of throughput maximization in the 5G system link using the noncontiguous multicarrier transmissions (NC-OFDM and NC-FBMC waveforms) in the presence of exemplary incumbent systems.

Example Results

Let us discuss some example results presented in [382]. The potential of the considered noncontiguous multicarrier transmission schemes have been considered as achievable throughput of the 5G spectral-opportunistic system using NC-OFDM and NC-FBMC waveforms in the presence of interference originating from the incumbent systems. The constraints on the allowable signal power and minimum SIRobserved by the PU receiver have been maintained. Two test scenarios have been considered: when the incumbent system is 2G cellular system Global System for Mobile Communications (GSM), and when the protected system is the 3G Universal Mobile Telecommunications System (UMTS).

The GSM and UMTS systems are assumed to operate at the carrier frequency of 940 MHz and 2130 MHz, respectively. The 5G (NC-OFDM or NC-FBMC) system is centered at 938.5 MHz or 2128.5 MHz in the presence of the earlier-generation systems and can utilize a maximum of 600 subcarriers separated by $\Delta f = 15$ kHz, that is, the maximal occupied bandwidth is 9 MHz. The IFFT size in the NC-OFDM-based systems is $N = 1024$. In the FBMC scheme,

4-times oversampling is assumed. The cyclic prefix in the NC-OFDM system consists of $N_{CP} = N/16$ samples. Additionally, for this kind of transmission, windowing has been considered to reduce the OOB sidelobe power, and the Hanning window extends the NC-OFDM symbol by $N_W = N/16$ samples. For the NC-FBMC transmission, the PHYDYAS filter with the overlapping factor of $K = 4$ has been used. The allowable maximum transmit power in the 5G link is 100 mW, and the noise figure at the 5G system User Equipment (UE) (terminal) receiver equals 12 dB. The transmit power of the incumbent system base station is assumed to be 33 dBm, and the required SIR is 9 dB, as defined in [390]. Furthermore, the antenna gains are 15 dBi and 0 dBi for the base station and for UE, respectively [390]. The path loss is calculated according to the Log-normal path-loss model [391] with the path-loss exponent equal to 3. The receiver-filter characteristic of the GSM receiver has been adopted from the Adjacent Channel Selectivity (ACS) metric provided in [390], and the PSD of the GSM transmission has been assumed to be equal to Spectrum Emission Mask (SEM) defined in [392], that is, the worst-case scenario has been considered. In case of UMTS, standards [199] and [393] have been used to obtain the Adjacent Channel Leakage Ratio (ACLR) and ACS values [382].

In case of the 5G and GSM systems-coexistence evaluation, the generated 5G noncontiguous multicarrier signal occupies subcarriers indexed as $\mathbf{I}_{DC} = \{-300, \ldots, -1\} \cup \{1, \ldots, 66\} \cup \{134, \ldots, 300\}$ (note the notch of 67 subcarriers corresponding to 1 MHz) with equally distributed power. In the experiment related to the 5G and UMTS systems coexistence, the multicarrier signal occupies subcarriers of indices $\mathbf{I}_{DC} = \{-300, \ldots, -67\} \cup \{267, \ldots, 300\}$ (note a notch of 5 MHz), with equally distributed power. Other details of the coexisting systems and their parameters can be found in [382].

In Figures 6.10 and 6.11, the achieved mean throughput of the 5G system link versus the distance of the 5G base station from the incumbent system terminal can be observed for all coexistence scenarios, that is, an opportunistic 5G system using either NC-OFDM or NC-FBMC waveforms, the incumbent system being either GSM or UMTS, and the Channel State Information (CSI) (channel characteristic) being either delayed or limited.

It has been assumed that the 5G base station has the knowledge of the CSI of the connected links. However, the measurements carried by the incumbent system terminal are relayed to the 5G base station by the base station of this system. Thus, in the considered model, the 5G base station has the delayed primary-link CSI. (It is assumed that this delay equals 5 ms.) Limited CSI knowledge means that only the power attenuation due to the path loss is known to the 5G base station. This approach needs much less control information to be collected by the 5G base station. Note that in the considered scenarios, both limited and delayed CSI utilizations provide similar results. Moreover, in Figure 6.11, it is visible that the achievable 5G system throughput using noncontiguous multicarrier waveforms is limited in the

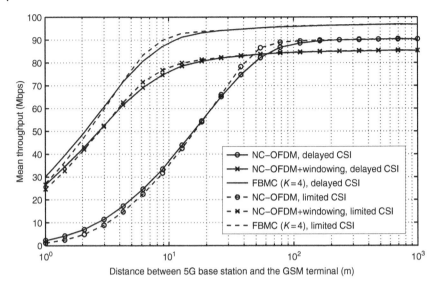

Figure 6.10 Throughput obtained with noncontiguous multicarrier schemes while protecting GSM downlink transmission. (Based on [382].)

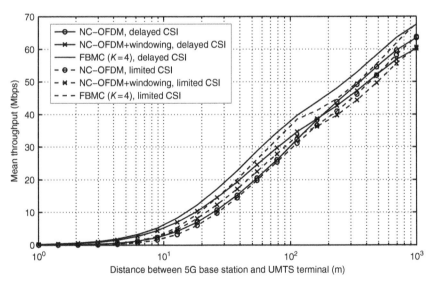

Figure 6.11 Throughput obtained with noncontiguous multicarrier schemes while protecting UMTS downlink transmission. (Based on [382].)

presence of the incumbent UMTS system more than in case of GSM. It is caused by the relatively small bandwidth of the spectrum available for the 5G transmission. As shown in [382], all considered 5G systems based on noncontiguous multicarrier waveforms achieve at least 75% of the maximal possible throughput. Moreover, the standard OFDM system achieves much smaller throughput.

Thus, feasibility of the small-scale spectrum aggregation has been confirmed. In contrast to the well-known carrier aggregation schemes already applied in 3GPP standards, even very narrow and noncontiguous frequency subbands can be utilized by a single link. This allows for smooth transition of the frequency assignment from legacy GSM and UMTS systems to new technologies while maintaining the electromagnetic compatibility with the incumbent systems.

6.5 Summary

To summarize this chapter, let us emphasize that advanced multicarrier transmission techniques using noncontiguous subcarriers such as NC-OFDM or NC-FBMC technique are capable of dynamic spectrum access and flexible spectrum aggregation. Cognitive spectrum sharing using these techniques in both licensed and unlicensed frequency bands of the future heterogeneous networks, frequency resources that have not been efficiently used so far, can be effectively used, and interference among coexisting systems can be avoided. Dynamic aggregation of potentially noncontiguous fragments of bands in a wide frequency range poses a number of challenges for the base-band processing, antenna and RF front-end transceiver design, particularly in changing radio environment [42]. Earlier in this chapter, it has been shown how to meet these challenges with noncontiguous multicarrier technologies and novel algorithms enhancing the spectral efficiency, interference robustness, and reception performance. Moreover, by applying efficient DSA methods based on spectrum pricing or the combination of cooperation and competition, it is possible to organize the dynamic access to the radio resources for multiple users of a system applying multicarrier techniques with orthogonal channels. This may be done in a semidistributed manner, requires only very limited exchange of necessary information, and results in the proper spectral efficiency and fairness.

7

Conclusions and Future Outlook

The future radio communication, fifth generation (5G) and beyond, is expected to provide an order of magnitude improvement in terms of more capacity, higher energy efficiency, lower latency, more mobility, more accuracy of terminal location, increased reliability, and availability [14]. The new system design shall ensure high flexibility in order to dynamicaly adapt to a broad range of usage requirements and deliver converged services. The development of cognitive features, spectral flexibility, spectrum aggregation and sharing capabilities are among the key directions in the definition of future radio interfaces and the Dynamic Spectrum Access (DSA) scheme. Multicarrier waveforms discussed in this book, with all the described enhancements, seem to have the required properties to address these requirements and 5G radio directions.

The enhanced Non-contiguous Orthogonal Frequency-Division Multiplexing (NC-OFDM) is one of such multicarrier techniques suitable for future applications, allowing for high transmission flexibility and spectral efficiency while preserving relatively low computational complexity. Moreover, it possesses the ability to flexibly shape and aggregate the available spectrum resources. As emphasized in Chapter 3, one very promising technique to reduce the Out-of-Band (OOB) power is the Optimized Cancellation Carriers Selection (OCCS) technique, as well as Extra Carriers (ECs) method that additionally accounts for the High Power Amplifier (HPA) nonlinear distortions.

Another merit of the enhanced NC-OFDM technique employed in the future cognitive radio-communication systems is the protection of the Primary Users (PU) obtained by means of the spectrum shaping algorithms and power control. However, the problem of the reduced reception quality of the NC-OFDM-based Cognitive Radio (CR) may occur, when the system operates in the presence of high-power PU-originating interference in-band of the Secondary User (SU) receiver. It has been shown that advanced signal detection and synchronization methods can be robust against this kind of interference and can enable high performance of a secondary NC-OFDM system, in the future systems coexistence scenarios.

Advanced Multicarrier Technologies for Future Radio Communication: 5G and Beyond, First Edition.
Hanna Bogucka, Adrian Kliks, and Paweł Kryszkiewicz.
© 2017 John Wiley & Sons, Inc. Published 2017 by John Wiley & Sons, Inc.

The Generalized Multicarrier (GMC) signal representation allows for the description of any signal waveform. Thus, all signals including various multicarrier waveforms, using orthogonal or nonorthogonal subcarriers, can be treated as a subclass of GMC signals. As a consequence, the transmission and reception algorithms presented in Chapter 4 for the GMC systems can be applied for the variations of the well-known multicarrier systems, including Orthogonal Frequency-Division Multiplexing (OFDM), filtered OFDM, and various fiter-bank-based multicarrier schemes; particularly, the GMC transceiver structure is based on the filter banks. Most of all, the parametric design of these filters (and the pulse shape) allows for controlling the signal properties, such as the OOB power and generated interference.

Appropriate modifications and refinements of the Peak-to-Average Power Ratio (PAPR) reduction methods, modulation and link adaptation techniques applied in a GMC transmitter improve the spectral features of the transmitted signals, increase the link throughput while protecting the PU transmissions in the coexisting systems. Moreover, even in case of lack of orthogonality between transmitted pulses (called atoms), it is possible to improve the reception performance by using the Successive Interference Cancellation (SIC) or Parallel Interference Cancellation (PIC) algorithms to remove the intrinsic self-interference.

Although GMC transceiver structure bases on the banks of filters, and the GMC description allows for the definition of any type of signal waveform, the Filter-Bank Multi-Carrier (FBMC) systems have been distinguished from all others as the very well researched technique with a great potential for the application in the future radio interfaces. The application of per-subcarrier or per-subband filtering allows for substantial reduction of unwanted emission out of the nominal band of the transmit signal, at the expense of increased computational complexity. This computational complexity, however, is usually scalable with the so-called overlapping factor. The FBMC systems using Offset Quadrature Amplitude Modulation (OQAM) mapping and advantageous Isotropic Orthogonal Transform Algorithm (IOTA) or PHYDYAS pulse shape are well-tailored for the future coexistence scenarios, DSA and spectrum-sharing strategies.

Finally, advanced multicarrier transmission techniques discussed in this book, particularly the ones using noncontiguous subcarriers such as NC-OFDM, or NC-FBMC technique, have been shown to be capable of dynamic spectrum access and flexible spectrum aggregation. Cognitive spectrum aggregation and sharing using these techniques in various coexisting systems scenarios have been well evaluated, showing their huge potential in 5G technologies.

Many issues and challenges of the advanced multicarrier schemes, addressed in this book, have been discussed in the vast literature. Some of these issues seem to be solved, some require further exploration. There are, however, still open issues that may be very interesting topics for further research. Among these topics, we believe, is the design of radio interfaces that take the most advantageous features of enhanced NC-OFDM, NC-FBMC, Universal Filtered Multicarrier (UFMC), or Generalized Frequency Division Multiplexing (GFDM) to be flexible enough, reasonably complex, scalable, transparent to the coexisting systems, and provide high-performance services even for the secondary users.

New systems applying advanced multicarrier waveforms with nonorthogonal subcarriers can be highly efficient in terms of the spectrum use; however, sophisticated algorithms are needed to increase their performance in the mobile radio environment, to combat the effect of intrinsic and external interference and to allow for their application in the newly developed communication concepts, such as Multiple Input, Multiple Output (MIMO) or full-duplex transmission. We hope that this book can help in understanding the currently offered and advanced multicarrier technologies and inspire research on their open issues.

References

1 Bhushan, N., Li, J., Malladi, D., and Gilmore, R. (2014) Network densification: the dominant theme for wireless evolution into 5G. *IEEE Communications Magazine*, **52** (2), 82–89, doi: 10.1109/MCOM.2014.6736747.

2 Parkvall, S., Furuskar, A., and Dahlman, E. (2011) Evolution of LTE toward IMT-advanced. *IEEE Communications Magazine*, **49** (2), 84–91, doi: 10.1109/MCOM.2011.5706315.

3 Bogucka, H., Wyglinski, A.M., Pagadarai, S., and Kliks, A. (2011) Spectrally agile multicarrier waveforms for opportunistic wireless access. *IEEE Communications Magazine*, **49** (6), 108–115, doi: 10.1109/MCOM.2011.5783994.

4 Kryszkiewicz, P., Bogucka, H., and Wyglinski, A. (2012) Protection of primary users in dynamically varying radio environment: practical solutions and challenges. *EURASIP Journal on Wireless Communications and Networking*, **2012** (1), 23, doi: 10.1186/1687-1499-2012-23.

5 Petersen, J.E. (2002) *The Telecommunications Illustrated Dictionary*, CRC Press.

6 Wesołowski, K. (2009) *Introduction to Digital Communication Systems*, John Wiley & Sons, Ltd.

7 IEEE Standard for Information Technology 802.11n 2009. (2009) Telecommunications and Information Exchange Between Systems; Local and Metropolitan Area Networks. Specific Requirements Part 11: Wireless LAN Medium Access Control (MAC) and Physical Layer (PHY) Specifications Amendment 5: Enhancement for Higher Throughputs.

8 Takagi, H. and Walke, B. (2008) *Spectrum Requirements Planning in Wireless Communications*, John Wiley & Sons, Inc., New York.

9 Cisco (2014) Cisco Visual Networking Index: Forecast and Methodology, 2014–2019. Tech. Rep., CISCO, http://www.cisco.com/c/en/us/solutions/collateral/service-provider/ip-ngn-ip-next-generation-network/white_paper_c11-481360.html (accessed 23 March 2017).

Advanced Multicarrier Technologies for Future Radio Communication: 5G and Beyond, First Edition.
Hanna Bogucka, Adrian Kliks, and Paweł Kryszkiewicz.
© 2017 John Wiley & Sons, Inc. Published 2017 by John Wiley & Sons, Inc.

10 Ericsson (2016) Ericsson mobility report on the pulse of the networked society. Tech. Rep., Ericsson, http://hugin.info/1061/R/1986667/728939.pdf (accessed 23 March 2017).

11 Pretz, T. (2014) *In the Works: Next-Generation Wireless*, IEEE, The Institute, http://theinstitute.ieee.org/technology-focus/technology-topic/in-the-works-nextgeneration-wireless (accessed 23 March 2017).

12 Huawei (2015) 5G: New air interface and radio access virtualization. White paper D5.1, Huawei Technologies Co. Ltd.

13 Ericsson (2016) 5G Radio Access. Ericsson White Paper, Tech. Rep., Ericsson, https://www.ericsson.com/res/docs/whitepapers/wp-5g.pdf (accessed 23 March 2017).

14 Association 5G Infrastructure (2015) 5G vision. The 5G infrastructure public private partnership: the next generation of communication networks and services 5G waveform candidate selection, Brochure 1, EU 5G Infrastructure Public Private Partnership.

15 Li, Q., Niu, H., Papathanassiou, A., and Wu, G. (2014) 5G network capacity: key elements and technologies. *IEEE Vehicular Technology Magazine*, **9** (**1**), 71–78,

16 3GPP (2012) Evolved Universal Terrestrial Radio Access (E-UTRA) and Evolved Universal Terrestrial Radio Access Network (E-UTRAN); Overall Description; Stage 2, TS 3GPP TS 36.300 v11.1.0, 3rd Generation Partnership Project (3GPP), https://www.etsi.org/deliver/etsi_ts/136300_136399/136300/11.03.00_60/ts_136300v110300p.pdf (accessed 23 March 2017).

17 Neira, E.M. (2014) IEEE Comsoc Technology News Special Issue on 5G, IEEE ComSoc Technology News, http://www.comsoc.org/ctn/ieee-comsoc-technology-news-special-issue-5g (accessed 23 March 2017).

18 Hu, R., Qian, Y., Kota, S., and Giambene, G. (2011) HetNets –a new paradigm for increasing cellular capacity and coverage. *IEEE Wireless Communications*, **18** (**3**), 8–9,

19 Thompson, J., Ge, X., Wu, H.C., Irmer, R., Jiang, H., Fettweis, G., and Alamouti, S. (2014) 5G wireless communication systems: prospects and challenges. *IEEE Communications Magazine*, **52** (**2**), 62–64.

20 Guvenc, I., Quek, T., Kountouris, M., and Lopez-Perez, D. (2013) Heterogeneous and small cell networks: Part 1 and 2. *IEEE Communications Magazine*, **51** (**5 and 6**), 34–35.

21 Bhushan, N. et al., (2014) Network densification: the dominant theme for wireless evolution into 5G. *IEEE Communications Magazine*, **52** (**2**), 82–89.

22 Andrews, J., Claussen, H., Dohler, M., Rangan, S., and Reed, M. (2012) Femtocells: past, present, and future. *IEEE Journal on Selected Areas in Communications*, **30** (**3**), 497–508.

23 Lopez-Perez, D. et al., (2011) Enhanced inter-cell interference coordination challenges in heterogeneous networks. *IEEE Wireless Communications*, **18** (3), 22–30.

24 Hwang, I., Song, B., and Soliman, S. (2013) A holistic view on hyper-dense heterogeneous and small cell networks. *IEEE Communications Magazine*, **51** (6), 20–27.

25 Fehske, A., Fettweis, G., Malmodin, J., and Biczok, G. (2011) The global footprint of mobile communications: the ecological and economic perspective. *IEEE Communications Magazine*, **49** (8), 55–62.

26 Shakir, M., Qaraqe, K., Tabassum, H., Alouini, M., Serpedin, E., and Imran, M. (2013) Green heterogeneous small-cell networks: toward reducing the CO_2 emissions of mobile communications industry using uplink power adaptation. *IEEE Communications Magazine*, **51** (6), 52–61.

27 Soh, Y., Quek, T., Kountouris, M., and Shin, H. (2013) Energy efficient heterogeneous cellular networks. *IEEE Journal on Selected Areas in Communications*, **31** (5), 840–850.

28 Aijaz, A., Holland, O., Pangalos, P., Aghvami, A., and Bogucka, H. (2012) Energy savings for mobile communication networks through dynamic spectrum and traffic load management, in *Green Communications: Theoretical Fundamentals, Algorithms, and Applications* (eds J. Wu, S. Rangan, and H. Zhang), Auerbach Publications, CRC Press, Taylor and Francis Group.

29 Holland, O., Dodgson, T., Aghvami, A., and Bogucka, H. (2012) Intra-operator dynamic spectrum management for energy efficiency. *IEEE Communications Magazine*, **50** (9), 178–184.

30 Ashraf, I., Boccardi, F., and Ho, L. (2011) Sleep mode techniques for small cell deployments. *IEEE Communications Magazine*, **49** (8), 73–79.

31 Kliks, A., Zalonis, A., Dimitrou, N., and Holland, O. (2013) WiFi Offloading for Energy Saving. *20th International Conference on Telecommunication, ICT 2013, Casablanca, Morocco*.

32 Bangerter, B., Talwar, S., Arefi, R., and Stewart, K. (2014) Networks and devices for the 5G era. *IEEE Communications Magazine*, **52** (2), 90–96,

33 Dimitrou, N., Zalonis, A., Polydoros, A., Kliks, A., and Holland, O. (2014) Context-Aware Radio Resource Management in HetNets. *Future HetNets workshop at IEEE Wireless Communications and Networking Conference (WCNC) 2014, Istanbul, Turkey.*

34 Triantafyllopoulou, D., Moessner, K., Bogucka, H., Zalonis, A., Dagres, I., and Polydoros, A. (2013) A Context-Aware Decision Making Framework for Cognitive and Coexisting Networking Environments. *European Wireless, 2013, Guildford, UK.*

35 Schaich, F., Sayrac, B., Schubert, M., Lin, H., Pedersen, K., Shaat, M., Wunder, G., and Georgakopoulos, A. (2015) Fantastic-5G: 5GPPP Project

on 5G Air Interface below 6 GHz. *European Conference on Networks and Communications, EuCNC 2015.*

36 Weiss, T. and Jondral, F. (2004) Spectrum pooling: an innovative strategy for the enhancement of spectrum efficiency. *IEEE Communications Magazine*, **42** (**3**), S8–S14, doi: 10.1109/MCOM.2004.1273768.

37 Pagadarai, S., Kliks, A., Bogucka, H., and Wyglinski, A. (2011) Non-contiguous multicarrier waveforms in practical opportunistic wireless systems. *IET Radar, Sonar and Navigation*, **5** (**6**), 674–680, doi: 10.1049/iet-rsn.2010.0332.

38 Mahmoud, H., Yucek, T., and Arslan, H. (2009) OFDM for cognitive radio: merits and challenges. *IEEE Wireless Communications*, **16** (**2**), 6–15, doi: 10.1109/MWC.2009.4907554.

39 Yamaguchi, H. (2004) Active Interference Cancellation Technique for MB-OFDM Cognitive Radio. *34th European Microwave Conference, vol. 2*, pp. 1105–1108.

40 Mahmoud, H. and Arslan, H. (2008) Spectrum Shaping of OFDM-Based Cognitive Radio Signals. *IEEE Radio and Wireless Symposium*, pp. 113–116, doi: 10.1109/RWS.2008.4463441.

41 Brandes, S., Cosovic, I., and Schnell, M. (2006) Reduction of out-of-band radiation in OFDM systems by insertion of cancellation carriers. *IEEE Communications Letters*, **10** (**6**), 420–422, doi: 10.1109/LCOMM.2006.1638602.

42 Bogucka, H., Kryszkiewicz, P., and Kliks, A. (2015) Dynamic spectrum aggregation for future 5G communications. *IEEE Communications Magazine*, **53** (**5**), 35–43, doi: 10.1109/MCOM.2015.7105639.

43 Kryszkiewicz, P. and Bogucka, H. (2013) Out-of-band power reduction in NC-OFDM with optimized cancellation carriers selection. *IEEE Communications Letters*, **17** (**10**), 1901–1904, doi: 10.1109/LCOMM.2013.081813.131515.

44 Fettweis, G., Krondorf, M., and Bittner, S. (2009) GFDM – Generalized Frequency Division Multiplexing. *Vehicular Technology Conference, 2009. VTC Spring 2009. IEEE 69th*, pp. 1–4, doi: 10.1109/VETECS.2009.5073571.

45 Michailow, N., Matthé, M., Gaspar, I.S., Caldevilla, A.N., Mendes, L.L., Festag, A., and Fettweis, G. (2014) Generalized frequency division multiplexing for 5th generation cellular networks. *IEEE Transactions on Communications*, **62** (**9**), 3045–3061, doi: 10.1109/TCOMM.2014.2345566.

46 Wunder, G., Jung, P., Kasparick, M., Wild, T., Schaich, F., Chen, Y., Brink, S.T., Gaspar, I., Michailow, N., Festag, A., Mendes, L., Cassiau, N., Ktenas, D., Dryjanski, M., Pietrzyk, S., Eged, B., Vago, P., and Wiedmann, F. (2014) 5GNOW: non-orthogonal, asynchronous waveforms for future mobile applications. *IEEE Communications Magazine*, **52** (**2**), 97–105, doi: 10.1109/MCOM.2014.6736749.

47 Matthé, M., Gaspar, I.S., Mendes, L.L., Zhang, D., Danneberg, M., Michailow, N., and Fettweis, G. (2017) Generalized FREQUENCY DIVISION MULTIPLEXING: A FLEXIBLE MULTI-CARRIER WAVEFORM for 5G, in *5G Mobile Communications* (eds W. Xiang, K. Zheng, and X.S. Shen), Springer International Publishing, Switzerland, pp. 223–259.

48 Kasparick, M. et al., (2013) 5G waveform candidate selection, Deliverable D3.1, EU 7th Framework Programme Project 5GNOW.

49 Scaglione, A., Barbarossa, S., and Giannakis, G. (1999) Filterbank transceivers optimizing information rate in block transmissions over dispersive channels. *IEEE Transactions on Information Theory*, **45** (3), 1019–1032, doi: 10.1109/18.761338.

50 Scaglione, A., Giannakis, G., and Barbarossa, S. (1999) Redundant filterbank precoders and equalizers Part I: unification and optimal designs. *IEEE Transactions on Signal Processing*, **47** (7), 1988–2005,

51 Stefanatos, S. and Polydoros, A. (2007) Gabor-Based Waveform Generation for Parametrically Flexible, Multi-Standard Transmitters. *European Signal Processing Conference, EUSIPCO'07, September 3–7, 2007, Poznan, Poland*, pp. 871–875.

52 Hunziker, T. and Dahlhaus, D. (2003) Iterative detection for multicarrier transmission employing time-frequency concentrated pulses. *IEEE Transactions on Communications*, **51** (4), 641–651, doi: 10.1109/TCOMM.2003.810811.

53 Proakis, J. and Salehi, M. (2008) *Digital Communications*, McGraw Hill Higher Education, New York.

54 ETSI EN 302 304 V1.1.1 (2004-11). (2004) ETSI Digital Video Broadcasting (DVB); Transmission System for Handheld Terminals (DVB-H).

55 ETSI EN (2009-09). (2009) ETSI Digital Video Broadcasting (DVB); Frame Structure Channel Coding and Modulation for a Second Generation Digital Terrestrial Television Broadcasting System (DVB-T2).

56 ETSI EN 301 958 – V1.1.1 (2002) Digital Video Broadcasting (DVB): Interaction Channel for Digital Terrestrial Television (RCT) Incorporating Multiple Access OFDM.

57 3GPP TS 36.331 V9.3.0 (2010-06). (2010) Technical Specification Group Radio Access Network; Evolved Universal Terrestrial Radio Access (E-UTRA); User Equipment (UE) Radio Transmission and Reception (Release 8).

58 IEEE 802.16-2009. (2009) IEEE Standard for Local and Metropolitan Area Networks Part 16: Air Interface for Broadband Wireless Access Systems.

59 ITU-R ITU-R M.2134. (2008) Requirements Related to Technical Performance for IMT-Advanced Radio Interface(s).

60 ITU-T Recommendation G.992.5. (2005) Transmission Systems and Media, Digital Systems and Networks; Digital Sections and Digital

Line System. Access Networks. Asymmetric Digital Subscriber Line Transceivers; Extended Bandwidth ADSL2 (ADSL2+).

61 ITU-T Recommendation G.993.1. (2004) Transmission Systems and Media, Digital Systems and Networks; Digital Sections and Digital Line System, Access Networks; Very High Speed Digital Subscriber Line Transceivers.

62 Bader, F. and Shaat, M. (2010) Pilot Pattern Adaptation and Channel Estimation in MIMO WiMAX-like FBMC System. *6th International Conference on Wireless and Mobile Communications (ICWMC)*, pp. 111–116, doi: 10.1109/ICWMC.2010.76.

63 Siohan, P. and Roche, C. (2000) Cosine-modulated filterbanks based on extended Gaussian functions. *IEEE Transactions on Signal Processing*, **48** (**11**), 3052–3061, doi: 10.1109/78.875463.

64 Vaidyanathan, P. (1993) *Multirate Systems and Filters Banks*, PTR Prentice Hall, Upper Saddle River, NJ.

65 Dagres, I., Miliou, N., Zalonis, A., Polydoros, A., and Kliks, A. (2010) Bit-Power Loading Algorithms Based on Effective SINR Mapping Techniques. *IEEE 21st International Symposium on Personal Indoor and Mobile Radio Communications (PIMRC)*, pp. 52–57, doi: 10.1109/PIMRC.2010.5671901.

66 Goldsmith, A. (2005) *Wireless Communication*, Cambridge University Press.

67 Kliks, A. and Bogucka, H. (2009) New Adaptive Bit and Power Loading Policies for Generalized Multicarrier Transmission. *17th European Signal Processing Conference (EUSIPCO 2009) Glasgow, Scotland, August 24–28, 2009*, pp. 1888–1892.

68 Kliks, A., Bogucka, H., and Stupia, I. (2009) On the Effective Adaptive Modulation Polices for Non-Orthogonal Multicarrier Systems. *6th International Symposium on Wireless Communication Systems. ISWCS 2009*, pp. 116–120, doi: 10.1109/ISWCS.2009.5285269.

69 Kliks, A. and Bogucka, H. (2010) Computationally-Efficient Bit-and-Power Allocation for Multicarrier Transmission. *Future Network and Mobile Summit 2010, Florence, Italy, June 16-18, 2010*.

70 Kliks, A., Sroka, P., and Debbah, M. (2010) Crystallized rate regions for MIMO transmission. *EURASIP Journal on Wireless Communications and Networking*, doi: 10.1155/2010/9190725.

71 Kliks, A., Sroka, P., and Debbah, M. (2010) MIMO Crystallized Rate Regions. *IEEE European Wireless Conference*, pp. 940–947, doi: 10.1109/EW.2010.5483430.

72 Song, G. and Li, Y. (2005) Cross-layer optimization for OFDM wireless networks—Part I: theoretical framework. *IEEE Transactions on Wireless Communications*, **4** (**2**), 614–624, doi: 10.1109/TWC.2004.843065.

73 Song, G. and Li, Y. (2005) Cross-layer optimization for OFDM wireless networks—Part II: algorithm development. *IEEE Transactions on Wireless Communications*, 4 (2), 625–634, doi: 10.1109/TWC.2004.843067.

74 Haykin, S. (2005) Cognitive radio: brain-empowered wireless communications. *IEEE Journal on Selected Areas in Communications*, 23 (2), 201–220, doi: 10.1109/JSAC.2004.839380.

75 Pagadarai, S., Kliks, A., Bogucka, H., and Wyglinski, A. (2010) On Non-Contiguous Multicarrier Waveforms for Spectrally Opportunistic Cognitive Radio Systems. *International Waveform Diversity and Design Conference (WDD)*, pp. 000–177–000–181, doi: 10.1109/WDD.2010.5592432.

76 Wyglinski, A. (2006) Effects of Bit Allocation on Non-Contiguous Multicarrier-Based Cognitive Radio Transceivers. *IEEE 64th Vehicular Technology Conference, VTC'06 Fall*, pp. 1–5, doi: 10.1109/VTCF.2006.159.

77 Zhou, Y. and Wyglinski, A. (2009) Cognitive Radio-Based OFDM Sidelobe Suppression Employing Modulated Filter Banks and Cancellation Carriers. *IEEE Military Communications Conference, MILCOM 2009*, pp. 1–5, doi: 10.1109/MILCOM.2009.5379927.

78 Han, H.S. and Lee, J.H. (2005) An overview of peak-to-average power ratio reduction techniques for multicarrier transmission. *IEEE Wireless Communications*, 12 (2), 56–65, doi: 10.1109/MWC.2005.1421929.

79 Kliks, A. and Bogucka, H. (2010) Improving effectiveness of the active constellation extension method for PAPR reduction in generalized multicarrier signals. *Wireless Personal Communications*, 61 (2), 323–334, doi: 10.1007/s11277-010-0025-5.

80 Han, F.M. and Zhang, X. (2009) Wireless multicarrier digital transmission via Weyl-Heisenberg frames over time-frequency dispersive channels. *IEEE Transactions on Communications*, 57 (6), 1721–1733, doi: 10.1109/TCOMM.2009.06.070406.

81 Jung, P. and Wunder, G. (2007) The WSSUS pulse design problem in multicarrier transmission. *IEEE Transactions on Communications*, 55 (10), 1918–1928, doi: 10.1109/TCOMM.2007.906426.

82 Kozek, W. and Molisch, A. (1998) Nonorthogonal pulseshapes for multicarrier communications in doubly dispersive channels. *IEEE Journal on Selected Areas in Communications*, 16 (8), 1579–1589, doi: 10.1109/49.730463.

83 Wang, Z. and Giannakis, G. (2000) Wireless multicarrier communications. *IEEE Signal Processing Magazine*, 17 (3), 29–48, doi: 10.1109/79.841722.

84 Cordis (2010) URANUS: Universal Radio-Link Platform for Efficient User-Centric Acc, http://cordis.europa.eu/fetch?CALLER=PROJ_ICT& ACTION=D&CAT=PROJ&RCN=86406 (accessed 26 July 2010).

85 Ju, Z. (2010) A filter bank based reconfigurable receiver architecture for universal wireless communications. PhD dissertation, University of Kassel.

86 Giannakis, G., Wang, Z., Scaglione, A., and Barbarossa, S. (1999) AMOUR-Generalized Multicarrier CDMA Irrespective of Multipath. *Global Telecommunications Conference, 1999. GLOBECOM '99, vol. 1B,* pp. 965–969, doi: 10.1109/GLOCOM.1999.830229.

87 Giannakis, G., Wang, Z., Scaglione, A., and Barbarossa, S. (1999) Mutually Orthogonal Transceivers for Blind Uplink CDMA Irrespective of Multipath Channel Nulls. *IEEE International Conference on Acoustics, Speech, and Signal Processing, ICASSP '99, vol. 5,* pp. 2741–2744, doi: 10.1109/ICASSP.1999.761311.

88 Newcom (2008) NEWCOM++: Network of Excellence in Wireless Communications, http://www.newcom-project.eu/ (accessed 15 December 2010).

89 PHYDYAS Physical Layer for Dynamic Spectrum Access and Cognitive Radio, http://www.ict-phydyas.org/ (accessed 15 December 2010).

90 ACROPOLIS Advanced Coexistence Technologies for Radio Optimisation in Licensed and Unlicensed Spectrum, http://www.ict-acropolis.eu/ (accessed 15 December 2010).

91 ICT-EMPhAtiC (2012) Enhanced Multicarrier Techniques for Professional Ad-hoc and Cell-Based Communications, http://www.ict-emphatic.eu/ (accessed 15 December 2010).

92 COST ACTION IC0902 Cognitive Radio and Networking for Cooperative Coexistence of Heterogeneous Wireless Networks, http://newyork.ing.uniroma1.it/IC0902/ (accessed 15 December 2010).

93 Matz, G., Schafhuber, D., Grochenig, K., Hartmann, M., and Hlawatsch, F. (2007) Analysis, optimization and implementation of low-interference wireless multicarrier systems. *IEEE Transactions on Wireless Communications,* **6** (5), 1921–1931, doi: 10.1109/TWC.2007.360393.

94 Wyglinski, A., Kliks, A., Kryszkiewicz, P., Sail, A., and Bogucka, H. (2015) Spectrally agile waveforms, in *Opportunistic Spectrum Sharing and White Space Access: The Practical Reality* (eds O. Holland, H. Bogucka, and A. Medeisis), John Wiley & Sons, Inc., New York.

95 Kliks, A., Zalonis, A., Dagres, I., Polydoros, A., and Bogucka, H. (2009) PHY abstraction methods for OFDM and NOFDM systems. *Journal of Telecommunications and Information Technology,* **2009** (3), 116–122.

96 Bingham, J. (1990) Multicarrier modulation for data transmission: an idea whose time has come. *IEEE Communications Magazine,* **28** (5), 5–14, doi: 10.1109/35.54342.

97 Prasad, R. (1998) *Universal Wireless Personal Communications,* Artech House Publishers, Norwood, MA.

98 Van Nee, R. and Prasad, R. (2000) *OFDM for Wireless Multimedia Communications*, Artech House Publishers, Norwood, MA.

99 Nassar, C., Natarajan, B., Wu, Z., Wiegandt, D., Zekavat, S., and Shattil, S. (2002) *Multi-Carrier Technologies for Wireless Communication*, Kluwer Academic Publishers, Norwell, MA.

100 Hanzo, L., Wong, C., and Yee, M. (2002) *Adaptive Wireless Transceivers: Turbo-Coded, Turbo-Equalised and Space-Time Coded TDMA, CDMA and OFDM Systems*, John Wiley & Sons, Inc., New York.

101 Hanzo, L., Yang, L.L., Kuan, E.L., and Yen, K. (2003) *Single- and Multi-Carrier DS-CDMA: Multi-User Detection, Space-Time Spreading, Synchronisation, Standards and Networking*, IEEE Press – John Wiley, New York.

102 Hanzo, L., Munster, M., Choi, B.J., and Keller, T. (2003) *OFDM and MC-CDMA for Broadband Multi-User Communications, WLANs and Broadcasting*, John Wiley & Sons.

103 Zou, W. and Wu, Y. (1995) COFDM: an overview. *IEEE Transactions on Broadcasting*, **41** (1), 1–8, doi: 10.1109/11.372015.

104 Sari, H., Karam, G., and Jeanclaude, I. (1995) Transmission techniques for digital terrestrial TV broadcasting. *IEEE Communications Magazine*, **33** (2), 100–109, doi: 10.1109/35.350382.

105 Bossert, M., Doner, A., and Zyablov, V. (1997) Coded Modulation for OFDM on Mobile Radio Channels. *European Personal Mobile Communication Conference*, pp. 109–116.

106 Kallenberg, O. (1997) *Foundations of Modern Probability*, Springer-Verlag, New York.

107 Behravan, A. and Eriksson, T. (2006) Some Statistical Properties of Multicarrier Signals and Related Measures. *IEEE 63rd Vehicular Technology Conference, VTC 2006-Spring, vol. 4*, pp. 1854–1858, doi: 10.1109/VETECS.2006.1683168.

108 Sezginer, S. and Sari, H. (2006) OFDM peak power reduction with simple amplitude predistortion. *IEEE Communications Letters*, **10** (2), 65–67, doi: 10.1109/LCOMM.2006.02015.

109 Sezginer, S. and Sari, H. (2007) Metric-based symbol predistortion techniques for peak power reduction in OFDM systems. *IEEE Transactions on Wireless Communications*, **6** (7), 2622–2629, doi: 10.1109/TWC.2007.05955.

110 Thompson, S., Proakis, J., and Zeidler, J. (2005) The Effectiveness of Signal Clipping for PAPR and Total Degradation Reduction in OFDM Systems. *IEEE Global Telecommunications Conference. GLOBECOM '05, vol. 5*, pp. 5–2811, doi: 10.1109/GLOCOM.2005.1578271.

111 Wang, L. and Tellambura, C. (2006) An Overview of Peak-to-Average Power Ratio Reduction Techniques for OFDM Systems. *IEEE*

International Symposium on Signal Processing and Information Technology, pp. 840–845, doi: 10.1109/ISSPIT.2006.270915.

112 Sharif, M., Gharavi-Alkhansari, M., and Khalaj, B. (2003) On the peak-to-average power of OFDM signals based on oversampling. *IEEE Transactions on Communications*, **51** (1), 72–78, doi: 10.1109/TCOMM.2002.807619.

113 Motorola R1-060023. (2006) Cubic Metric in 3GPP-LTE.

114 R1-040642. (2004) Comparison of PAR and Cubic Metric for Power De-rating.

115 Skrzypczak, A., Siohan, P., and Javaudin, J.-P. (2006) Power Spectral Density and Cubic Metric for the OFDM/OQAM Modulation. *IEEE International Symposium on Signal Processing and Information Technology*, pp. 846–850, doi: 10.1109/ISSPIT.2006.270916.

116 Ciochina, C., Buda, F., and Sari, H. (2006) An Analysis of OFDM Peak Power Reduction Techniques for WiMAX Systems. *IEEE International Conference on Communications, ICC '06, vol. 10*, pp. 4676–4681, doi: 10.1109/ICC.2006.255378.

117 Saleh, A.A.M. (1981) Frequency-independent and frequency-dependent nonlinear models of TWT amplifiers. *IEEE Transactions on Communications*, **COM-29**, 1715–1720,

118 Rapp, C. (1991) Effects on HPA-Nonlinearity on a 4-DPSK / OFDM Signal for a Digital Sound Broadcasting System. *2nd European Conference on Satellite Communications, ECSC-2, Liége, Belgium.*

119 Gharaibeh, K.M. (2011) *Nonlinear Distortion in Wireless Systems: Modeling and Simulation with MATLAB*, John Wiley & Sons, Inc., New York.

120 Weekley, J. and Mangus, B. (2005) TWTA versus SSPA: a comparison of on-orbit reliability data. *IEEE Transactions on Electron Devices*, **52** (5), 650–652, doi: 10.1109/TED.2005.845864.

121 Bogucka, H. (2006) Directions and Recent Advances in PAPR Reduction Methods. *IEEE International Symposium on Signal Processing and Information Technology*, pp. 821–827, doi: 10.1109/ISSPIT.2006.270912.

122 Jiang, T. and Wu, T. (2008) An overview: peak-to-average power ratio reduction techniques for OFDM signals. *IEEE Transactions on Broadcasting*, **54** (2), 257–268, doi: 10.1109/TBC.2008.915770.

123 Louet, Y. and Palicot, J. (2008) A classification of methods for efficient power amplification of signals. *Annals of Telecommunications*, **63**, 351–368, doi: 10.1007/s12243-008-0035-4.

124 Ermolova, N. (2006) Nonlinear amplifier effects on clipped-filtered multicarrier signals. *IEE Proceedings – Communications*, **153** (2), 213–218, doi: 10.1049/ip-com:20045178.

125 Guel, D. and Palicot, J. (2009) Clipping Formulated as an Adding Signal Technique for OFDM Peak Power Reduction. *IEEE 69th*

Vehicular Technology Conference, VTC'09 Spring, pp. 1–5, doi: 10.1109/VETECS.2009.5073442.

126 Li, X. and Cimini, L.J.J. (1998) Effects of clipping and filtering on the performance of OFDM. *IEEE Communications Letters*, **2** (5), 131–133, doi: 10.1109/4234.673657.

127 Cha, S., Park, M., Lee, S., Bang, K.J., and Hong, D. (2008) A new PAPR reduction technique for OFDM systems using advanced peak windowing method. *IEEE Transactions on Consumer Electronics*, **54** (2), 405–410, doi: 10.1109/TCE.2008.4560106.

128 Chen, G., Ansari, R., and Yao, Y. (2009) Improved Peak Windowing for PAPR Reduction in OFDM. *IEEE 69th Vehicular Technology Conference, VTC'09 Spring*, pp. 1–5, doi: 10.1109/VETECS.2009.5073593.

129 van Nee, R. and de Wild, A. (1998) Reducing the Peak-to-Average Power Ratio of OFDM. *48th IEEE Vehicular Technology Conference, VTC 98, vol. 3*, pp. 2072–2076, doi: 10.1109/VETEC.1998.686121.

130 Kliks, A. and Bogucka, H. (2007) A Modified Method of Active Constellation Extension for Generalized Multicarrier Signals. *12th International OFDM-Workshop InOWo2007*, Hamburg, Germany, August 29-30, 2007, pp. 16–20.

131 Krongold, B. and Jones, D. (2003) PAR reduction in OFDM via active constellation extension. *IEEE Transactions on Broadcasting*, **49** (3), 258–268, doi: 10.1109/TBC.2003.817088.

132 Armstrong, J. (2002) Peak-to-average power reduction for OFDM by repeated clipping and frequency domain filtering. *Electronics Letters*, **38** (5), 246–247, doi: 10.1049/el:20020175.

133 Wang, L. and Tellambura, C. (2005) A simplified clipping and filtering technique for PAR reduction in OFDM systems. *IEEE Signal Processing Letters*, **12** (6), 453–456, doi: 10.1109/LSP.2005.847886.

134 Bauml, R., Fischer, R., and Huber, J. (1996) Reducing the peak-to-average power ratio of multicarrier modulation by selected mapping. *Electronics Letters*, **32** (22), 2056–2057, doi: 10.1049/el:19961384.

135 Baxley, R. and Zhou, G. (2007) Comparing selected mapping and partial transmit sequence for PAR reduction. *IEEE Transactions on Broadcasting*, **53** (4), 797–803, doi: 10.1109/TBC.2007.908335.

136 Jayalath, A. and Tellambura, C. (2005) SLM and PTS peak-power reduction of OFDM signals without side information. *IEEE Transactions on Wireless Communications*, **4** (5), 2006–2013, doi: 10.1109/TWC.2005.853916.

137 ETSI EN 303 035-1 V1.2.1 (2001-12). (2001) Terrestrial Trunked Radio (TETRA); Harmonized EN for TETRA Equipment Covering Essential Requirements Under Article 3.2 of the RTTE Directive; Part 1: Voice plus Data (V+D).

138 ETSI EN 303 035-2 V1.2.2 (2003-01). (2003) Terrestrial Trunked Radio (TETRA); Harmonized EN for TETRA Equipment Covering Essential Requirements Under Article 3.2 of the RTTE Directive; Part 2: Direct Mode Operation (DMO).

139 Duel-Hallen, A., Hu, S., and Hallen, H. (2000) Long-range prediction of fading signals – enabling adapting transmission for mobile radio channels. *IEEE Signal Processing Magazine*, **17** (3), 62–75.

140 Conti, A., Win, M., and Chiani, M. (2007) Slow adaptive M-QAM with diversity in fast fading and shadowing. *IEEE Transactions on Communications*, **55** (5), 895–905,

141 Bogucka, H. and Conti, A. (2010) Utility-Based QAM Adaptation with Diversity and Ambiguous CSI Under Energy Constraints. *IEEE International Communications Conference (ICC 2010)*, pp. 1–5, doi: 10.1109/ICC.2010.5502087.

142 Zalonis, A., Miliou, N., Dagres, I., Polydoros, A., and Bogucka, H. (2010) Trends in adaptive modulation and coding. *Advances in Electronics and Telecommunications*, **1** (1), 104–111,

143 Fantacci, R., Marabissi, D., Tarchi, D., and Habib, I. (2009) Adaptive modulation and coding techniques for ofdma systems. *IEEE Transactions on Wireless Communications*, **8** (9), 4876–4883, doi: 10.1109/TWC.2009.090253.

144 Meng, J. and Yang, E.H. (2014) Constellation and rate selection in adaptive modulation and coding based on finite blocklength analysis and its application to lte. *IEEE Transactions on Wireless Communications*, **13** (10), 5496–5508, doi: 10.1109/TWC.2014.2350974.

145 Ramakrishnan, S., Balakrishnan, J., and Ramasubramanian, K. (2010) Exploiting Signal and Noise Statistics for Fixed Point FFT Design Optimization in OFDM Systems. *Communications (NCC), 2010 National Conference on*, pp. 1–5, doi: 10.1109/NCC.2010.5430229.

146 Grimm, M., Allen, M., Marttila, J., Valkama, M., and Thoma, R. (2014) Joint mitigation of nonlinear RF and baseband distortions in wideband direct-conversion receivers. *IEEE Transactions on Microwave Theory and Techniques*, **62** (1), 166–182, doi: 10.1109/TMTT.2013.2292603.

147 Morelli, M., Kuo, C.C., and Pun, M.O. (2007) Synchronization techniques for orthogonal frequency division multiple access (OFDMA): a tutorial review. *Proceedings of the IEEE*, **95** (7), 1394–1427, doi: 10.1109/JPROC.2007.897979.

148 Schmidl, T. and Cox, D. (1997) Robust frequency and timing synchronization for OFDM. *IEEE Transactions on Communications*, **45** (12), 1613–1621, doi: 10.1109/26.650240.

149 van de Beek, J.J., Sandell, M., and Borjesson, P. (1997) ML estimation of time and frequency offset in OFDM systems. *IEEE Transactions on Signal Processing*, **45** (7), 1800–1805, doi: 10.1109/78.599949.

150 Ozdemir, M.K. and Arslan, H. (2007) Channel estimation for wireless OFDM systems. *IEEE Communication Surveys and Tutorials*, **9** (2), 18–48, doi: 10.1109/COMST.2007.382406.

151 Yee, N. and Linnartz, J.P. (1994) Controlled Equalization of Multi-Carrier CDMA in an Indoor Rician Fading Channel. *Vehicular Technology Conference, 1994 IEEE 44th*, pp. 1665–1669, doi: 10.1109/VETEC.1994.345379.

152 Slimane, S. (2000) Partial Equalization of Multi-Carrier CDMA in Frequency Selective Fading Channels. *2000 IEEE International Conference on Communications, 2000, ICC 2000*, pp. 26–30, doi: 10.1109/ICC.2000.853057.

153 Fazel, K. and Kaiser, S. (2003) *Multi-Carrier and Spread Spectrum Systems*, John Wiley & Sons, Ltd.

154 Brandes, S., Cosovic, I., and Schnell, M. (2005) Reduction of Out-of-Band Radiation in OFDM Based Overlay Systems. *IEEE International Symposium on New Frontiers in Dynamic Spectrum Access Networks (DySPAN)*, pp. 662–665, doi: 10.1109/DYSPAN.2005.1542691.

155 Yuan, Z. and Wyglinski, A. (2010) On sidelobe suppression for multicarrier-based transmission in dynamic spectrum access networks. *IEEE Transactions on Vehicular Technology*, **59** (4), 1998–2006, doi: 10.1109/TVT.2010.2044428.

156 Li, D., Dai, X., and Zhang, H. (2008) Sidelobe Suppression in NC-OFDM Systems Using Phase Shift. *4th International Conference on Wireless Communications, Networking and Mobile Computing (WiCOM)*, pp. 1–4, doi: 10.1109/WiCom.2008.297.

157 Van De Beek, J. (2009) Sculpting the multicarrier spectrum: a novel projection precoder. *IEEE Communications Letters*, **13** (12), 881–883, doi: 10.1109/LCOMM.2009.12.091614.

158 Zhang, J., Huang, X., Cantoni, A., and Guo, Y. (2012) Sidelobe suppression with orthogonal projection for multicarrier systems. *IEEE Transactions on Communications*, **60** (2), 589–599, doi: 10.1109/TCOMM.2012.012012.110115.

159 Zielinski, T. (2005) *Cyfrowe Przetwarzanie Sygnalow*, WKL, Warszawa.

160 Faulkner, M. (2000) The effect of filtering on the performance of OFDM systems. *IEEE Transactions on Vehicular Technology*, **49** (5), 1877–1884, doi: 10.1109/25.892590.

161 Weiss, T., Hillenbrand, J., Krohn, A., and Jondral, F. (2004) Mutual Interference in OFDM-Based Spectrum Pooling Systems. *Vehicular Technology Conference (VTC), vol. 4*, pp. 1873–1877, doi: 10.1109/VETECS.2004.1390598.

162 El-Saadany, M., Shalash, A., and Abdallah, M. (2009) Revisiting Active Cancellation Carriers for Shaping the Spectrum of OFDM-Based Cognitive Radios. *IEEE Sarnoff Symposium (SARNOFF)*, pp. 1–5, doi: 10.1109/SARNOF.2009.4850359.

163 Yu, L., Rao, B., Milstein, L., and Proakis, J. (2010) Reducing Out-of-Band Radiation of OFDM-Based Cognitive Radios. *IEEE 11th International Workshop on Signal Processing Advances in Wireless Communications (SPAWC)*, pp. 1–5, doi: 10.1109/SPAWC.2010.5670975.

164 Sutton, P., Ozgul, B., Macaluso, I., and Doyle, L. (2010) OFDM Pulse-Shaped Waveforms for Dynamic Spectrum Access Networks. *IEEE Symposium on New Frontiers in Dynamic Spectrum*, pp. 1–2, doi: 10.1109/DYSPAN.2010.5457921.

165 Mahmoud, H. and Arslan, H. (2008) Sidelobe suppression in OFDM-based spectrum sharing systems using adaptive symbol transition. *IEEE Communications Letters*, **12** (**2**), 133–135, doi: 10.1109/LCOMM.2008.071729.

166 Pagadarai, S., Rajbanshi, R., Wyglinski, A., and Minden, G. (2008) Sidelobe Suppression for OFDM-Based Cognitive Radios Using Constellation Expansion. *IEEE Wireless Communications and Networking Conference (WCNC)*, pp. 888–893, doi: 10.1109/WCNC.2008.162.

167 Cosovic, I., Brandes, S., and Schnell, M. (2005) A Technique for Sidelobe Suppression in OFDM Systems. *IEEE Global Telecommunications Conference (GLOBECOM)*, vol. 1, p. 5, doi: 10.1109/GLOCOM.2005.1577381.

168 Cosovic, I., Brandes, S., and Schnell, M. (2006) Subcarrier weighting: a method for sidelobe suppression in OFDM systems. *IEEE Communications Letters*, **10** (**6**), 444–446, doi: 10.1109/LCOMM.2006.1638610.

169 Cosovic, I. and Mazzoni, T. (2006) Suppression of sidelobes in OFDM systems by multiple-choice sequences. *European Transactions on Telecommunications*, **17** (**6**), 623–630, doi: 10.1002/ett.1162.

170 Li, D., Dai, X., and Zhang, H. (2009) Sidelobe suppression in NC-OFDM systems using constellation adjustment. *IEEE Communications Letters*, **13** (**5**), 327–329, doi: 10.1109/LCOMM.2009.090031.

171 Ahmed, S., Rehman, R., and Hwang, H. (2008) New Techniques to Reduce Sidelobes in OFDM System. *3rd International Conference on Convergence and Hybrid Information Technology (ICCIT)*, vol. 2, pp. 117–121, doi: 10.1109/ICCIT.2008.157.

172 Panta, K. and Armstrong, J. (2003) Spectral analysis of OFDM signals and its improvement by polynomial cancellation coding. *IEEE Transactions on Consumer Electronics*, **49** (**4**), 939–943, doi: 10.1109/TCE.2003.1261178.

173 Noreen, S. and Azeemi, N. (2010) A Technique for Out-of-Band Radiation Reduction in OFDM-Based Cognitive Radio. *IEEE 17th International Conference on Telecommunications (ICT)*, pp. 853–856, doi: 10.1109/ICTEL.2010.5478875.

174 Zhou, X., Li, G., and Sun, G. (2011) Low-Complexity Spectrum Shaping for OFDM-Based Cognitive Radios. *IEEE Wireless Communications and Networking Conference (WCNC)*, pp. 1471–1475, doi: 10.1109/WCNC.2011.5779347.

175 Cosovic, I. and Mazzoni, T. (2007) Sidelobe Suppression in OFDM Spectrum Sharing Systems via Additive Signal Method. *IEEE 65th Vehicular Technology Conference (VTC)*, pp. 2692–2696, doi: 10.1109/VETECS.2007.553.

176 Jiang, W. and Schellmann, M. (2012) Suppressing the Out-of-Band Power Radiation in Multi-Carrier Systems: A Comparative Study. *IEEE Global Communications Conference (GLOBECOM)*, pp. 1477–1482, doi: 10.1109/GLOCOM.2012.6503322.

177 Xu, R. and Chen, M. (2009) A precoding scheme for DFT-based OFDM to suppress sidelobes. *IEEE Communications Letters*, **13** (10), 776–778, doi: 10.1109/LCOMM.2009.091339.

178 Ma, M., Huang, X., Jiao, B., and Guo, Y. (2011) Optimal orthogonal precoding for power leakage suppression in DFT-based systems. *IEEE Transactions on Communications*, **59** (3), 844–853, doi: 10.1109/TCOMM.2011.121410.100071.

179 Zhou, X., Li, G., and Sun, G. (2013) Multiuser spectral precoding for OFDM-based cognitive radio systems. *IEEE Journal on Selected Areas in Communications*, **31** (3), 345–352, doi: 10.1109/JSAC.2013.130302.

180 van de Beek, J. (2010) Orthogonal multiplexing in a subspace of frequency well-localized signals. *IEEE Communications Letters*, **14** (10), 882–884, doi: 10.1109/LCOMM.2010.081610.100997.

181 van de Beek, J. and Berggren, F. (2009) EVM-Constrained OFDM Precoding for Reduction of Out-of-Band Emission. *IEEE Vehicular Technology Conference (VTC)*, pp. 1–5, doi: 10.1109/VETECF.2009.5378740.

182 Van De Beek, J. and Berggren, F. (2009) N-continuous OFDM. *IEEE Communications Letters*, **13** (1), 1–3, doi: 10.1109/LCOMM.2009.081446.

183 Zheng, Y., Zhong, J., Zhao, M., and Cai, Y. (2012) A precoding scheme for N-continuous OFDM. *IEEE Communications Letters*, **16** (12), 1937–1940, doi: 10.1109/LCOMM.2012.102612.122168.

184 Wei, P., Dan, L., Xiao, Y., and Li, S. (2013) A Low-Complexity Time-Domain Signal Processing Algorithm for N-Continuous OFDM. *IEEE International Conference on Communications (ICC)*, pp. 5754–5758, doi: 10.1109/ICC.2013.6655513.

185 Lizarraga, E., Sauchelli, V., and Maggio, G. (2011) N-Continuous OFDM Signal Analysis of FPGA-Based Transmissions. *VII Southern Conference on Programmable Logic (SPL)*, pp. 13–18, doi: 10.1109/SPL.2011.5782618.

186 Qu, D., Wang, Z., and Jiang, T. (2010) Extended active interference cancellation for sidelobe suppression in cognitive radio OFDM systems with cyclic prefix. *IEEE Transactions on Vehicular Technology*, **59** (4), 1689–1695, doi: 10.1109/TVT.2010.2040848.

187 Qu, D., Wang, Z., Jiang, T., and Daneshmand, M. (2009) Sidelobe Suppression Using Extended Active Interference Cancellation with Self-Interference Constraint for Cognitive OFDM System. *International*

Conference on Communications and Networking in China, ChinaCOM 2009, pp. 1–5, doi: 10.1109/CHINACOM.2009.5339921.

188 Wang, Z., Qu, D., Jiang, T., and He, Y. (2008) Spectral Sculpting for OFDM Based Opportunistic Spectrum Access by Extended Active Interference Cancellation. *IEEE Global Telecommunications Conference (GLOBECOM)*, pp. 1–5, doi: 10.1109/GLOCOM.2008.ECP.852.

189 Golub, G.H. and Van Loan, C.F. (1996) *Matrix Computations*, Johns Hopkins University Press, Baltimore, MD.

190 Yuan, Z., Pagadarai, S., and Wyglinski, A. (2008) Cancellation Carrier Technique Using Genetic Algorithm for OFDM Sidelobe Suppression. *IEEE Military Communications Conference (MILCOM)*, pp. 1–5, doi: 10.1109/MILCOM.2008.4753557.

191 Pagadarai, S., Wyglinski, A.M., and Rajbanshi, R. (2008) A Sub-Optimal Sidelobe Suppression Technique for OFDM-Based Cognitive Radios. *IEEE Military Communications Conference (MILCOM)*, pp. 1–6, doi: 10.1109/MILCOM.2008.4753556.

192 Huang, S.G. and Hwang, C.H. (2009) Improvement of active interference cancellation: avoidance technique for OFDM cognitive radio. *IEEE Transactions on Wireless Communications*, **8** (**12**), 5928–5937, doi: 10.1109/TWC.2009.12.081277.

193 Pagadarai, S., Wyglinski, A.M., and Rajbanshi, R. (2008) A Novel Sidelobe Suppression Technique for OFDM-Based Cognitive Radio Transmission. *IEEE Symposium on New Frontiers in Dynamic Spectrum Access Networks (DySPAN)*, pp. 1–7, doi: 10.1109/DYSPAN.2008.10.

194 Sokhandan, N. and Safavi, S. (2010) Sidelobe Suppression in OFDM-Based Cognitive Radio Systems. *International Conference on Information Sciences Signal Processing and their Applications (ISSPA)*, pp. 413–417, doi: 10.1109/ISSPA.2010.5605455.

195 Wang, Z. and Giannakis, G. (2003) Complex-field coding for OFDM over fading wireless channels. *IEEE Transactions on Information Theory*, **49** (**3**), 707–720, doi: 10.1109/TIT.2002.808101.

196 Kryszkiewicz, P. (2010) *On the Improvement of the OFDM-Based Cognitive-Radio Performance Using Cancellation Carriers*, Poznanskie Warsztaty Telekomunikacyjne (PWT), Poland, pp. 1–4.

197 IEEE Std 802.11-2012 (Revision of IEEE Std 802.11-2007). (2012) IEEE Standard for Information Technology–Telecommunications and Information Exchange Between Systems Local and Metropolitan Area Networks–Specific Requirements Part 11: Wireless LAN Medium Access Control (MAC) and Physical Layer (PHY) Specifications, pp. 1–2793, doi: 10.1109/IEEESTD.2012.6178212.

198 3GPP TS 36.101. (2008) Evolved Universal Terrestrial Radio Access (E-UTRA); User Equipment (UE) Radio Transmission and Reception, 3rd Generation Partnership Project (3GPP), http://www.3gpp.org/ftp/Specs/html-info/36101.htm (accessed 24 March 2017).

199 3GPP TS 25.101. (2008) User Equipment (UE) Radio Transmission and Reception (FDD), 3rd Generation Partnership Project (3GPP), http://www.3gpp.org/ftp/Specs/html-info/25101.htm.

200 Kryszkiewicz, P. and Bogucka, H. (2012) Flexible Quasi-Systematic Precoding for the Out-of-Band Energy Reduction in NC-OFDM. *IEEE Wireless Communications and Networking Conference (WCNC)*, pp. 209–214, doi: 10.1109/WCNC.2012.6214138.

201 Cho, K. and Yoon, D. (2002) On the general BER expression of one- and two-dimensional amplitude modulations. *IEEE Transactions on Communications*, **50** (7), 1074–1080, doi: 10.1109/TCOMM.2002.800818.

202 Ettus Research http://www.ettus.com (accessed 12 December 2014).

203 Bogucka, H., Kryszkiewicz, P., Kliks, A., and Holland, O. (2012) Holistic approach to green wireless communications based on multicarrier technologies, in *Green Communications: Theoretical Fundamentals, Algorithms, and Applications* (eds J. Wu, S. Rangan, and H. Zhang), CRC Press.

204 ETSI EN 302 755 v1.3.1. (2012) Digital Video Broadcasting (DVB); Frame Structure Channel Coding and Modulation for a Second Generation Digital Terrestrial Television Broadcasting System (DVB-T2).

205 Senst, M., Jordan, M., Dorpinghaus, M., Farber, M., Ascheid, G., and Meyr, H. (2007) Joint Reduction of Peak-to-Average Power Ratio and Out-of-Band Power in OFDM Systems. *IEEE Global Telecommunications Conference (GLOBECOM)*, pp. 3812–3816, doi: 10.1109/GLOCOM.2007.724.

206 Ghassemi, A., Lampe, L., Attar, A., and Gulliver, T. (2010) Joint Sidelobe and Peak Power Reduction in OFDM-Based Cognitive Radio. *IEEE Vehicular Technology Conference (VTC)*, pp. 1–5, doi: 10.1109/VETECF.2010.5594133.

207 Kryszkiewicz, P., Kliks, A., and Louet, Y. (2013) Reduction of Subcarriers Spectrum Sidelobes and Intermodulation in NC-OFDM Systems. *IEEE International Conference on Wireless and Mobile Computing, Networking and Communications (WiMob)*, pp. 124–129, doi: 10.1109/WiMOB.2013.6673350.

208 Tellado-Mourelo, J. (1999) Peak to average power reduction for multicarrier transmission. PhD thesis, Stanford University.

209 Sanguinetti, L., D'Amico, A., and Cosovic, I. (2008) On the Performance of Cancellation Carrier-Based Schemes for Sidelobe Suppression in OFDM Networks. *IEEE Vehicular Technology Conference (VTC)*, pp. 1691–1696, doi: 10.1109/VETECS.2008.389.

210 Grant, M. and Boyd, S. (2014) CVX: Matlab Software for Disciplined Convex Programming, Version 2.1, http://cvxr.com/cvx (accessed 24 March 2017).

211 Hussain, S., Guel, D., Louet, Y., and Palicot, J. (2009) Performance Comparison of PRC Based PAPR Reduction Schemes for

WiLAN Systems. *European Wireless Conference (EW)*, pp. 167–172, doi: 10.1109/EW.2009.5358005.

212 Cabric, D. and Brodersen, R. (2005) Physical Layer Design Issues Unique to Cognitive Radio Systems. *IEEE 16th International Symposium on Personal, Indoor and Mobile Radio Communications (PIMRC)*, vol. 2, pp. 759–763, doi: 10.1109/PIMRC.2005.1651545.

213 Minn, H., Bhargava, V., and Letaief, K. (2003) A robust timing and frequency synchronization for OFDM systems. *IEEE Transactions on Wireless Communications*, **2** (**4**), 822–839, doi: 10.1109/TWC.2003.814346.

214 Nickel, P., Gerstacker, W., Jonietz, C., Kilian, G., Heuberger, A., and Koch, W. (2006) Window Design for Non-Orthogonal Interference Reduction in OFDM Receivers. *2006 IEEE 7th Workshop on Signal Processing Advances in Wireless Communications*, pp. 1–5, doi: 10.1109/SPAWC.2006.346334.

215 Dahlman, E., Parkvall, S., and Skold, J. (2013) *4G: LTE/LTE-Advanced for Mobile Broadband*, Elsevier Science, https://books.google.pl/books?id=AbkPAAAAQBAJ (accessed 24 March 2017).

216 Kryszkiewicz, P. and Bogucka, H. (2016) In-band-interference robust synchronization algorithm for an NC-OFDM system. *IEEE Transactions on Communications*, **64** (**5**), 2143–2154.

217 Brandes, S., Schnell, M., Berthold, U., and Jondral, F. (2007) OFDM Based Overlay Systems – Design Challenges and Solutions. *IEEE 18th International Symposium on Personal, Indoor and Mobile Radio Communications (PIMRC)*, pp. 1–5, doi: 10.1109/PIMRC.2007.4394130.

218 Morelli, M. and Mengali, U. (1999) An improved frequency offset estimator for OFDM applications. *IEEE Communications Letters*, **3** (**3**), 75–77, doi: 10.1109/4234.752907.

219 Coulson, A. (2004) Narrowband interference in pilot symbol assisted OFDM systems. *IEEE Transactions on Wireless Communications*, **3** (**6**), 2277–2287, doi: 10.1109/TWC.2004.837471.

220 Marey, M. and Steendam, H. (2007) Analysis of the narrowband interference effect on OFDM timing synchronization. *IEEE Transactions on Signal Processing*, **55** (**9**), 4558–4566, doi: 10.1109/TSP.2007.896020.

221 Zivkovic, M. and Mathar, R. (2011) Performance Evaluation of Timing Synchronization in OFDM-Based Cognitive Radio Systems. *IEEE Vehicular Technology Conference (VTC)*, pp. 1–5, doi: 10.1109/VETECF.2011.6092909.

222 Kryszkiewicz, P. and Bogucka, H. (2013) *Performance of NC-OFDM Autocorrelation-Based Synchronization Under Narrowband Interference*, Poznanskie Warsztaty Telekomunikacyjne (PWT), Poland, pp. 1–5.

223 Weiss, T., Krohn, A., Capar, F., Martoyo, I., and Jondral, F. (2003) Synchronization Algorithms and Preamble Concepts for Spectrum Pooling Systems. *IST Mobile & Wireless Telecommunications Summit*, pp. 1–5.

224 Sun, P. and Zhang, L. (2010) Timing Synchronization for OFDM Based Spectrum Sharing System. *International Symposium on Wireless Communication Systems (ISWCS)*, pp. 951–955, doi: 10.1109/ISWCS.2010.5624337.

225 Fort, A., Weijers, J.W., Derudder, V., Eberle, W., and Bourdoux, A. (2003) A Performance and Complexity Comparison of Auto-Correlation and Cross-Correlation for OFDM Burst Synchronization. *IEEE International Conference on Acoustics, Speech, and Signal Processing, (ICASSP), vol. 2*, pp. II–341–4, doi: 10.1109/ICASSP.2003.1202364.

226 Saha, D., Dutta, A., Grunwald, D., and Sicker, D. (2011) Blind Synchronization for NC-OFDM When Channels are Conventions, Not Mandates. *IEEE Symposium on New Frontiers in Dynamic Spectrum Access Networks (DySPAN)*, pp. 552–563, doi: 10.1109/DYSPAN.2011.5936246.

227 Huang, B., Wang, J., Tang, W., and Li, S. (2010) An Effective Synchronization Scheme for NC-OFDM Systems in Cognitive Radio Context. *IEEE International Conference on Wireless Information Technology and Systems (ICWITS)*, pp. 1–4, doi: 10.1109/ICWITS.2010.5611980.

228 Awoseyila, A., Kasparis, C., and Evans, B. (2009) Robust time-domain timing and frequency synchronization for OFDM systems. *IEEE Transactions on Consumer Electronics*, **55** (2), 391–399, doi: 10.1109/TCE.2009.5174399.

229 Abdzadeh-Ziabari, H. and Shayesteh, M. (2011) Robust timing and frequency synchronization for OFDM systems. *IEEE Transactions on Vehicular Technology*, **60** (8), 3646–3656, doi: 10.1109/TVT.2011.2163194.

230 Ren, G., Chang, Y., Zhang, H., and Zhang, H. (2007) An efficient frequency offset estimation method with a large range for wireless OFDM systems. *IEEE Transactions on Vehicular Technology*, **56** (4), 1892–1895, doi: 10.1109/TVT.2006.878560.

231 Wei, S., Goeckel, D., and Kelly, P. (2010) Convergence of the complex envelope of bandlimited OFDM signals. *IEEE Transactions on Information Theory*, **56** (10), 4893–4904, doi: 10.1109/TIT.2010.2059550.

232 Gouba, O. and Louet, Y. (2012) Predistortion Performance Considering Peak to Average Power Ratio Reduction in OFDM Context. *IEEE Wireless Communications and Networking Conference (WCNC)*, pp. 204–208, doi: 10.1109/WCNC.2012.6214128.

233 Yucek, T. and Arslan, H. (2007) MMSE noise plus interference power estimation in adaptive OFDM systems. *IEEE Transactions on Vehicular Technology*, **56** (6), 3857–3863, doi: 10.1109/TVT.2007.901883.

234 Candan, C. (2013) Analysis and further improvement of fine resolution frequency estimation method from three DFT samples. *IEEE Signal Processing Letters*, **20** (9), 913–916, doi: 10.1109/LSP.2013.2273616.

235 Sorensen, H. and Burrus, C. (1993) Efficient computation of the DFT with only a subset of input or output points. *IEEE Transactions on Signal Processing*, **41** (3), 1184–1200, doi: 10.1109/78.205723.

236 Rajbanshi, R., Wyglinski, A.M., and Minden, G. (2006) An Efficient Implementation of NC-OFDM Transceivers for Cognitive Radios. *1st International Conference on Cognitive Radio Oriented Wireless Networks and Communications (CROWNCOM)*, pp. 1–5, doi: 10.1109/CROWNCOM.2006.363452.

237 Qian, S. and Chen, D. (1999) Joint time-frequency analysis. *IEEE Signal Processing Magazine*, 16 (2), 52–67, doi: 10.1109/79.752051.

238 Qian, S. and Chen, D. (1996) *Joint Time-Frequency Analysis. Methods and Applications*, Prentice Hall PTR, Upper Saddle River, NJ.

239 Feichtinger, H. and Strohmer, T. (1998) *Gabor Analysis and Algorithms. Theory and Applications*, Birkhäser, Basel.

240 Hunziker, T. and Dahlhaus, D. (2000) Iterative Symbol Detection for Bandwidth Efficient Nonorthogonal Multicarrier Transmission. *IEEE 51st Vehicular Technology Conference Proceedings. VTC'00, vol.1*, Spring, Tokyo, pp. 61–65, doi: 10.1109/VETECS.2000.851418.

241 Bialasiewicz, J.T. (2004) *Falki i aproksymacje*, WNT, Warszawa.

242 Janssen, A. (1998) The duality condition for Weyl-Heisenberg frames, in *Gabor Analysis and Algorithms: Theory and Applications*, 2nd edn, vol. 2 (eds H.G. Feichtinger and T. Strohmer), Birkhäuser, Boston, MA, pp. 33–84, doi: 10.1007/978-1-4612-2016-9_2.

243 Gabor, D. (1946) Theory of communication. *Journal of IEE: Part III: Radio and Communication Engineering*, 93 (26), 429–457.

244 Strohmer, T. and Beaver, S. (2003) Optimal OFDM design for time-frequency dispersive channels. *IEEE Transactions on Communications*, 51 (7), 1111–1122, doi: 10.1109/TCOMM.2003.814200.

245 Kovacevic, J., Dragotti, P., and Goyal, V. (2002) Filter bank frame expansions with erasures. *IEEE Transactions on Information Theory*, 48 (6), 1439–1450, doi: 10.1109/TIT.2002.1003832.

246 Daubechies, I., Grossmann, A., and Meyer, Y. (1986) Painless nonorthogonal expansions. *Journal of Mathematical Physics*, 27 (5), 1271–1283, doi: 10.1063/1.527388.

247 Casazza, P. (2000) The art of frame theory. *Taiwanese Journal of Mathematics*, 4 (2), 129–202.

248 Christensen, O. (2002) *An Introduction to Frames and Riesz Basis*, Birkhaser, Boston, MA.

249 Casazza, P.G., Christensen, O., and Janssen, A.J.E.M. (2001) Weyl–Heisenberg frames, translation invariant systems and the Walnut representation. *Journal of Functional Analysis*, 180 (1), 85–147.

250 Heil, C. (2007) History and evolution of the density theorem for Gabor frames. *Journal of Fourier Analysis and Applications*, 13, 113–166, doi: 10.1007/s00041-006-6073-2.

251 Conway, J. and Sloane, N. (1993) *Sphere Packings, Lattices and Groups*, Grundlehren in Mathematischen Wissenschaften, Springer, New York.

252 Haas, R. and Belfiore, J.C. (1997) A time-frequency well-localized pulse for multiple carrier transmission. *Wireless Personal Communications*, **5**, 1–18, doi: 10.1023/A:1008859809455.

253 Benedetto, J. and Zimmermann, G. (1997) Sampling multipliers and the poisson summation formula. *Journal of Fourier Analysis and Applications*, **3**, 505–523, doi: 10.1007/BF02648881.

254 Qian, S. and Chen, D. (1993) Discrete Gabor transform. *IEEE Transactions on Signal Processing*, **41** (7), 2429–2438, doi: 10.1109/78.224251.

255 Prinz, P. (1996) Calculating the dual Gabor window for general sampling sets. *IEEE Transactions on Signal Processing*, **44** (8), 2078–2082, doi: 10.1109/78.533729.

256 Subbanna, N. and Eldar, Y. (2004) A Fast Algorithm for Calculating the Dual Gabor Window with Integer Oversampling. *23rd IEEE Convention of Electrical and Electronics Engineers in Israel*, pp. 368–371.

257 Bolcskei, H., Hlawatsch, F., and Feichtinger, H.G. (1995) Equivalence of DFT Filter Banks and Gabor Expansions. *Proceedings of SPIE 2569, Wavelet Applications in Signal and Image Processing III, vol. 2569*, pp. 128–139, doi: 10.1117/12.217569.

258 Bolcskei, H. and Hlawatsch, F. (1997) Discrete Zak transforms, polyphase transforms, and applications. *IEEE Transactions on Signal Processing*, **45** (4), 851–866, doi: 10.1109/78.564174.

259 Slimane, S. (2002) Peak-to-Average Power Ratio Reduction of OFDM Signals Using Broadband Pulse Shaping. *IEEE 56th Vehicular Technology Conference, VTC'02-Fall, vol. 2*, pp. 889–893, doi: 10.1109/VETECF.2002.1040728.

260 Hughes-Hartogs, D. (1989) Ensemble modem structure for imperfect transmission media. US Patents Nos. 4679227, July 1987, 4731816 March 1988 and 4833796 May 1989.

261 Kliks, A. and Bogucka, H. (2008) The Application of Water-Filling Principle for Generalized Multicarrier Signal. *13th International OFDM-Workshop InOWo2008, Hamburg, Germany, August 27-28, 2008*, pp. 211–215.

262 Campello, J. (1999) Practical Bit Loading for DMT. *IEEE International Conference on Communications, ICC '99, vol. 2*, pp. 801–805, doi: 10.1109/ICC.1999.765384.

263 Kliks, A., Bogucka, H., Stupia, I., and Lottici, V. (2009) A Pragmatic Bit and Power Allocation Algorithm for NOFDM Signalling. *IEEE Wireless Communications and Networking Conference, WCNC 2009*, pp. 1–6, doi: 10.1109/WCNC.2009.4917534.

264 Fischer, R. and Huber, J. (1996) A New Loading Algorithm for Discrete Multitone Transmission. *Global Telecommunications Conference, 1996. GLOBECOM '96. 'Communications: The Key to Global Prosperity, vol. 1*, pp. 724–728, doi: 10.1109/GLOCOM.1996.594456.

265 Chow, P., Cioffi, J., and Bingham, J. (1995) A practical discrete multitone transceiver loading algorithm for data transmission over spectrally shaped channels. *IEEE Transactions on Communications*, **43** (**234**), 773–775, doi: 10.1109/26.380108.

266 Glover, I. and Grant, P. (2004) *Digital Communications*, Prentice Hall.

267 Speth, M., Fechtel, S., Fock, G., and Meyr, H. (2001) Optimum receiver design for OFDM-based broadband transmission .II. A case study. *IEEE Transactions on Communications*, **49** (**4**), 571–578, doi: 10.1109/26.917759.

268 Speth, M., Fechtel, S., Fock, G., and Meyr, H. (1999) Optimum receiver design for wireless broad-band systems using OFDM. I. *IEEE Transactions on Communications*, **47** (**11**), 1668–1677, doi: 10.1109/26.803501.

269 Rabiei, A. and Beaulieu, N. (2009) Cochannel interference mitigation using whitening receiver designs in bandlimited microcellular wireless systems. *IEEE Transactions on Wireless Communications*, **8** (**3**), 1284–1294, doi: 10.1109/TWC.2008.071046.

270 Winters, J. (1984) Optimum combining in digital mobile radio with cochannel interference. *IEEE Journal on Selected Areas in Communications*, **2** (**4**), 528–539, doi: 10.1109/JSAC.1984.1146095.

271 Wang, T., Proakis, J., and Zeidler, J. (2007) Interference analysis of filtered multitone modulation over time-varying frequency- selective fading channels. *IEEE Transactions on Communications*, **55** (**4**), 717–727, doi: 10.1109/TCOMM.2007.892455.

272 Kasdin, N. (1995) Discrete simulation of colored noise and stochastic processes and 1/f alpha; power law noise generation. *Proceedings of the IEEE*, **83** (**5**), 802–827, doi: 10.1109/5.381848.

273 Mochizuki, K. and Uchino, M. (2001) Efficient digital wide-band coloured noise generator. *Electronics Letters*, **37** (**1**), 62–64, doi: 10.1049/el:20010026.

274 Kay, S. (1981) Efficient generation of colored noise. *Proceedings of the IEEE*, **69** (**4**), 480–481, doi: 10.1109/PROC.1981.12000.

275 Golub, G.H. and Van Loan, C.F. (1996) *Matrix Computation*, The Johns Hopkins University Press, Baltimore, MD and London.

276 Davis, M., Monk, A., and Milstein, L. (1996) A noise whitening approach to multiple-access noise rejection .II. Implementation issues. *IEEE Journal on Selected Areas in Communications*, **14** (**8**), 1488–1499, doi: 10.1109/49.539403.

277 Monk, A., Davis, M., Milstein, L., and Helstrom, C. (1994) A noise whitening approach to multiple access noise rejection .I. Theory and background. *IEEE Journal on Selected Areas in Communications*, **12** (**5**), 817–827, doi: 10.1109/49.298055.

278 Frohlich, F. and Martin, U. (2004) Frequency-Domain MIMO Interference Cancellation Technique for Space-Time Block-Coded

Single-Carrier Systems. *ITG Workshop on Smart Antennas*, pp. 30–34, doi: 10.1109/WSA.2004.1407644.

279 Manohar, S., Tikiya, V., Annavajjala, R., and Chockalingam, A. (2007) BER-optimal linear parallel interference cancellation for multicarrier DS-CDMA in Rayleigh fading. *IEEE Transactions on Communications*, **55** (**6**), 1253–1265, doi: 10.1109/TCOMM.2007.898860.

280 Patel, P. and Holtzman, J. (1994) Analysis of a simple successive interference cancellation scheme in a DS/CDMA system. *IEEE Journal on Selected Areas in Communications*, **12** (**5**), 796–807, doi: 10.1109/49.298053.

281 Shankar Kumar, K. and Chockalingam, A. (2004) Parallel Interference Cancellation in Multicarrier DS-CDMA systems. *IEEE International Conference on Communications, vol. 5*, pp. 2874–2878, doi: 10.1109/ICC.2004.1313054.

282 Xi, S., Song, M., Zhao, Y., Ren, L., and Song, J. (2003) Co-Channel Interference Cancellation for MIMO CDMA Wireless Communications. *14th IEEE Proceedings on Personal, Indoor and Mobile Radio Communications. PIMRC 2003, vol. 2*, pp. 1405–1409, doi: 10.1109/PIMRC.2003.1260344.

283 Foschini, C. (1996) Layered space-time architecture for wireless communications in fading when using multiple antennas. *Bell Laboratories Technical Journal*, **1** (**2**), 41–59.

284 Golden, G., Foschini, G., Velenzuela, R., and Wolnianski, I. (1999) Detection algorithm and initial laboratory results using the V-BLAST space-time communication architecture. *Electronics Letters*, **35** (**1**), 14–15.

285 Kim, J.H., Jeong, J.Y., Yeom, S.J., Choi, B.G., and Park, Y.W. (1999) Performance Analysis of the Hybrid Interference Canceller for Multiple Access Interference Cancellation. *IEEE Region 10 Conference TENCON'99, vol. 2*, pp. 1236–1239, doi: 10.1109/TENCON.1999.818651.

286 Sun, S., Rasmussen, L., Lim, T., and Sugimoto, H. (1998) A Matrix-Algebraic Approach to Linear Hybrid Interference Cancellation in CDMA. *IEEE 1998 International Conference on Universal Personal Communications, ICUPC '98, vol. 2*, pp. 1319–1323, doi: 10.1109/ICUPC.1998.733707.

287 Farhang-Boroujeny, B. (2011) OFDM versus filter bank multicarrier. *IEEE Signal Processing Magazine*, **28** (**3**), 92–112, doi: 10.1109/MSP.2011.940267.

288 Sahin, A., Guvenc, I., and Arslan, H. (2014) A survey on multicarrier communications: prototype filters, lattice structures, and implementation aspects. *IEEE Communication Surveys and Tutorials*, **16** (**3**), 1312–1338, doi: 10.1109/SURV.2013.121213.00263.

289 Plimmer, S.A., David, J.P.R., Ong, D.S., and Li, K.F. (1999) A simple model for avalanche multiplication including deadspace effects. *IEEE Transactions on Electron Devices*, **46** (**4**), 769–775, doi: 10.1109/16.753712.

290 Saeedi-Sourck, H., Wu, Y., Bergmans, J.W.M., Sadri, S., and Farhang-Boroujeny, B. (2011) Complexity and performance comparison of filter bank multicarrier and OFDM in uplink of multicarrier multiple access networks. *IEEE Transactions on Signal Processing*, **59** (4), 1907–1912, doi: 10.1109/TSP.2010.2104148.

291 Bellanger, M. (2010) Physical Layer for Future Broadband Radio Systems. *2010 IEEE Radio and Wireless Symposium (RWS)*, pp. 436–439, doi: 10.1109/RWS.2010.5434093.

292 Medjahdi, Y., Terre, M., Ruyet, D.L., Roviras, D., and Dziri, A. (2011) Performance analysis in the downlink of asynchronous OFDM/FBMC based multi-cellular networks. *IEEE Transactions on Wireless Communications*, **10** (8), 2630–2639, doi: 10.1109/TWC.2011.061311.101112.

293 Waldhauser, D.S., Baltar, L.G., and Nossek, J.A. (2008) MMSE Subcarrier Equalization for Filter Bank Based Multicarrier Systems. *2008 IEEE 9th Workshop on Signal Processing Advances in Wireless Communications*, pp. 525–529, doi: 10.1109/SPAWC.2008.4641663.

294 Schaich, F., Wild, T., and Chen, Y. (2014) Waveform Contenders for 5G – Suitability for Short Packet and Low Latency Transmissions. *2014 IEEE 79th Vehicular Technology Conference (VTC Spring)*, pp. 1–5, doi: 10.1109/VTCSpring.2014.7023145.

295 Caus, M. and Neira, A.I. (2012) Transmitter-receiver designs for highly frequency selective channels in MIMO FBMC systems. *IEEE Transactions on Signal Processing*, **60** (12), 6519–6532, doi: 10.1109/TSP.2012.2217133.

296 Chang, R.W. (1966) High-speed multichannel data transmission with bandlimited orthogonal signals. *Bell System Technical Journal*, **45**, 1775–1796.

297 Saltzberg, B. (1967) Performance of an efficient parallel data transmission system. *IEEE Transactions on Communication Technology*, **15** (6), 805–811, doi: 10.1109/TCOM.1967.1089674.

298 Bellanger, M. and Daguet, J. (1974) TDM-FDM transmultiplexer: digital polyphase and FFT. *IEEE Transactions on Communications*, **22** (9), 1199–1205, doi: 10.1109/TCOM.1974.1092391.

299 Hirosaki, B. (1981) An orthogonally multiplexed QAM system using the discrete fourier transform. *IEEE Transactions on Communications*, **29** (7), 982–989, doi: 10.1109/TCOM.1981.1095093.

300 Cherubini, G., Eleftheriou, E., Oker, S., and Cioffi, J.M. (2000) Filter bank modulation techniques for very high speed digital subscriber lines. *IEEE Communications Magazine*, **38** (5), 98–104, doi: 10.1109/35.841832.

301 Cherubini, G., Eleftheriou, E., and Olcer, S. (2002) Filtered multi tone modulation for very high-speed digital subscriber lines. *IEEE Journal on Selected Areas in Communications*, **20** (5), 1016–1028, doi: 10.1109/JSAC.2002.1007382.

302 Bolcskei, H., Duhamel, P., and Hleiss, R. (1999) Design of Pulse Shaping OFDM/OQAM Systems for High Data-Rate Transmission Over Wireless

Channels. *IEEE International Conference on Communications, vol. 1*, pp. 559–564, doi: 10.1109/ICC.1999.768001.

303 Du, J. and Signell, S. (2007) Time Frequency Localization of Pulse Shaping Filters in OFD/OQAM Systems. *6th International Conference on Information, Communications Signal Processing*, pp. 1–5, doi: 10.1109/ICICS.2007.4449830.

304 Le Floch, B., Alard, M., and Berrou, C. (1995) Coded orthogonal frequency division multiplex [TV broadcasting]. *Proceedings of the IEEE*, **83** (6), 982–996, doi: 10.1109/5.387096.

305 Farhang-Boroujeny, B. and Yuen, C.G. (2010) Cosine modulated and offset QAM filter bank multicarrier techniques: a continuous-time prospect. *EURASIP Journal on Advances in Signal Processing*, **2010**, 16.

306 Farhang-Boroujeny, B. and Lin, L. (2005) Cosine Modulated Multitone for Very High-Speed Digital Subscriber Lines. *Proceedings. (ICASSP '05). IEEE International Conference on Acoustics, Speech, and Signal Processing, 2005, vol. 3*, pp. iii/345–iii/348, doi: 10.1109/ICASSP.2005.1415717.

307 Pérez-Neira, A.I., Caus, M., Zakaria, R., Ruyet, D.L., Kofidis, E., Haardt, M., Mestre, X., and Cheng, Y. (2016) MIMO signal processing in offset-QAM based filter bank multicarrier systems. *IEEE Transactions on Signal Processing*, **64** (**21**), 5733–5762, doi: 10.1109/TSP.2016.2580535.

308 Caus, M. and Pérez-Neira, A.I. (2014) Multi-stream transmission for highly frequency selective channels in MIMO-FBMC/OQAM systems. *IEEE Transactions on Signal Processing*, **62** (**4**), 786–796, doi: 10.1109/TSP.2013.2293973.

309 Viholainen, A., Bellanger, M., and Huchard, M. (2009) Prototype Filter and Structure Optimization. Deliverable D5.1, EU ICT – 211887 Project, PHYDYAS.

310 Bellanger, M.G. e.a. (2010) FBMC Physical Layer: A Primer, online. PHYDYAS FP7 Project Document, http://www.ict-phydyas.org/teamspace/internal-folder/FBMC-Primer_06-2010.pdf (accessed 30 June 2016).

311 Bellanger, M.G. (2001) Specification and Design of a Prototype Filter for Filter Bank Based Multicarrier Transmission. *Acoustics, Speech, and Signal Processing, 2001. Proceedings. (ICASSP '01). 2001 IEEE International Conference on, vol. 4*, pp. 2417–2420, doi: 10.1109/ICASSP.2001.940488.

312 Farhang-Boroujeny, B. (2014) Filter bank multicarrier modulation: a waveform candidate for 5G and beyond. *Advances in Electrical Engineering*, 2014, **25**, doi: 10.1155/2014/482805.

313 Amini, P., Yuen, C.H., Chen, R.R., and Farhang-Boroujeny, B. (2010) Isotropic Filter Design for MIMO Filter Bank Multicarrier Communications. *Sensor Array and Multichannel Signal Processing Workshop (SAM), 2010 IEEE*, pp. 89–92, doi: 10.1109/SAM.2010.5606775.

314 Jackowski, T. (2014) Filter-bank based multicarrier transmission. MSc thesis, Poznan University of Technology, Faculty of Electronics and Telecommunication.

315 Siohan, P. and Roche, C. (2004) Derivation of extended Gaussian functions based on the Zak transform. *IEEE Signal Processing Letters*, **11** (3), 401–403, doi: 10.1109/LSP.2003.821727.

316 Vahlin, A. and Holte, N. (1996) Optimal finite duration pulses for OFDM. *IEEE Transactions on Communications*, **44** (1), 10–14, doi: 10.1109/26.476088.

317 Slepian, D. and Pollak, H. (1961) Prolate spheroidal wave functions, Fourier analysis and uncertainty. *Bell System Technical Journal*, **40**, 43–64.

318 Chen, C.Y. and Vaidyanathan, P. (2008) MIMO radar space time adaptive processing using prolate spheroidal wave functions. *IEEE Transactions on Signal Processing*, **56** (2), 623–635, doi: 10.1109/TSP.2007.907917.

319 Walter, G.G. and Shen, X. (2004) Wavelets based on prolate spheroidal wave functions. *Journal of Fourier Analysis and Applications*, **10**, 1–26, doi: 10.1007/s00041-004-8001-7.

320 Zhao, H., Ran, Q.W., Ma, J., and Tan, L.Y. (2010) Generalized prolate spheroidal wave functions associated with linear canonical transform. *IEEE Transactions on Signal Processing*, **58** (6), 3032–3041, doi: 10.1109/TSP.2010.2044609.

321 Zakaria, R. and Ruyet, D.L. (2010) On Maximum Likelihood MIMO Detection in QAM-FBMC Systems. *21st Annual IEEE International Symposium on Personal, Indoor and Mobile Radio Communications*, pp. 183–187, doi: 10.1109/PIMRC.2010.5671632.

322 Zakaria, R. and Ruyet, D.L. (2013) On Interference Cancellation in Alamouti Coding Scheme for Filter Bank Based Multicarrier Systems. *Wireless Communication Systems (ISWCS 2013), Proceedings of the 10th International Symposium on*, pp. 1–5.

323 Zakaria, R. and Ruyet, D.L. (2012) A novel filter-bank multicarrier scheme to mitigate the intrinsic interference: application to MIMO systems. *IEEE Transactions on Wireless Communications*, **11** (3), 1112–1123, doi: 10.1109/TWC.2012.012412.110607.

324 Moret, N., Tonello, A., and Weiss, S. (2011) MIMO Precoding for Filter Bank Modulation Systems Based on PSVD. *Vehicular Technology Conference (VTC Spring), 2011 IEEE 73rd*, pp. 1–5, doi: 10.1109/VETECS.2011.5956567.

325 Ihalainen, T., Ikhlef, A., Louveaux, J., and Renfors, M. (2011) Channel equalization for multi-antenna FBMC/OQAM receivers. *IEEE Transactions on Vehicular Technology*, **60** (5), 2070–2085, doi: 10.1109/TVT.2011.2145424.

326 Cheng, Y., Baltar, L.G., Haardt, M., and Nossek, J.A. (2015) Precoder and Equalizer Design for Multi-User MIMO FBMC/OQAM with Highly Frequency Selective Channels. *2015 IEEE International Conference on Acoustics, Speech and Signal Processing (ICASSP)*, pp. 2429–2433, doi: 10.1109/ICASSP.2015.7178407.

327 Rottenberg, F., Mestre, X., and Louveaux, J. (2016) Optimal Zero Forcing Precoder and Decoder Design for Multi-User MIMO FBMC Under Strong Channel Selectivity. *2016 IEEE International Conference on Acoustics, Speech and Signal Processing (ICASSP)*, pp. 3541–3545, doi: 10.1109/ICASSP.2016.7472336.

328 Rottenberg, F., Mestre, X., Horlin, F., and Louveaux, J. (2016) Single-tap precoders and decoders for multi-user MIMO FBMC-OQAM under strong channel frequency selectivity. *IEEE Transactions on Signal Processing*, **PP** (**99**), 1, doi: 10.1109/TSP.2016.2621722.

329 Estella, I., Pascual-Iserte, A., and Payaró, M. (2010) OFDM and FBMC Performance Comparison for Multistream MIMO Systems. *2010 Future Network Mobile Summit*, pp. 1–8.

330 Tonello, A. and Pecile, F. (2008) Analytical results about the robustness of FMT modulation with several prototype pulses in time-frequency selective fading channels. *IEEE Transactions on Wireless Communications*, **7** (**5**), 1634–1645, doi: 10.1109/TWC.2008.060528.

331 Farhang-Boroujeny, B. and Yuen, C.H.G. (2010) Cosine modulated and offset QAM filter bank multicarrier techniques: a continuous-time prospect. *EURASIP Journal on Advances in Signal Processing*, **2010**, doi: 10.1155/2010/165654.

332 Lin, L. and Farhang-Boroujeny, B. (2006) Cosine-modulated multitone for very-high-speed digital subscriber lines. *EURASIP Journal on Applied Signal Processing*, **2006**, 1–17.

333 Vakilian, V., Wild, T., Schaich, F., ten Brink, S., and Frigon, J.F. (2013) Universal-Filtered Multi-Carrier Technique for Wireless Systems Beyond LTE. *Proceedings. (GLOBECOM '13). IEEE Global Communications Conference, 2013*, pp. 223–228, doi: 10.1109/GLOCOMW.2013.6824990.

334 Chen, Y., Schaich, F., and Wild, T. (2014) Multiple Access and Waveforms for 5G: IDMA and Universal Filtered Multi-Carrier. 2014 IEEE 79th Vehicular Technology Conference (VTC Spring), pp. 1–5, doi: 10.1109/VTCSpring.2014.7022995.

335 Schaich, F. and Wild, T. (2014) Waveform Contenders for 5G: OFDM vs. FBMC vs. UFMC. *Communications, Control and Signal Processing (ISCCSP), 2014 6th International Symposium on*, pp. 457–460, doi: 10.1109/ISCCSP.2014.6877912.

336 Mitola, J. and Maguire, G.Q. (1999) Cognitive radio: making software radios more personal. *IEEE Personal Communications*, **6** (**4**), 13–18, doi: 10.1109/98.788210.

337 Mitola, J. (2000) Cognitive radio: an integrated agent architecture for software defined radio. PhD thesis, Royal Institute of Technology (KTH), Sweden.

338 Mitola, J. (2006) *Cognitive Radio Architecture: The Engineering Foundation of Radio XML*, John Wiley & Sons, Inc., New York.

339 Hossain, E., Niyato, D., and Han, Z. (2009) *Dynamic Spectrum Access and Management in Cognitive Radio Networks*, Cambridge University Press, Cambridge.

340 Fette, B.A. (2009) *Cognitive Radio Technology*, Academic Press, New York.

341 Biglieri, E., Goldsmith, A., Greenstein, L., Mandayam, N., and Poor, H. (2013) *Principles of Cognitive Radio*, Cambridge University Press, Cambridge.

342 Di Benedetto, M.G. and Bader, F. (eds) (2014) *Cognitive Communication and Cooperative HetNet Coexistence*, Springer International Publishing, Switzerland.

343 Holland, O., Bogucka, H., and Medeisis, A. (eds) (2015) *Opportunistic Spectrum Sharing and White Space Access: The Practical Reality*, John Wiley & Sons, Inc., New York.

344 Akyildiz, I.F., Lee, W.Y., Vuran, M.C., and Mohanty, S. (2006) Next generation/dynamic spectrum access/cognitive radio wireless networks: a survey. *Computer Networks (Elsevier)*, **50** (**13**), 2127–2159.

345 Zhao, Q. and Sadler, B.M. (2007) A survey of dynamic spectrum access. *IEEE Signal Processing Magazine*, **24** (**3**), 79–89, doi: 10.1109/MSP.2007.361604.

346 Wang, B. and Liu, K. (2011) Advances in cognitive radio networks: a survey. *IEEE Journal on Selected Topics in Signal Processing*, **5** (**1**), 5–23, doi: 10.1109/JSTSP.2010.2093210.

347 Haykin, S., Thomson, D., and Reed, J. (2010) Spectrum sensing for cognitive radio. *Proceedings of IEEE*, **97** (**5**), 849–877.

348 Lu, L., Zhou, X., Onunkwo, U., and Li, G. (2012) Ten years of research in spectrum sensing and sharing in cognitive radio. *EURASIP Journal on Wireless Communications and Networking*, **2012** (**28**), 1–16.

349 Pucker, L. (2010) Review of contemporary spectrum sensing technologies, Ts, IEEE-SA P1900.6 Standards Group.

350 Yucek, T. and Arslan, H. (2009) A survey of spectrum sensing algorithms for cognitive radio applications. *IEEE Communication Surveys and Tutorials*, **11** (**1**), 116–130, doi: 10.1109/SURV.2009.090109.

351 Cichon, K., Kliks, A., and Bogucka, H. (2016) Energy-efficient cooperative spectrum sensing: a survey. *IEEE Communication Surveys and Tutorials*, **18** (**3**), 1861–1886.

352 Jouini, W., Moy, C., and Palicot, J. (2012) Decision making for cognitive radio equipment: analysis of the first 10 years of exploration. *EURASIP Journal on Wireless Communications and Networking*, **2012** (**26**), 1–16.

353 Hardin, B. (1968) The tragedy of the commons. *Science*, **162** (**3859**), 1243–1248.

354 Matinmikko, M., Mustonen, M., Roberson, D., Paavola, J., Höyhtyä, M., Yrjölä, S., and Röning, J. (2014) Overview and Comparison of Recent Spectrum Sharing Approaches in Regulation and Research: From Opportunistic Unlicensed Access Towards Licensed Shared Access.

Dynamic Spectrum Access Networks (DYSPAN), 2014 IEEE International Symposium on, pp. 92–102, doi: 10.1109/DySPAN.2014.6817783.

355 Report to the President's Council of Advisors on Science and Technology (PCAST) (2012) Realizing the Full Potential of Government-Held Spectrum to Spur Economic Growth, https://www.whitehouse.gov/sites/default/files/microsites/ostp/pcast_spectrum_report_final_july_20_2012.pdf (accessed 08 November 2016).

356 Sohul, M.M., Yao, M., Yang, T., and Reed, J.H. (2015) Spectrum access system for the citizen broadband radio service. *IEEE Communications Magazine*, **53** (7), 18–25, doi: 10.1109/MCOM.2015.7158261.

357 Holland, O., Nardis, L.D., Nolan, K., Medeisis, A., Anker, P., Minervini, L.F., Velez, F., Matinmikko, M., and Sydor, J. (2012) Pluralistic Licensing. *Dynamic Spectrum Access Networks (DYSPAN), 2012 IEEE International Symposium on*, pp. 33–41, doi: 10.1109/DYSPAN.2012.6478113.

358 Kliks, A., Holland, O., Basaure, A., and Matinmikko, M. (2015) Spectrum and license flexibility for 5G networks. *IEEE Communications Magazine*, **53** (7), 42–49, doi: 10.1109/MCOM.2015.7158264.

359 LTE–U Forum (2012) Online Web Page of the LTE–U Forum, http://www .lteuforum.org/ (accessed 08 November 2016).

360 Chen, Q., Yu, G., Yin, R., Maaref, A., Li, G.Y., and Huang, A. (2016) Energy efficiency optimization in licensed-assisted access. *IEEE Journal on Selected Areas in Communications*, **34** (4), 723–734, doi: 10.1109/JSAC.2016.2544605.

361 Galanopoulos, A., Foukalas, F., and Tsiftsis, T.A. (2016) Efficient coexistence of LTE with wifi in the licensed and unlicensed spectrum aggregation. *IEEE Transactions on Cognitive Communications and Networking*, **2** (2), 129–140, doi: 10.1109/TCCN.2016.2594780.

362 Li, Y., Baccelli, F., Andrews, J.G., Novlan, T.D., and Zhang, J. (2015) Modeling and Analyzing the Coexistence of Licensed-Assisted Access LTE and Wi-Fi. *2015 IEEE Globecom Workshops (GC Wkshps)*, pp. 1–6, doi: 10.1109/GLOCOMW.2015.7414197.

363 Singh, B., Hailu, S., Koufos, K., Dowhuszko, A.A., Tirkkonen, O., Jäntti, R., and Berry, R. (2015) Coordination protocol for inter-operator spectrum sharing in co-primary 5G small cell networks. *IEEE Communications Magazine*, **53** (7), 34–40, doi: 10.1109/MCOM.2015.7158263.

364 Irnich, T., Kronander, J., Selén, Y., and Li, G. (2013) Spectrum Sharing Scenarios and Resulting Technical Requirements for 5G Systems. *Personal, Indoor and Mobile Radio Communications (PIMRC Workshops), 2013 IEEE 24th International Symposium on*, pp. 127–132, doi: 10.1109/PIMRCW.2013.6707850.

365 Zhang, Z., He, Y., and Chong, E. (2008) Opportunistic scheduling for OFDM systems with fairness constraints. *EURASIP Journal on Wireless Communications and Networking*, **2008** (1), 1–12.

366 Han, Z., Ji, Z., and Liu, K. (2005) Fair multiuser channel allocation for OFDMA networks using nash bargaining solutions and coalitions. *IEEE Transactions on Communications*, **53** (8), 1366–1376.

367 Sacchi, C., Granelli, F., and Schlegel, C. (2011) A QoE-oriented strategy for OFDMA radio resource allocation based on min-MOS maximization. *IEEE Communication Letters*, **15** (5), 494–496.

368 Chen, J. and Swindlehurst, A. (2012) Applying bargaining solutions to resource allocation in multiuser MIMO-OFDMA broadcast systems. *IEEE Journal on Selected Topics in Signal Processing*, **6** (2), 127–139.

369 Han, Z., Ji, Z., and Liu, K. (2007) Non-cooperative resource competition game by virtual referee in multi-cell OFDMA networks. *IEEE Journal on Selected Areas in Communications*, **25** (6), 1079–1089.

370 Buzzi, S., Colavolpe, G., Saturnino, D., and Zappone, A. (2012) Potential games for energy-efficient power control and subcarrier allocation in uplink multicell OFDMA systems. *IEEE Journal on Selected Topics in Signal Processing*, **6** (2), 89–103.

371 Wu, D., Yu, D., and Cai, Y. (2008) Subcarrier and Power Allocation in Uplink OFDMA Systems Based on Game Theory. *International Conference on Neural Networks and Signal Processing*, pp. 522–526.

372 Yu, D., Wu, D., Cai, Y., and Zhong, W. (2008) Power Allocation Based on Power Efficiency in Uplink OFDMA Systems: A Game Theoretic Approach. *IEEE Singapore International Conference on Communication Systems, (ICCS) 2008*, pp. 92–97.

373 Chen, F., Xu, L., Mei, S., Zhenhui, T., and Huan, L. (2007) OFDM Bit and Power Allocation Based on Game Theory. *International Symposium on Microwave, Antenna, Propagation and EMC Technologies for Wireless Communications*, pp. 1147–1150.

374 Wang, F., Krunz, M., and Cui, S. (2008) Price-based spectrum management in cognitive radio networks. *IEEE Journal on Selected Topics in Signal Processing*, **2** (1), 74–87.

375 Kwon, H. and Lee, B. (2006) Distributed Resource Allocation Through Noncooperative Game Approach in Multi-Cell OFDMA Systems. *IEEE International Conference on Communications*, pp. 4345–4350.

376 Wang, L., Xue, Y., and Schulz, E. (2006) Resource Allocation in Multi-cell OFDM Systems Based on Noncooperative Game. *IEEE International Symposium on Personal, Indoor and Mobile Radio Communications*, pp. 1–5.

377 Liang, Z., Chew, Y., and Ko, C. (2008) Decentralized Bit, Subcarrier and Power Allocation with Interference Avoidance in Multicell OFDMA Systems Using Game Theoretic Approach. *IEEE Military Communications Conference, MILCOM 2008*, pp. 1–7.

378 Niyato, D. and Hossain, E. (2008) Competitive pricing for spectrum sharing in cognitive radio networks: dynamic game, inefficiency of nash

equilibrium, and collusion. *IEEE Journal on Selected Areas in Communications*, **26** (**1**), 192–202.

379 Bogucka, H. (2012) Optimal pricing of spectrum resources in wireless opportunistic access. *Journal of Computer Networks and Communications*, **2012** (**1**), 1–14, doi: 10.1155/2012/794572.

380 Brandenburger, A. and Nalebuff, B. (1997) *Co-opetition: A Revolutionary Mindset that Combines Competition and Co-operation: The Game Theory Strategy That's Changing the Game of Business*, Currency Doubleday, New York.

381 Parzy, M. and Bogucka, H. (2014) Coopetition methodology for resource sharing in distributed OFDM-based cognitive radio networks. *IEEE Transactions on Communications*, **62** (**5**), 1518–1529, doi: 10.1109/TCOMM.2014.031214.130451.

382 Kryszkiewicz, P., Kliks, A., and Bogucka, H. (2016) Small-scale spectrum aggregation and sharing. *IEEE Journal on Selected Areas in Communications*, **34** (**10**), 2630–2641.

383 Siohan, P., Siclet, C., and Lacaille, N. (2002) Analysis and design of OFDM/OQAM systems based on filterbank theory. *IEEE Transactions on Signal Processing*, **50** (**5**), 1170–1183.

384 Sanguinetti, L., Morelli, M., and Poor, H. (2010) Frame detection and timing acquisition for OFDM transmissions with unknown interference. *IEEE Transactions on Wireless Communication*, **9** (**3**), 1226–1236.

385 Kryszkiewicz, P. and Bogucka, H. (2016) Low complex, narrowband-interference robust synchronization for NC-OFDM cognitive radio. *IEEE Transactions on Communications*, **64** (**9**), 3644–3654.

386 Stitz, T., Ihalainen, T., Viholainen, A., and Renfors, M. (2010) Pilot-based synchronization and equalization in filter bank multicarrier communications. *EURASIP Journal on Advances in Signal Processing*, **2010**, 1–19.

387 Fusco, T., Petrella, A., and Tanda, M. (2009) Data-aided symbol timing and CFO synchronization for filter bank multicarrier systems. *IEEE Transactions on Wireless Communications*, **8** (**5**), 2705–2715.

388 Thein, C., Fuhrwerk, M., and Peissig, J. (2013) About the Use of Different Processing Domains for Synchronization in Non-Contiguous FBMC Systems. *IEEE 24th International Symposium on Personal Indoor and Mobile Radio Communications (PIMRC)*, pp. 791–795, doi: 10.1109/PIMRC.2013.6666244.

389 Rahimi, S. and Champagne, B. (2014) Joint channel and frequency offset estimation for oversampled perfect reconstruction filter bank transceivers. *IEEE Transactions on Communications*, **62** (**6**), 2009–2021.

390 40, C.R. (2010) Compatibility study for LTE and WiMAX operating within the bands 880–915 MHz / 925–960 MHz and 1710–1785 MHz /1805–1880 MHz (900/1800 MHz bands), Ts, CEPT.

391 Rappaport, T. (2001) *Wireless Communications: Principles and Practice*, 2nd edn, Prentice Hall PTR, Upper Saddle River, NJ.

392 ETSI TS 5.05. (1996) Digital Cellular Telecommunications System (Phase 2+); Radio Transmission and Reception, ETSI.

393 3GPP TS 25.104. (2008) Base Station (BS) Radio Transmission and Reception (FDD), 3rd Generation Partnership Project (3GPP), http://www.3gpp .org/ftp/Specs/html-info/25104.htm (accessed 27 March 2017).

394 Bogucka, H. (2013) *Technologie radia kognitywnego*, Wydawnictwo Naukowe PWN, Warszawa (in Polish).

Index

Advanced Multicarrier Technologies for Future Radio Communication: 5G and Beyond, First Edition.
Hanna Bogucka, Adrian Kliks, and Paweł Kryszkiewicz.
© 2017 John Wiley & Sons, Inc. Published 2017 by John Wiley & Sons, Inc.